Aiden Cawley.

T & D

Heinhold/Stubbe
Power Cables and their Application
Part 2

Authors	Clause
Wilfried Glaubitz	4.1, 5 – 9
Herbert Postler	4.1, 5 – 9
Dirk Rittinghaus	1.2, 4.2
Günter Seel	1.1, 2, 3
Frank Sengewald	7 – 9
(RXS GmbH, Hagen)	

Scientific and technological fundamentals:
Franz Winkler

Translation:
Natascha Wacker

Power Cables and their Application

Part 2

Tables Including
Project Planning Data for Cables
and Accessories.
Details for the Determination
of the Cross-Sectional Area

Editors: Lothar Heinhold and Reimer Stubbe

3rd revised edition, 1993

Siemens Aktiengesellschaft

Die Deutsche Bibliothek – CIP-Einheitsaufnahme

Power cables and their application / ed.: Lothar Heinhold and Reimer Stubbe. – Berlin;
Munich: Siemens-Aktienges., [Abt. Verl.].
  Einheitssacht.: Kabel und Leitungen für Starkstrom < engl. >
  Teilw. hrsg. von Lothar Heinhold
NE: Heinhold, Lothar [Hrsg.]; EST
Pt. 2. Tables including project planning data for cables and accessories:
  details for the determination of the cross sectional area /
  [authors: Wilfried Glaubitz... Transl.: Natascha Wacker]. – 3., rev. ed. – 1993
  ISBN 3-8009-1575-8
NE: Glaubitz, Wilfried

The product names referred to in this book are registered trademarks of Siemens AG

ARCOFLEX, CORDAFLEX, FLEXIPREN, OZOFLEX, PLANOFLEX, PROTODUR, PROTOFLEX,
PROTOLIN, PROTOLON, PROTOMONT, PROTOTHEN, SIENOPYR, SIFLA, SINOTHERM,
SUPROMONT

ISBN 3-8009-1575-8

3rd edition, 1993

Published by Siemens Aktiengesellschaft, Berlin and Munich
© 1970 by Siemens Aktiengesellschaft, Berlin and Munich

Preface

The 3rd edition of the book "Power Cables and their Application" covers many topics. Therefore, it was split into two parts in order to facilitate the handling. Part 1 includes the topics "Materials, Construction, Criteria for Selection, Project Planning of Installations, Laying and Installation, Accessories, Measuring and Testing". Besides explanations and instructions for project planning, Part 2 mainly contains tables with actual constructional and electrical data of insulated wires and flexible cables, power cables and accessories. In conjunction with Part 1, it facilitates an optimum selection and a safe project planning of cable installations.

As already done in the supplement to the 3rd German edition, the tables for power cables were drawn up by means of data processing. However, it was important that Part 2 forms a self-contained guide so that it can be used for project planning without Part 1. Therefore, each chapter includes the clause "Selection" followed by the tables for insulated wires and flexible cables and power cables. Good knowledge of the technical fundamentals, as described in Part 1, is necessary for the correct application.

The tables for selection which are included in Part 1, Clause 8.2 and 13.3, can also be found in the clauses "Selection" together with a summary of the contents of tables. The summary shows the designs and conductor cross-sectional areas taken into account. The following clause includes brief instructions for project planning for the determination of the conductor cross-sectional areas.

For accessories the application can be derived from the designation (e.g. indoor and outdoor terminations) so that a simplified summary of tables is sufficient.

Detailed explanations to tabulated values conclude the clauses for insulated wires and flexible cables, power cables and accessories. The corresponding explanation can be found via the numbering of items. Furthermore, reference is given to clauses in Part 1 of the book and important VDE specifications.

We wish to thank all those who have participated in the preparation of this book for their efforts and advice and for documents made available. Our special gratitude is due to Franz Winkler whose scientific and technological fundamentals were the basis for this Part 2.

Erlangen, February 1993

Lothar Heinhold
Reimer Stubbe

Introduction

In contrast to the 2nd edition, the tables were considerably enlarged in order to consider the requirements for the selection and determination of insulated wires and flexible cables and power cables. The most important details are explained below.

Insulated Wires and Flexible Cables

The instructions for project planning are completed by information on the maximum permissible cable lengths and the limitation of the voltage drop as well as by information on the rules given in DIN VDE 0100 Part 430 for the protection of insulated wires and flexible cables against overcurrents. The application of these rules and diagrams is explained by the use of selected examples.

In contrast to the 2nd edition, the tables were considerably enlarged in order to meet the present state of cable engineering. Besides the harmonized designs, HYDROFIRM cables for application in water and the range of products of SIENOPYR cables with improved performance in the case of fire are included.

Power Cables

With regard to the contents the tables for cables were also enlarged; tables for cables which have an insulation of thermoplastic PE were omitted, i.e. the details in the supplement to the 3rd German edition shall be used.

The permissible pulling forces were incorporated into the clause "Mechanical Properties" for the first time. The current-carrying capacities were calculated again in order to consider the new modifications in the constructional rules of VDE.

These modifications mainly concern the new d.c. resistances of conductors according to DIN VDE 0295 and the PE outer sheaths normally used for XLPE-insulated medium-voltage cables. The calculation was carried out on the basis of the calculation methods agreed for DIN VDE 0298 Part 2 and the values for materials for insulation and outer sheath. Differences are therefore observed compared with the present values according to DIN VDE 0298 Part 2. However, they do not go too far. This procedure was necessary in order to get proof and comprehensible results on the basis of the calculation methods given in Part 1. For this purpose, the fictitious thermal resistances of the cables were given for most designs.

For single-core cables with $U_0/U \geq 1$ kV laid in ground, the resistances per unit length and reactances per unit length in the zero phase-sequence system for the determination of the short-circuit currents in the system as well as the reduction factors for the interference with telecommunications cables and control cables can be taken from the tables. Furthermore, the current-carrying capacities of screens in case of short-circuit as well as the minimum time values of all cables installed in air are given. The clause "Determination of the Cross-Sectional Area" is a summary of the most important rules for project planning taken from DIN VDE 0298 Part 2. A particular clause deals with the rules for the current-carrying capacity of ship cables specified in DIN IEC 92 Part 201.

Accessories for Cables

Clause 8 includes the most important joints and terminations for power cables. Accessories for insulated wires and flexible cables and FRNC cables could not be considered yet.

In contrast to the 2nd edition, the structure of the tables was modified for reasons of clarity. Besides measures and weights the tables also include a brief description of the designs of the relevant accessories as well as information on the permissible peak short-circuit current, the mounting position and the installation.

Contents

Insulated Wires and Flexible Cables

1 Selection of Cables [1]

1.1 Summary and Application

Designation of Cables

Cables are designated with:

▷ type designation code,
▷ number of cores × nominal cross-sectional area of conductor in mm² and nominal cross-sectional area of the split PE conductor or of additional control and pilot cores, if necessary,
▷ rated voltages of cables in V or kV.

Cables according to DIN VDE 0250 and cables for which no VDE specifications exist

For cables according to the specification DIN VDE 0250, the type designation code always begins with the letter "N". For cables with core numbers which deviate from the specification or nominal cross-sectional areas which differ from the specification but which fulfil the other requirements, the letter N is put in parantheses (N).

Example for a cable according to DIN VDE 0250:

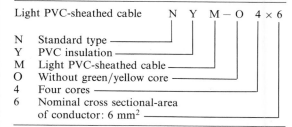

Light PVC-sheathed cable	N Y M – O 4 × 6
N	Standard type
Y	PVC insulation
M	Light PVC-sheathed cable
O	Without green/yellow core
4	Four cores
6	Nominal cross sectional-area of conductor: 6 mm²

For cables for which no VDE specifications exist, the letter N is omitted in the type designation code. These cables are designed in line with existing specifications and fulfil the safety requirements laid down in these specifications. The following letters describe the design and the construction, based on the insulation material.

Cables with a green/yellow marked core are identified by "J" after the type designation code and cables without a green/yellow marked core by "O".

Example for a cable
for which no VDE specification exists:

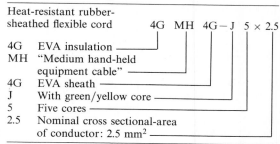

Heat-resistant rubber-sheathed flexible cord	4G MH 4G – J 5 × 2.5
4G	EVA insulation
MH	"Medium hand-held equipment cable"
4G	EVA sheath
J	With green/yellow core
5	Five cores
2.5	Nominal cross sectional-area of conductor: 2.5 mm²

Cables according to DIN VDE 0281 and 0282

For cables according to the harmonized specifications DIN VDE 0281 and 0282, a new type designation code is used; it is split into three parts: The first part identifies the standard and the rated voltage of the cable. The second part describes the constructional elements. The third part includes the number of cores and the nominal cross-sectional area and indicates whether or not a green/yellow core exists.

The letter "H" at the beginning of the type designation code indicates that the cables in all respects comply with the harmonized specifications.

The letter "A" at the beginning of the type designation code indicates that the cables basically comply with the harmonized specifications but are only approved in a specific country (recognized national types).

Example for a cable according to DIN VDE 0281:

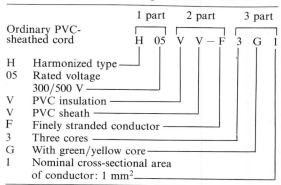

	1 part 2 part 3 part
Ordinary PVC-sheathed cord	H 05 V V – F 3 G 1
H	Harmonized type
05	Rated voltage 300/500 V
V	PVC insulation
V	PVC sheath
F	Finely stranded conductor
3	Three cores
G	With green/yellow core
1	Nominal cross-sectional area of conductor: 1 mm²

---

[1] In this chapter the term "cable" is used for "insulated wires and flexible cables" (so-called "Leitungen", see Introduction, Part 1)

1.1 Summary and Application

Design	Type designation code	Number of cores	Nominal cross-sectional area of conductors q_n mm²	Rated voltage U_0/U V	Highest permissible operating voltage $U_{b\,max}$				a.c. test voltage kV	Project planning data	
					~ L-E V	3 ~ L-L V	⎓ L-E V	⎓ L-L V		Table	Page

1.1.1 Building wires and cables

Design	Type designation code	Number of cores	Nominal cross-sectional area of conductors q_n mm²	Rated voltage U_0/U V	~ L-E V	3 ~ L-L V	⎓ L-E V	⎓ L-L V	a.c. test voltage kV	Table	Page
Light PVC-sheathed cables	NYM	1 2 3	1.5 – 16 1.5, 2.5 1.5 – 10	300/500	318	550	413	825	2	2.1.1 a	56
		4 5	1.5 – 35 1.5 – 25							2.1.1 b	56
		7 10	1.5, 2.5 1.5							2.1.1 c	58
Metal-clad PVC-sheathed cables	NHYRUZY	3 – 5 3, 5	1.5 2.5							2.1.2	59
Lead-covered PVC-sheathed cables	NYBUY	3, 4 4 5	1.5 2.5, 6 2.5, 4							2.1.3	60
SIFLA flat building wires	NYIF	2, 5 3, 4	1.5, 2.5 1.5 – 4	220/380	241	418	314	627		2.1.4	60

Permissible operating temperature ϑ_{Lr} °C	Permissible short-circuit temperature ϑ_e °C	Suitable for Class II equipment	Application	
			Location	Permissible stress and/or installation
70	160	yes	In dry, damp and wet locations, in operating areas and storerooms potentially subject to fire hazard, and outdoors, if not exposed to direct solar radiation, – but not in ground – (e.g. in dwellings, industrial and administrative buildings, in agricultural premises, dairies and laundries) According to DIN VDE 0165 also in potentially explosive areas. According to DIN VDE 0100 Part 520, single-core cables may also be used without protection against overcurrents for the connection of switchgear and distribution boards.	On, in and under plaster In masonry and in concrete, but not for direct bedding in heaped, vibrated and compressed concrete
70	160	yes	In rooms containing high-frequency equipment, but not in potentially explosive areas and not in bathrooms and shower rooms in dwellings and hotels	
70	160	yes	Mainly in chemical works, in heavy industry and in mining installations where high safety is required, also in potentially explosive areas according to DIN VDE 0165, but not in bathrooms and shower rooms in dwellings and hotels	
70	160	no	In dry locations, also in bathrooms and shower rooms in dwellings and hotels. Not permitted in wooden houses and buildings used for agricultural purposes, or in adjacent sections of buildings not separated from them by fire-proof walls	In and under plaster. Without plaster covering in voids of ceilings and walls consisting of non-flammable materials

1.1 Summary and Application

Design	Type designation code	Number of cores	Nominal cross-sectional area of conductors q_n mm^2	Rated voltage U_0/U V	Highest permissible operating voltage $U_{b\,max}$				a.c. test voltage kV	Project planning data	
					$\sim$ L-E V	$3\sim$ L-L V	$---$ L-E V	$---$ L-L V		Table	Page

1.1.2 Single-core cables for internal wiring, general and special purposes

Design	Type designation code	Number of cores	Nominal cross-sectional area of conductors q_n mm^2	Rated voltage U_0/U V	$\sim$ L-E V	$3\sim$ L-L V	$---$ L-E V	$---$ L-L V	a.c. test voltage kV	Table	Page
PVC-insulated single-core non-sheathed cables for internal wiring	H05V-U	1	0.5 – 1	300/500	318	550	413	825	2	2.2.1	62
	H05V-K									2.2.2	62
PVC-insulated single-core non-sheathed cables for general purposes	H07V-U		1.5 – 10	450/750	476	825	619	1238	2.5	2.2.3	63
	H07V-R		16 – 300							2.2.4	64
	H07V-K		1.5 – 150							2.2.5	64
Rubber-insulated single-core cables for special purposes	NSGAFÖU		1.5 – 240	1.8/3 kV	2.1 kV	3.6 kV	2.7 kV	5.4 kV	6	2.2.6	66

Permissible operating temperature ϑ_{Lr} °C	Permissible short-circuit temperature ϑ_e °C	Suitable for Class II equipment	Application		
			Location		Permissible stress and/or installation
70	160	no	In dry locations, in and on electronic equipment in power installations for measuring, controlling and regulating; for internal wiring of equipment, in and on luminaires		For protected fixed installation. For installation in conduits on, in and under plaster, but only for signalling systems, e. g. bell system
70	160	no	In dry locations, in and on electronic equipment in power installations for measuring, controlling and regulating. They may be used for internal wiring of equipment, switchgear and distribution boards as well as for protected installation in and on luminaires with a rated voltage up to 1 000 V a. c. or up to 750 d. c. to earth. For use in rail vehicles the d. c. operating voltage may be 900 V to earth.		In conduits on, in and under plaster as well as in cable trunking. Not for direct installation on ladders or trays. As equipotential bonding cables also directly on, in and under plaster as well as on ladders etc.
			In dry, damp and wet locations		Open installation on insulators over plaster beyond arm's reach
			In operating areas and storerooms potentially subject to fire hazard		In plastic conduits on, in and under plaster
			In potentially explosive areas according to DIN VDE 0165		In switchgear and distribution boards
90	200	Not applicable	In dry locations; in rail vehicles and trolley buses according to DIN VDE 0115. According to DIN VDE 0100 Part 520, these cables may also be used without protection against overcurrents for the connection of switchgear and distribution boards.		In equipment as well as in conduits on, in and under plaster and in cable trunking

1.1 Summary and Application

Design	Type designation code	Number of cores	Nominal cross-sectional area of conductors q_n mm²	Rated voltage U_0/U V	Highest permissible operating voltage $U_{b\,max}$ ~ L-E V	3~ L-L V	--- L-E V	--- L-L V	a.c. test voltage kV	Project planning data Table	Page

1.1.3 Heat-resistant cables and cords

Design	Type designation code	Number of cores	Nominal cross-sectional area of conductors q_n mm²	Rated voltage U_0/U V	~ L-E V	3~ L-L V	--- L-E V	--- L-L V	a.c. test voltage kV	Table	Page
Heat-resistant rubber-insulated single-core non-sheathed cables	N4GA	1	1.5, 2.5	450/750	476	825	619	1238	2.5	2.3.1	68
	N4GAF		0.75 – 35							2.3.2	68
Heat-resistant rubber-sheathed flexible cord	4GMH4G	3 4 5, 7 5	0.75, 1.5 1.5 0.75, 1.5 2.5	300/500	318	550	413	825	2	2.3.3	70
SINOTHERM heat-resistant silicone rubber-insulated single-core non-sheathed cables	SIA	1	1.5 – 10							2.3.4	70
	SIAF		0.75 – 120							2.3.5	72
			1.5 – 95	0.6/1 kV	0.7 kV	1.2 kV	0.9 kV	1.8 kV	2.5	2.3.6	72
	H05SJ-K		0.75 – 16	300/500	318	550	413	825	2	2.3.7	74
	A05SJ-K		25 – 95								
SINOTHERM heat-resistant silicone rubber-insulated and -sheathed cords	N2GMH2G	2 – 4 5	0.75 – 4 1.5							2.3.8	74

Permissible operating temperature ϑ_{Lr} °C	Permissible short-circuit temperature ϑ_e °C	Suitable for Class II equipment	Application	
			Location	Permissible stress and/or installation
120	200	no	At ambient temperatures above 55 °C In dry locations for internal wiring of luminaires, heating appliances, electric machines, switchgear and distribution boards. Cables with nominal cross-sectional areas ≥ 1.5 mm^2 may be used for internal wiring of appliances with a rated voltage up to 1000 V a.c. or up to 750 V d.c. to earth.	For protected fixed installation. Not for direct installation on ladders or trays. Cables with nominal cross-sectional areas ≥ 1.5 mm^2 may also be installed in conduits on, in and under plaster as well as in cable ducting.
120	200	yes	In dry, damp and wet locations as well as outdoors	Used as flexible connecting cable for cooking and heating appliances with medium mechanical stresses, e.g. cookers, electric storage heaters
180	350	no	In dry locations for internal wiring of high-intensity luminaires, spotlights, electric motors, heating appliances of all kinds, in chemical and ceramic industry, in nuclear power plants, foundries, in iron and steel works etc.	For protected fixed installation. Not for direct installation on ladders or trays
180	350	no		In switchgear and distribution boards with difficult installation conditions and increased mechanical stresses. Cables with nominal cross-sectional areas ≥ 1.5 mm^2 may also be used in conduits on, in and under plaster. It shall be ensured, however, that the conduit system is open at the ends and ventilated.
			In operating areas and storerooms potentially subject to fire hazard	In plastic conduits on, in and under plaster
			In potentially explosive areas according to DIN VDE 0165	In switchgear and distribution boards
180	350	yes	In dry, damp and wet locations as well as outdoors	Used as flexible connecting cables for cooking and heating appliances with low mechanical stresses, e.g. cookers, melting furnaces, glass extruders etc. Also for fixed installation, but only in open and ventilated conduit systems or ducts

1.1 Summary and Application

Design	Type designation code	Number of cores	Nominal cross-sectional area of conductors q_n mm²	Rated voltage U_0/U V	Highest permissible operating voltage $U_{b\,max}$				a.c. test voltage kV	Project planning data	
					~ L-E V	3~ L-L V	--- L-E V	--- L-L V		Table	Page

1.1.4 Cables for measuring, controlling and regulating

Design	Type designation code	Number of cores	Nominal cross-sectional area of conductors q_n mm²	Rated voltage U_0/U V	~ L-E V	3~ L-L V	--- L-E V	--- L-L V	a.c. test voltage kV	Table	Page
PROTOFLEX PVC control cables	NYSLYÖ	3 – 60	0.75	300/500	318	550	413	825	2	2.4.1 a	76
			1							2.4.1 b	76
			1.5							2.4.1 c	78
		3 – 18	2.5							2.4.1 d	78
	(N)YSLYÖ	2	1, 1.5							2.4.2	80
	YSLYÖ	4, 5, 7	4, 6							2.4.3	80
Screened PROTOFLEX PVC control cables	NYSLYCYÖ	3 – 25	0.75, 1							2.4.4 a	82
			1.5, 2.5							2.4.4 b	82
	YSLYCYÖ	4, 5, 7	4, 6							2.4.5	84
	YSLYCY	3, 4 3, 5 5	0.75 1.5, 2.5 4, 6							2.4.6	84
PVC control cables	SYSL	3 – 60	0.5							2.4.7 a	86
			0.75							2.4.7 b	86
		2 – 60	1							2.4.7 c	88
			1.5							2.4.7 d	88
		3 – 18 4, 5, 7	2.5 4, 6							2.4.7 e	90

Permissible operating temperature ϑ_{Lr} °C	Permissible short-circuit temperature ϑ_e °C	Suitable for Class II equipment	Application	
			Location	Permissible stress and/or installation
70	150	yes	In dry, damp and wet locations but not outdoors	Used as connecting cables subject to medium mechanical stresses in and on processing machines, conveyor systems, production lines etc. For fixed installation and flexible application with free movement without tensile stress and without forced guiding over rollers. Not for operational reeling and dereeling
70	150	yes		For installations where interference suppression is required
70	150	yes		For fixed installation and temporary but non-periodical bendings, for medium mechanical stresses, as connecting cables in and on processing machines, conveyor systems, production lines etc.

1.1 Summary and Application

| Design | Type designation code | Number of cores | Nominal cross-sectional area of conductors q_n mm² | Rated voltage U_0/U V | Highest permissible operating voltage $U_{b\,max}$ | | | | a.c. test voltage kV | Project planning data | |
					~ L-E V	3~ L-L V	$=$ L-E V	$=$ L-L V		Table	Page
1.1.5 Flexible PVC-sheathed cords											
Flat non-sheathed cord	H03VH-H	2	0.75	300/300	330	330	495	495	2	2.5.1	92
Light PVC-sheathed cords	H03VV-F	2								2.5.2 a	93
		3, 4								2.5.2 b	93
	H05VV-F	2	1.5	300/500	318	550	413	825		2.5.3 a	94
		3	0.75 − 2.5							2.5.3 b	94
		4, 5	1.5, 2.5								
1.1.6 Flexible rubber-sheathed cords and cables											
Ordinary rubber-sheathed cords	H05RR-F	2	0.75 − 2.5	300/500	318	550	413	825	2	2.6.1 a	96
		3, 4	0.75 − 2.5							2.6.1 b	96
		5	1.5, 2.5								
THERMOSTABIL ordinary tough rubber-sheathed cord	H05RR-F	3	0.75							2.6.1 b	96
Ordinary rubber-sheathed cords	H05RN-F	2	0.75, 1							2.6.2 a	98
		3								2.6.2 b	98
	A05RN-F	4	0.75							2.6.3	99
Highly flexible rubber-sheathed cord	H05RN-F	3	0.75							2.6.4	99

Permissible operating temperature ϑ_{Lr} °C	Permissible short-circuit temperature ϑ_e °C	Suitable for Class II equipment	Application	
			Location	Permissible stress and/or installation
70	150	yes	In dry locations, e.g. in households, kitchens and offices, but not outdoors and not in industrial and agricultural premises	For connecting light electrical equipment with very low mechanical stresses, e.g. radio sets, table lamps, clocks etc., if permitted in the relevant VDE specifications. This cord is not suitable for connecting cooking or heating appliances or electric tools.
70	150	yes		For connecting light electrical equipment with low mechanical stresses, e.g. table lamps, floor lamps, kitchen machines, vacuum cleaners, office machines, radio sets etc., if permitted in the relevant VDE specifications. These cords are not suitable for connecting cooking or heating appliances if they can get in contact with hot parts and are exposed to other heat effects. They are not suitable for connecting industrial electric tools either.
70	150	yes	However, permitted in tailors' shops and similar premises. For domestic appliances also in damp and wet locations	For connecting electrical appliances with medium mechanical stresses (e.g. washing machines, spin-driers and refrigerators) if permitted in the relevant VDE specifications. These cords are not suitable for connecting cooking or heating appliances if they can get in contact with hot parts and are exposed to other heat effects. They are not suitable for connecting industrial electric tools either. These cords may be used for fixed installation (e.g. in furniture and panels).
60	200	yes	In dry locations, e.g. in households, kitchens and offices, but not outdoors and not in industrial or agricultural premises. Permitted in tailors' shops and similar premises	For low mechanical stresses. The cords may be installed permanently (e.g. in furniture and panels and hollow spaces of pre-fabricated building sections). For connecting electrical appliances (e.g. vacuum cleaners, kitchen appliances, cookers, soldering irons)
60	200	yes		For connecting electric heating appliances where the cords are exposed to special mechanical and thermal stresses (e.g. irons)
60	200	yes	In dry, damp and wet locations as well as outdoors, but not in industrial and agricultural premises. Permitted in tailors' shops and similar premises and potentially explosive areas according to DIN VDE 0165	For connecting garden equipment (e.g. lawn-mowers). The cords may get in contact with fats and oils (e.g. deep fat friers)
60	200	yes		For connecting electrical hand-held equipment with stresses where cords are subject to stresses by kinking and twisting (e.g. electric tools, such as power drills, hand grinders, circular saws, do-it-yourself equipment)

1.1 Summary and Application

1.1.6 Flexible rubber-sheathed cords and cables (continued)

Design	Type designation code	Number of cores	Nominal cross-sectional area of conductors q_n mm²	Rated voltage U_0/U V	Highest permissible operating voltage $U_{b\,max}$ ~ L-E V	3 ~ L-L V	$=$ L-E V	$=$ L-L V	a.c. test voltage kV	Project planning data Table	Page
OZOFLEX rubber-sheathed flexible cables	H07RN-F	1 2	10 – 300 1.5 – 4	450/750	476	825	619	1238	2.5	2.6.5 a	100
		3	1 – 16							2.6.5 b	100
		4	1.5 – 120							2.6.5 c	102
		5	1 – 25							2.6.5 d	102
	A07RN-F	3 4	1.5 – 95 10 – 35							2.6.6 a	104
		7 8 – 24	1.5, 2.5 2.5							2.6.6 b	104
Highly flexible OZOFLEX rubber-sheathed cords	H07RN-F	3 – 5	1, 1.5							2.6.7	106
Heavy duty PROTOMONT rubber-sheathed flexible cables	NSSHÖU	1	16 – 150	0.6/1 kV	0.7 kV	1.2 kV	0.9 kV	1.8 kV	3	2.6.8 a	108
		2 3 4	1.5 1.5 – 6 1.5 – 50							2.6.8 b	108
		3	70/35, 95/50							2.6.8 c	110
		5 7, 12, 18	1.5 – 25 2.5							2.6.8 d	110

Permissible operating temperature ϑ_{Lr} °C	Permissible short-circuit temperature ϑ_e °C	Suitable for Class II equipment	Application	
			Location	Permissible stress and/or installation
60	200[1]	yes	In dry, damp and wet locations as well as outdoors. In agricultural premises and operating areas potentially subject to fire hazard. In potentially explosive areas according to DIN VDE 0165, on sites and in industrial premises For application in open-cast mines and quarries see DIN VDE 0168	For connecting electrical appliances with medium mechanical stresses, e.g. portable motors or machines, electric tools, large water heaters, hot-plates, hand lamps and traction motors. The cables may be used for fixed installation, e.g. on plaster, in temporary buildings and huts and for direct installation on components of hoisting gear, machines etc.. For protected fixed installation in conduits or appliances and as connecting cable to rotors of electrical machines, the cables may be used with a rated voltage up to 1 000 V a.c. or up to 750 V d.c. to earth. For use in rail vehicles the d.c. operating voltage may be 900 V to earth.
60	200	yes		For connecting electrical hand-held appliances where cords are subject to stresses by kinking and twisting, e.g. electric tools, such as power drills, hand grinders, circular saws, do-it-yourself equipment
90	200	yes	In open-cast mines, in quarries as well as in underground mines	For connecting heavy electrical equipment with very high mechanical stresses. These cables may also be used for fixed installation, e.g. on plaster. For forced guiding on festoon and trolley systems, only single-core cables are suitable.

[1] For untinned conductor 250 °C

1.1 Summary and Application

Design	Type designation code	Number of cores	Nominal cross-sectional area of conductors q_n mm²	Rated voltage U_0/U V	Highest permissible operating voltage $U_{b\,max}$				a.c. test voltage kV	Project planning data	
					~ L-E V	3~ L-L V	⚌ L-E V	⚌ L-L V		Table	Page

1.1.6 Flexible rubber-sheathed cords and cables (continued)

Design	Type designation code	Number of cores	Nominal cross-sectional area	Rated voltage	~ L-E	3~ L-L	⚌ L-E	⚌ L-L	a.c. test voltage	Table	Page
HYDROFIRM(T) rubber-sheathed flexible cables	TGKT	4	1.5 – 16	450/750	476	825	619	1238	2.5	2.6.9	112
HYDROFIRM rubber-sheathed flexible cables	TGK	3	1 – 70							2.6.10 a	114
		4 5	1.5 – 70 1.5, 2.5							2.6.10 b	114
	TGW	1	1.5 – 70							2.6.11	116
	TGFLW	3								2.6.12 a	118
		4								2.6.12 b	118

1.1.7 Welding cables

Design	Type designation code	Number of cores	Nominal cross-sectional area	Rated voltage	~ L-E	3~ L-L	⚌ L-E	⚌ L-L	a.c. test voltage	Table	Page
ARCOFLEX welding cables	NSLFFÖU	1	16 – 120	–	–	–	–	–	1	2.7.1	120
FLEXIPREN welding cables	NSLFFÖU		25 – 70							2.7.2	120

Permissible operating temperature ϑ_{Lr} °C	Permissible short-circuit temperature ϑ_e °C	Suitable for Class II equipment	Application	
			Location	Permissible stress and/or installation
90	250	yes	In water (also in salt water and brackish water), in water at depths up to 500 m and in dry, damp and wet locations as well as outdoors, but not in potentially explosive areas. On sites, in agricultural and industrial premises	For connecting electrical appliances in water subject to medium mechanical stresses, e.g. submersible pump-motors. For protected fixed installation in conduits or equipment, e.g. in well installations, and as connecting cable to rotors of electrical machines, the cables may be applied with a rated voltage up to 1000 V a.c. or up to 750 V d.c. to earth. Permissible a.c. test voltage in connection with motor tests 3 kV (maximum test period 3 min)
			In drinking water	For water temperatures up to 40 °C
90	250	yes	In service water	
90	250	yes		For water temperatures up to 60 °C
80	–	Not applicable	In dry, damp and wet locations as well as outdoors and in industrial premises	For connecting electrodes for arc welding with high mechanical stresses and no-load voltages approximately up to 100 V
				For machine and hand-welding operation
80	–			For hand-welding operation

1.1 Summary and Application

| Design | Type designation code | Number of cores | Nominal cross-sectional area of conductors q_n mm² | Rated voltage U_0/U kV | Highest permissible operating voltage $U_{b\,max}$ | | | | a.c. test voltage kV | Project planning data | |
					~ L-E kV	3~ L-L kV	$===$ L-E kV	$===$ L-L kV		Table	Page

1.1.8 Flexible cables for forced guiding and trailing

Design	Type designation code	Number of cores	Nominal cross-sectional area mm²	Rated voltage U_0/U kV	~ L-E kV	3~ L-L kV	$===$ L-E kV	$===$ L-L kV	a.c. test voltage kV	Table	Page
Rubber-sheathed flexible cables											
CORDAFLEX	NSHTÖU	5, 7, 12, 24	1.5	0.6/1	0.7	1.2	0.9	1.8	2.5	2.8.1	122
		7, 12, 24	2.5								
		4	2.5 – 50								
CORDAFLEX (K)	NSHTÖU	5, 7, 12, 24	1.5							2.8.2 a	124
		4, 7, 12, 18, 24, 30	2.5								
		19+5, 25+5	2.5+1(C)							2.8.2 b	124
		4	4 – 120							2.8.2 c	126
CORDAFLEX (SM)	NSHTÖU	46	1							2.8.3	128
		12, 24	1.5								
		7, 12, 24, 30	2.5								
		4	10 – 50								
SPREADERFLEX basket cables	YSLTÖ	48	1	300/500 V	318 V	550 V	413 V	825 V	2	2.8.4	128
		30, 36, 42	2.5								
PLANOFLEX rubber-sheathed flat flexible cables	NGFLGÖu	4, 7, 8, 10, 12, 24	1.5							2.8.5 a	130
		4, 7, 8, 12, 24	2.5								
		4	4 – 95							2.8.5 b	130
		7	4 – 35							2.8.5 c	132
PVC lift control cables	YSLTK	7, 12, 18, 24, 30	1							2.8.6	132
	YSLYTK	28+2	1+0.5 FM (C)							2.8.7	133
	YSLYCYTK									2.8.8	134

Permissible operating temperature ϑ_{Lr} °C	Permissible short-circuit temperature ϑ_e °C	Suitable for Class II equipment	Application	
			Location	Permissible stress and/or installation
90	200	yes	In dry, damp and wet locations as well as outdoors. In areas subject to fire hazard, on sites and in industrial premises. In potentially explosive areas according to DIN VDE 0165. In open-cast mines and in quarries	For connecting hoisting gear, transportation and conveyor systems for forced guiding with high mechanical stresses, e.g. operational reeling and dereeling On cable reels without guide rollers for apparatus with travel speeds up to 60 m/min
90	200	yes		On cable reels with and without guide rollers for apparatus with travel speeds up to 120 m/min
90	200	yes		On cable reels with and without guide rollers and on trolley systems, for apparatus with travel speeds up to 120 m/min and with very high dynamic stresses, e.g. electro-hydraulic grab cranes, crane lifting magnets
70	150	yes	In dry, damp and wet locations as well as outdoors	For cables which are suspended vertically and are laid into a spreader basket with travel speeds up to 120 m/min; e.g. for connecting and controlling supporting gear of containers (spreader)
90	250	yes	In potentially explosive areas according to DIN VDE 0165	For connecting hoisting gear, transportation and conveyor systems with medium mechanical stresses and operational bendings only in one plane, e.g. for operation on festoon systems on cable tender systems etc.
70	150	yes	In dry, damp and wet locations	For operation where the cables are suspended vertically and moved up and down during operation; e.g. for lifts, hoisting gear, transportation and conveyor systems. The cables are not suitable for forced guiding or guide rollers, e.g. on reels. Greatest suspension length 50 m
70	150	yes		Greatest suspension length 150 m
70	150	yes		For installations where interference suppression is required according to DIN VDE 0875

1.1 Summary and Application

Design	Type designation code	Number of cores	Nominal cross-sectional area of conductors q_n mm²	Rated voltage U_0/U kV	Highest permissible operating voltage $U_{b\,max}$				a.c. test voltage kV	Project planning data	
					~ L-E kV	3~ L-L kV	═ L-E kV	═ L-L kV		Table	Page

1.1.8 Flexible cables for forced guiding and trailing (continued)

Reeling and trailing cables

Design	Type designation code	Number of cores	Nominal cross-sectional area of conductors q_n mm²	Rated voltage U_0/U kV	~ L-E kV	3~ L-L kV	═ L-E kV	═ L-L kV	a.c. test voltage kV	Table	Page
PROTOLON	NTSWÖU	3+3	120+70/3 150+70/3 185+95/3	0.6/1	0.7	1.2	0.9	1.8	4	2.8.9	135
	NTSCGEWÖU		25+25/3 35+25/3 50+25/3 70+35/3 95+50/3 120+70/3	3.6/6, 6/10	4.2, 6.9	7.2, 12	5.4, 9	10.8, 18	11, 17	2.8.10, 2.8.11	136, 138
PROTOLON (SM)	NTSCGEWÖU		25+25/3 35+25/3 50+25/3 70+35/3							2.8.12, 2.8.13	140, 141
PROTOLON (SB)	NTSCGEWÖU		35+50/3 50+50/3 70+50/3 95+50/3	6/10, 12/20	6.9, 13.9	12, 24	9, 18	18, 36	17, 29	2.8.14, 2.8.15	142, 143
PROTOLON (SF)	NTSCGEWÖU		35+25/3 50+25/3 70+35/3 95+50/3 120+70/3							2.8.16, 2.8.17	144, 145
PROTOLON (ST)	NTSCGEWÖU		35+25/3 50+25/3 70+35/3 95+50/3	6/10	6.9	12	9	18	17	2.8.18	146
			25+25/3 35+25/3 50+25/3 70+35/3	12/20	13.9	24	18	36	29	2.8.19	147
PROTOLON	NTMCGCWÖU	1	25, 50, 95							2.8.20	148
	NTSCGERLWÖU	3+3	25+16/3E 35+16/3E 50+25/3E 70+35/3E 95+50/3E	6/10, 12/20	6.9, 13.9	12, 24	9, 18	18, 36	17, 29	2.8.21, 2.8.22	149, 150

Permissible operating temperature ϑ_{Lr} °C	Permissible short-circuit temperature ϑ_e °C	Suitable for Class II equipment	Application	
			Location	Permissible stress and/or installation
90	200	yes	In dry, damp and wet locations as well as outdoors. In areas potentially subject to fire hazard, on sites and in industrial premises. In potentially explosive areas according to DIN VDE 0165. In open-cast mines, quarries and underground mines according to DIN VDE 0118 and on sites below ground (tunnelling)	For connecting hoisting gear, transportation and conveyor systems with very high mechanical stresses For forced guiding, e.g. operational reeling and dereeling On cable reels with and without guide rollers for apparatus with travel speeds up to 60 m/min
		Not applicable		
90	200			On cable reels with multiple guide rollers and very high dynamic stresses for apparatus with travel speeds up to 120 m/min
90	200			For trailing operation where cables are stressed by grinding and chafing
90	200			For fixed installation also suitable for occasional moving, e.g. transportation purposes
90	200		In water up to 40 °C, even under the influence of oil	For application with free movement and fixed installation, e.g. for dredgers and floating docks
90	200			For connecting portable transformer load-centres, for interconnecting cells etc. with high mechanical stresses
90	200			For very high mechanical stresses for interconnecting distribution cables, for site supply etc.

27

1.1 Summary and Application

Design	Type designation code	Number of cores	Nominal cross-sectional area of conductors q_n mm²	Rated voltage U_0/U kV	Highest permissible operating voltage $U_{b\,max}$				a.c. test voltage kV	Project planning data	
					$\sim$ L-E kV	$3\sim$ L-L kV	$\overline{}$ L-E kV	$\overline{}$ L-L kV		Table	Page

1.1.9 Mining cables

Design	Type designation code	Number of cores	Nominal cross-sectional area of conductors q_n mm²	Rated voltage U_0/U kV	$\sim$ L-E kV	$3\sim$ L-L kV	$\overline{}$ L-E kV	$\overline{}$ L-L kV	a.c. test voltage kV	Table	Page
Heavy duty PROTOMONT rubber-sheathed flexible cables	NSSHÖU 3×..+3×../3E	3+3	2.5+2.5 6+ 6 10+10 16+16 25+16 35+16 50+25 70+35 95+50 120+70	0.6/1	0.7	1.2	0.9	1.8	3	2.9.1 a	152
	NSSHÖU 3×..+3×../3E +3×..ST	3+3+3	2.5+2.5+1.5 6+ 6 +1.5 10+10 +2.5 16+16 +2.5 25+16 +2.5 35+16 +2.5 50+25 +2.5 70+35 +2.5 95+50 +2.5							2.9.1 b	154
	NSSHÖU ..×../..KON	3, 5 5 5	2.5/2.5 4/4 6/6							2.9.1 c	156
	NSSHCGEÖU... 3×..+3×.. (1.5 ST KON+.. 3/KON) SM	3+3	35+16 50+25 70+35 95+50							2.9.2	157
	NSSHCGEÖU... 3×../..KON +3×(1.5 ST KON/ 1.5 ÜL KON) V		25/16 35/16 50/25 70/35 95/50							2.9.3	158
	NSSHCGEÖU... 3×../..KON +3×(1.5 ST KON/ 1.5 ÜL KON) Z									2.9.4	159

Permissible operating temperature ϑ_{Lr} °C	Permissible short-circuit temperature ϑ_e °C	Suitable for Class II equipment	Location	Application	
					Permissible stress and/or installation
90	200	yes	In underground mines at any locations according to the relevant clauses of BVOE[1] and DIN VDE 0118, in open-cast mines and in quarries as well as on sites above and below ground (tunnelling). In dry, damp and wet locations as well as outdoors. Furthermore in agricultural premises and areas potentially subject to fire hazard and in industrial premises. In potentially explosive areas according to DIN VDE 0165	For connecting electric machines, apparatus and distribution boards with very high mechanical stresses, also in combination with protection and monitoring equipment, e. g. N-type monitors	
90	200	yes		For forced guiding for mobile equipment, e. g. coal cutters	
				On cable reels, in cable trailers and power track systems	
90	200	yes		In cable trailers and power track systems	
90	200	yes		On cable reels and for trailing operation	

German mining rules
BVOE = Bergverordnung des Landesoberbergamtes Nordrhein-Westfalen für elektrische Anlagen

1.1 Summary and Application

Design	Type designation code	Number of cores	Nominal cross-sectional area of conductors q_n mm²	Rated voltage U_0/U kV	Highest permissible operating voltage $U_{b\,max}$				a.c. test voltage kV	Project planning data	
					~ L-E kV	3 ~ L-L kV	$=$ L-E kV	$=$ L-L kV		Table	Page

1.1.9 Mining cables (continued)

Design	Type designation code	Number of cores	Nominal cross-sectional area of conductors q_n mm²	Rated voltage U_0/U kV	~ L-E kV	3 ~ L-L kV	$=$ L-E kV	$=$ L-L kV	a.c. test voltage kV	Table	Page
Heavy duty PROTOMONT rubber-sheathed flexible cables	NTMTWÖU .. × 2.5 ST + .. × 1 FM(C)	8+2 14+6	2.5+1	0.6/1	0.7	1.2	0.9	1.8	3	2.9.5	160
Screened SUPROMONT PVC cables	NYHSSYCY 3 × .. + 3 × ../3E + 3 × 2.5ST + ÜL	3+3	25+16 35+16 50+25 70+35 95+50	3.6/6	4.2	7.2	5.4	10.8	11	2.9.6	161

Permissible operating temperature ϑ_{Lr} °C	Permissible short-circuit temperature ϑ_e °C	Suitable for Class II equipment	Application	
			Location	Permissible stress and/or installation
90	200	yes	In underground mines at any locations according to the relevant clauses of BVOE[1] and DIN VDE 0118, in open-cast mines and in quarries as well as on sites above and below ground (tunnelling). In dry, damp and wet locations as well as outdoors. Furthermore in agricultural premises and areas potentially subject to fire hazard and in industrial premises. In potentially explosive areas according to DIN VDE 0165	For cables which are suspended vertically and moved up and down during operation, e.g. for intrinsically safe cage-controlled winches in blind shafts and similar systems with telephone communication to the cage. Greatest suspension length 200 m with a 5fold safety factor
70	150	Not applicable	In underground mines and on sites below ground (tunnelling). According to the relevant clauses of BVOE[1] and DIN VDE 0118, these cables must not be used in mining operation.	For connecting permanently installed and mobile high voltage equipment, e.g. pressure-resistant transformers, also in connection with protection and monitoring equipment, e.g. H-type monitors

German mining rules

BVOE = Bergverordnung des Landesoberbergamtes Nordrhein-Westfalen für elektrische Anlagen

1.1 Summary and Application

Design	Type designation code	Number of cores	Nominal cross-sectional area of conductors q_n mm²	Rated voltage U_0/U kV	Highest permissible operating voltage $U_{b\,max}$				a.c. test voltage kV	Project planning data	
					~ L-E kV	3~ L-L kV	--- L-E kV	--- L-L kV		Table	Page

1.1.10 Halogen-free cables with improved performance in the case of fire

Design	Type designation code	Number of cores	Nominal cross-sectional area of conductors q_n mm²	Rated voltage U_0/U kV	~ L-E kV	3~ L-L kV	--- L-E kV	--- L-L kV	a.c. test voltage kV	Table	Page
Light-sheathed SIENOPYR cables	NHXMH	1 2 3	1.5 – 16 1.5 1.5 – 10	300/500 V	318 V	550 V	413 V	825 V	2	2.10.1 a	162
		4 5 7	1.5 – 35 1.5 – 16 1.5, 2.5							2.10.1 b	162
SIENOPYR rubber-insulated single-core cables for special purposes	(N)HXSGAFHXÖ	1	4 – 240	1.8/3	2.1	3.6	2.7	5.4	6	2.10.2	164
SIENOPYR heat-resistant single-core non-sheathed cables	(N)HX4GA		0.75 – 4	450/750 V	476 V	825 V	619 V	1238 V	2.5	2.10.3	164
	(N)HX4GAF		0.75 – 95							2.10.4	166
SIENOPYR (X) rubber-sheathed flexible cables	(N)HXSHXÖ	3	1.5 – 50	0.6/1	0.7	1.2	0.9	1.8	2.5	2.10.5 a	166
		4 5	1.5 – 50 1.5 – 25							2.10.5 b	168
		7 8,16, 20,36 33	1.5 – 6 1.5 2.5							2.10.5 c	168
		4 + 2, 7 + 2 4 + 2, 7 + 2 4 + 2	2.5 + 1.5 4 + 1.5 6 + 1.5							2.10.5 d	170

Permissible operating temperature ϑ_{Lr} °C	Permissible short-circuit temperature ϑ_e °C	Suitable for Class II equipment	Application	
			Location	Permissible stress and/or installation
70	250	yes	In buildings, installations and apparatus with high concentration of people and valuable contents for preventive fire protection, e.g. in hospitals, hotels, department stores, electronic data processing equipment etc. In dry, damp and wet locations as well as outdoors if protected against direct solar radiation – but not in ground. In potentially explosive areas according to DIN VDE 0165, in areas and storerooms potentially subject to fire hazard and in agricultural premises. According to DIN VDE 0100 Part 520, single-core cables may also be used without protection against overcurrents for the connection of switchgear and distribution boards.	On, in and under plaster, in masonry and in concrete, except for direct bedding in heaped, vibrated and compressed concrete
90	200	Not applicable	e.g. in means of transport according to DIN VDE 0115, such as undergrounds, buses etc. In dry locations. According to DIN VDE 0100 Part 520, these cables may also be used without protection against overcurrents for the connection of switchgear and distribution boards.	In apparatus and conduits on, in and under plaster and in cable ducting
120	200	no	e.g. in switchgear and distribution boards, equipment and luminaires, in this case especially for through-wiring of ceiling luminaires and, according to DIN VDE 0115, in means of transport. Furthermore for internal wiring of apparatus with a nominal voltage up to 1000 V a.c. or up to 750 V d.c. to earth. For use in rail vehicles the d.c. operating voltage may be 900 V to earth. In dry locations	For protected fixed installation. Not for direct installation on ladders or trays. Cables with nominal cross-sectional areas ≥ 1.5 mm^2 may also be laid in conduits in, on and under plaster and in cable ducting.
90	200	yes	e.g. in hospitals, theatres, computer centres and nuclear power plants. In dry, damp and wet locations as well as outdoors	For connecting electric machines and appliances with medium mechanical stresses. The cables are also suitable and type-tested for the special conditions of application in the containment of nuclear power plants. Also suitable for fixed installation.

1.2 Determination of the Cross-Sectional Area of Cables

1.2.1 General

In DIN VDE 0298 Part 4 of February 1988 the current-carrying capacities of cables for fixed installation in buildings and of flexible cables were summarized in one specification for the first time. The previous classification of the installation methods was substituted by the installation methods internationally used:

Installation method A: Installation in thermally insulated walls

Installation method B: Installation in conduits or trunkings (on or in the wall or under plaster)

Installation method C: Direct installation (on or in the wall or under plaster)

Installation method E: Installation in free air.

The current-carrying capacities of cables recommended in DIN VDE 0298 Part 4 replace the data specified in DIN VDE 0100 Part 523.

In order to achieve a safe project planning of cable installations the cross-sectional area of conductor shall be selected such that the requirement

current-carrying capacity $I_z \geq$ loading I_b

is fulfilled for all operating conditions which can occur. Thus it is ensured that the permissible operating and short-circuit temperatures according to Table 1.2.16 are not exceeded.

The current-carrying capacity shall be calculated for

▷ normal operation (Clause 1.2.2) and for

▷ short-circuit (operation under fault conditions) (Clause 1.2.3).

Especially in low-voltage systems, the cross-sectional area of conductor shall be additionally determined in respect of the permitted voltage drop ΔU (Clause 1.2.4) and in order to avoid thermal overloading of the cable a suitable protective device is to be selected (Clause 1.2.5). Besides that the respective installation rules shall be observed.

Furthermore it shall be checked according to DIN VDE 0100 Supplement 5 that the maximum cable length is not exceeded with regard to the

▷ protection against indirect contact,
▷ protection against short-circuit,
▷ permitted voltage drop.

For the selection of a cable, the details according to Table 1.2.1 (planning aid) are necessary.

For the determination of the cross-sectional area of cables, the terms and definitions specified in DIN VDE 0298 Part 4 shall basically be used. These details and further ones shall be analogously taken from Part 1, Clauses 18.1 and 18.2, where they are summarized for cables.

The current-carrying capacities specified in DIN VDE 0298 Part 4 are rated values. The current-carrying capacities given in Clause 2 are based on these values. Clause 2 also includes values for cable designs which are not given in DIN VDE 0298 Part 4 as well as values for designs for which no VDE specifications exist.

Table 1.2.1   Planning aid: Necessary data for cable installations

1 Cable design	1.1 Type designation 1.2 Insulation material (e.g. NR/SR, PVC, EPR) 1.3 Number of cores (single-core, multi-core) 1.4 Nominal cross-sectional area of conductor q_n
2 Voltage	2.1 Rated voltage of the system U_n 2.2 Maximum operating voltage U_{bmax} 2.3 System frequency f 2.4 Type of current (three-phase, alternating or direct current)

3 Earthing,
   PE conductor
	3.1 Protection against electric shock (DIN VDE 0100 Part 410, HD 384.4.41 S1)
	3.2 Consideration of
	        earthing,
	        PE conductor,
	        equipotential bonding conductor
	        (DIN VDE 0100 Part 540)

4 Current-carrying capacity
   for normal operation,
   operating conditions
	4.1 Type of operation (continuous, short-time or intermittent operation)
	4.2 Installation conditions
	        Installation
	        – in thermally insulated walls (A),
	        – on or in walls,
	          in conduits or trunking (B),
	        – directly in the wall or under plaster,
	          on a wall (C),
	        – in free air (E)
	4.3 Ambient conditions
	        Ambient temperature ϑ_u

5 Current-carrying capacity
   for short-circuit
   (thermal stress)
	5.1 Rating with values taken from the network calculation
	        Initial symmetrical short-circuit current I_k''
	        Steady-state short-circuit current I_k
	        Short-circuit duration t_k

6 Voltage drop	6.1 System frequency
	        Transmitted power S or operating current I_b
	        Power factor of load $\cos \varphi$
	        Length of cable run l
	        Type of current (three-phase, alternating or direct current)
	        Permitted voltage drop ΔU or Δu

7 Protection against overcurrents	7.1 Allocation of protective devices

8 Maximum cable length	8.1 Maximum permissible length of cables with regard to
	        the protection against indirect contact,
	        the protection against short-circuit,
	        the permitted voltage drop

1.2.2 Determination of the Cross-Sectional Area for Normal Operation

Cables for Fixed Installation

The calculation method for the determination of the cross-sectional area of cables for fixed installation is given in Table 1.2.2 for installation methods A, B, C and E. The "reference operating conditions", given in this table, in most cases represent the basic project planning data of cables for fixed installation in buildings. The rated current-carrying capacities I_r for these "reference operating conditions" are completely given for a certain installation method in Tables 1.2.3 and 1.2.4 as well as in the tables of Clause 2.

For "site operating conditions" the rated current-carrying capacity I_r shall be multiplied by the product of the respective rating factors f given in Tables 1.2.10 to 1.2.13:

$$I_z = I_r \cdot \Pi f.$$

Instructions for this procedure shall be taken from the right column of Table 1.2.2.

Example for Installation on the Wall

50 PVC-insulated single-core non-sheathed cables for general purposes H07V-U with a conductor cross-sectional area of 6 mm² are installed in a trunking on the wall in a closed operating area. These cables form ten three-phase circuits together with the relevant PE and neutral conductors; the ambient temperature is 40 °C. The current-carrying capacity is calculated as follows:

Rated current I_r for installation method B1 (Table 1.2.3):	36 A
Rating factor for	
– ambient temperature 40 °C (Table 1.2.10):	0.87
– grouping (Table 1.2.11):	0.48
Total rating factor Πf:	0.42
Current carrying capacity I_z:	15.1 A

Example for Installation under the Ceiling

Two light PVC-sheathed cables NYM-J 3 × 2.5 (2 loaded cores) and two light PVC-sheathed cables NYM-J 4 × 1.5 (3 loaded cores) are installed under the ceiling of a building.

They are arranged side by side in flat formation. The ambient temperature is up to 35 °C. The current-carrying capacity is calculated as follows:

Cross-sectional area of conductor:	2.5 mm²	1.5 mm²
Rated current I_r for installation method C (Table 2.1.1 a and b, Item 39):	26 A	17.5 A
Rating factor for		
– ambient temperature 35 °C (Table 1.2.10):	0.94	0.94
– grouping (Table 1.2.11):	0.68	0.68
Total rating factor Πf:	0.64	0.64
Current-carrying capacity I_z	16.6 A	11.2 A

Example for Installation on Cable Ladders

A three-phase power of 60 kVA shall be transmitted with not more than six lead-covered PVC cables NY-BUY at a nominal voltage of $U_n = 380$ V. The cables shall be laid side by side on a cable ladder. They are in a large room having an ambient temperature of 45 °C. Which cross-sectional area of conductor shall be selected?

Load current:

$$I_b = 60000 \text{ VA}/(\sqrt{3} \cdot 380 \text{ V})$$
$$= 91.2 \text{ A}$$

Rating factor for	
– ambient temperature 45 °C (Table 1.2.10):	0.79
– grouping (Table 1.2.12):	0.79
Total rating factor Πf:	0.62

Thus the "fictitious load current" I_{bf} converted to the "reference operating conditions" is at least

$$I_{bf} = 91.2 \text{ A}/0.62 = 147.1 \text{ A.}$$

For six cables laid in parallel, a minimum current-carrying capacity of 24.5 A is obtained per cable. Table 1.2.4 shows that a cross-sectional area of 2.5 mm² is necessary for installation method E. The rated current for this is $I_r = 25$ A.

Therefore, six lead-covered PVC-sheathed cables NYBUY 4 × 2.5 are required.

Table 1.2.2 Operating conditions for cables for fixed installation

Reference operating conditions for the determination of rated currents I_r	Site operating conditions and calculation of the current-carrying capacity $I_z = I_r \, \Pi f$
Type of operation	
Continuous operation with the current-carrying capacities according to Tables 1.2.3 and 1.2.4 and Clause 2	Current-carrying capacity at intermittent operation according to Figure 1.2.1 and Part 1, Clause 18.6
Installation conditions	
Installation method A[1] Installation in thermally insulated walls according to Table 1.2.3 – single-core non-sheathed cables in conduit (reference installation method) – multi-core cable in conduit – multi-core cable in the wall	Rating factors for grouping according to Table 1.2.11, for multi-core cables according to Table 1.2.13
Installation method B1, B2[2] Installation in conduits or trunking according to Table 1.2.3 – single-core non-sheathed cables in conduit on the wall (reference installation method, B1) – single-core non-sheathed cables in trunking on the wall (B1) – single-core non-sheathed cables, single-core light-sheathed cables or multi-core cable in trunking in the wall or under plaster (B1) – multi-core cable in conduit on the wall or on the floor (B2) – multi-core cable in trunking on the wall or on the floor (B2)	Rating factors for grouping according to Table 1.2.11, for multi-core cables according to Table 1.2.13
Installation method C[3] Direct installation according to Table 1.2.3 – multi-core cable on the wall or on the floor (reference installation method) – single-core light-sheathed cables on the wall or on the floor – multi-core cable in the wall or under plaster – flat building wires under plaster	Rating factors for grouping according to Table 1.2.11, for multi-core cables according to Table 1.2.13
Installation method E Installation in free air according to Table 1.2.4, i.e. unhindered heat dissipation is ensured – for a cable spaced from the wall according to Table 1.2.4; – for cables in flat formation with a horizontal spacing of at least twice the cable diameter; – for cables in flat formation with a vertical spacing of at least twice the cable diameter	Rating factors for grouping according to Table 1.2.12, for multi-core cables according to Table 1.2.13
Ambient conditions (see Part 1, Clause 18.2.2)	
Ambient temperature 30 °C	Rating factors
Sufficiently large or ventilated rooms in which the ambient temperature is not noticeably increased by the heat losses from the cables	for deviating ambient temperatures according to Table 1.2.10
Protection from solar radiation etc.	see Part 1, Clause 18.4.2
System frequency 50 to 60 Hz	

[1] The thermally insulated wall consists of an outer weatherproof skin, a thermal insulation and an inner skin of wooden or wood-like material having a thermal resistivity of 0.1 $m^2 \cdot K/W$. The conduit or the multi-core cable is fixed so as to be close to, but not necessarily touching, the inner skin. Heat from the cable is assumed to escape through the inner skin only. The conduit can be metal or plastic.

[2] Conduit mounted so that the gap between conduit and surface is less than 0.3 times the conduit diameter.

[3] Cable mounted so that the gap between cable and surface is less than 0.3 times the cable diameter.

Table 1.2.3 Current-carrying capacity, cables for fixed installation, installation method A, B1, B2 and C

Insulation material	PVC							
Type designation codes[1]	NYM, NYBUY, NHYRUZY, NYIF, NYIFY, H07V-U, H07V-R, H07V-K, NHXMH[8]							
Permissible operating temperature	70 °C							
Number of loaded cores	2	3	2	3	2	3	2	3
Installation method	A		B1		B2		C	
	In thermally insulated walls		On or in walls or under plaster in conduits or trunkings				directly installed	

A — In thermally insulated walls:
- Single-core non-sheathed cables in conduit[2][5]
- Multi-core cable in conduit[5]
- Multi-core cable in the wall

B1 — On or in walls or under plaster in conduits or trunkings:
- Single-core non-sheathed cables in conduit on the wall[3]
- Single-core non-sheathed cables in trunking on the wall
- Single-core non-sheathed cables, single-core light-sheathed cables, multi-core cable in conduit in masonry[6]

B2 — On or in walls or under plaster in conduits or trunkings:
- Multi-core cable in conduit on the wall or on the floor
- Multi-core cable in trunking on the wall or on the floor

C — directly installed:
- Multi-core cable on the wall or on the floor[4]
- Single-core light-sheathed cables on the wall or on the floor
- Multi-core cable, flat building wires in the wall or under plaster[7]

Copper conductor Nominal cross-sectional area mm²	Current-carrying capacity in A							
1.5	15.5	13	17.5	15.5	15.5	14	19.5	17.5
2.5	19.5	18	24	21	21	19	26	24
4	26	24	32	28	28	26	35	32
6	34	31	41	36	37	33	46	41
10	46	42	57	50	50	46	63	57
16	61	56	76	68	68	61	85	76
25	80	73	101	89	90	77	112	96
35	99	89	125	111	110	95	138	119
50	119	108	151	134	—	—	—	—
70	151	136	192	171	—	—	—	—
95	182	164	232	207	—	—	—	—
120	210	188	269	239	—	—	—	—

Table 1.2.4 Current-carrying capacity, cables for fixed installation, installation method E, in free air

Insulation material	PVC	
Type designation codes[1]	NYM, NYMZ, NYMT, NYBUY, NHYRUZY, NHXMH[2], NHYSSYCY[3]	
Permissible operating temperature	70 °C	
Number of loaded cores	2	3
Installation method	E	E
	≥0,3d	≥0,3d
Copper conductor Nominal cross-sectional area mm²	Current-carrying capacity in A	
1.5	20	18.5
2.5	27	25
4	37	34
6	48	43
10	66	60
16	89	80
25	118	101
35	145	126
50[4]	–	153
70[4]	–	196
95[4]	–	238

[1] Type designation codes and further details are given in Clause 1.1
[2] Not included in DIN VDE 0298 Part 4, insulation
consists of cross-linked polyolefin compound
[3] Rated voltage 3.6/6 kV
[4] These cross-sectional areas are not included in DIN VDE 0298 Part 4

◁ [1] Type designation codes and further details are given in Clause 1.1
[2] Also applies to single-core non-sheathed cables in conduit
in flush cable trunking in a floor
[3] Also applies to single-core non-sheathed cables in conduit
in ventilated cable channels in a floor
[4] Also applies to multi-core cable in open or ventilated ducts
[5] Also applies to single-core non-sheathed cables, single-core light
PVC-sheathed cables, multi-core cable in trunking in the floor
[6] Also applies to single-core non-sheathed cables in conduit in the ceiling
[7] Also applies to multi-core cable in the ceiling
[8] Not included in DIN VDE 0298 Part 4,
insulation consists of cross-linked polyolefin compound

Flexible Cables with Rated Voltages up to 1000 V

Table 1.2.5 shows the calculation method for the determination of the cross-sectional area of flexible cables with rated voltages up to 1000 V for installation in free air and on a wall. The current-carrying capacities I_r for the "reference operating conditions" (see Table 1.2.5) are shown in Table 1.2.6 as well as in the tables of Clause 2.

For "site operating conditions" the rated current I_r shall be multiplied by the product of the respective rating factors f given in Tables 1.2.10 to 1.2.13:

$$I_z = I_r \cdot \Pi f.$$

Instructions for this procedure shall be taken from the right column of Table 1.2.5.

Before applying grouping factors to the *single-core cables given in Table 1.2.6, Columns 2 and 3*, the current-carrying capacity shall be multiplied by the factor 0.95. For touching or bundled cables the tabulated values shall be additionally multiplied by the factor 0.8 for one-phase circuits or by the factor 0.7 for three-phase circuits.

At high ambient temperatures the rating factors according to Table 1.2.7 shall be used *for cables with increased heat resistance.*

At low temperatures such cables shall not be loaded with higher operating currents than with the rated currents allocated to them in the above-mentioned tables although high operating temperatures are permitted. Therefore, only rating factors ≤ 1.0 are given in Table 1.2.7.

Example for a Multi-Core Cable

Only 18 out of 25 cores of a PVC control cable NYSLYÖ-J 25 × 1.5 are always loaded at the same time. The cable lies on a surface at an ambient temperature of 30 °C. The current-carrying capacity is calculated as follows:

Rated current I_r (Table 2.4.1c, Item 41):	18 A
Rating factor for 18 loaded cores of multi-core cables (Table 1.2.13):	0.45
Current-carrying capacity I_z:	8.1 A

Table 1.2.5 Operating conditions for flexible cables with rated voltages up to 1000 V

Reference operating conditions for the determination of rated currents I_r	*Site operating conditions* and calculation of the current-carrying capacity $I_z = I_r \, \Pi f$
Type of operation	
Continuous operation with the current-carrying capacity according to Table 1.2.6	Current-carrying capacity at intermittent operation according to Figure 1.2.1 and Part 1, Clause 18.6
Cable types and installation conditions	
Single-core rubber-insulated cables, in free air according to Table 1.2.6, Column 2 Single-core PVC-insulated cables, in free air according to Table 1.2.6, Column 3 Multi-core rubber-insulated cables with **two** loaded conductors on a surface, e.g. for domestic appliances or hand-held equipment according to Table 1.2.6, Column 4 Multi-core rubber-insulated cables with **three** loaded conductors on a surface, e.g. for domestic appliances or hand-held equipment according to Table 1.2.6, Column 5	Rating factors for grouping according to Table 1.2.11
Multi-core rubber-insulated cable with **three** loaded conductors, on a surface (apart from use for domestic appliances or hand-held equipment) according to Table 1.2.6, Column 6	Rating factors for grouping according to Table 1.2.11, for multi-core cables acc. to Table 1.2.13
Multi-core PVC-insulated cable with **two** loaded conductors on a wall, e.g. for domestic appliances or hand-held equipment according to Table 1.2.6, Column 7 Multi-core PVC-insulated cable with **three** loaded conductors on a surface, e.g. for domestic appliances or hand-held equipment according to Table 1.2.6, Column 8	Rating factors for grouping according to Table 1.2.11
Multi-core PVC-insulated cable with **three** loaded conductors, on a wall (apart from use for domestic appliances or hand-held equipment) according to Table 1.2.6, Column 9	Rating factors for grouping according to Table 1.2.11, for multi-core cables acc. to Table 1.2.13
Ambient conditions (see Part 1, Clause 18.2.2)	
Ambient temperature 30 °C	Rating factors
Sufficiently large or ventilated rooms in which the ambient temperature is not noticeably increased by the heat losses of the cables	for deviating ambient temperatures according to Table 1.2.7 or 1.2.10
Protection from solar radiation etc.	see Part 1, Clause 18.4.2
System frequency 50 to 60 Hz	

Table 1.2.6 Current-carrying capacity, flexible cables with rated voltages up to 1000 V (HYDROFIRM cables TG... with rated voltages 450/750 V are given in Table 1.2.9)

Column 1	2	3	4	5	6	7	8	9
Insulation material	NR/SR	PVC	NR/SR			PVC		
Type designation codes[1]	A05RN-F H07RN-F	H05V-U[2] H05V-K[2] H07V-U[2] H07V-R[2] H07V-K[2] NFYW[2]	H03RT-F H05RR-F A05RR-F A05RRT-F H05RN-F A05RN-F H07RN-F A07RN-F		NPL[2] NIFLÖU[2] NMHVÖU NGFLGÖU NSHCÖU NSHTÖU[3] H05RND5-F H05RT2D5-F H07RND5-F H07RT2D5-F H07RN-F A07RN-F	H03VH-Y[4] H03VH-H H03VV-F H03VVH2-F H05VV-F H05VVH2-F	H03VV-F H05VV-F	NYMHYV NYSLYÖ NYSLYCYÖ H05VVH6-F H05VVD3H6-F H07VVH6-F H07VVD3H6-F YSLYÖ[5] (N)YSLYÖ[5] YSLYCY[5] YSLYCÖ[5] SYSL[5] YSLTK[5] YSLYTK[5] YSLYCYTK[5]
Permissible operating temperature	60 °C	70 °C	60 °C			70 °C		
Number of loaded cores	1		2	3	2 or 3	2	3	2 or 3
Installation method								

Copper conductor Nominal cross-sectional area mm²	Current-carrying capacity in A							
	2–3		4	5	6	7	8	9
0.5	12[5]		3	3	—	3	3	9.5[5]
0.75	15		6	6	12	6	6	12
1	19		10	10	15	10	10	15
1.5	24		16	16	18	16	16	18
2.5	32		25	20	26	25	20	26
4	42		32	25	34	—	—	34
6	54		40	—	44	—	—	44
10	73		63	—	61	—	—	61
16	98		—	—	82	—	—	82
25	129		—	—	108	—	—	108
35	158		—	—	135	—	—	—
50	198		—	—	168	—	—	—
70	245		—	—	207	—	—	—
95	292		—	—	250	—	—	—
120	344		—	—	292	—	—	—
150	391		—	—	335	—	—	—
185	448		—	—	382	—	—	—
240	528		—	—	453	—	—	—
300	608		—	—	523	—	—	—
400	726		—	—	—	—	—	—
500	830		—	—	—	—	—	—

[1] Type designation codes and further details are given in Clause 1.1
[2] Cables for fixed installation
[3] Rating factors for reeled cables: Table 1.2.14
[4] Nominal cross-sectional area 0.1 mm², loadable with 0.2 A, independent of the ambient temperature
[5] Not included in DIN VDE 0298 Part 4

Table 1.2.7 Rating factors for the current-carrying capacity of cables with increased heat resistance, applicable to the values in Table 1.2.6

Insulation material	PVC		EVA		ETFE	SiR	
Type designation codes[1]	NYFAW NYFAFW NYFAZW	NYPLYW	N4GA N4GAF (N)HX4GA[2] (N)HX4GAF[2]	4GMH4G[2]	N7YA N7YAF	N2GFA N2GFAF H05SJ-K A05SJ-K A05SJ-U SIA[2] SIAF[2]	N2GSA N2GMH2G
Permissible operating temperature	90 °C		120 °C		135 °C	180 °C	
Number of loaded cores	1	2 or 3	1	2 or 3	1	1	2 or 3
Installation method							

Ambient temperature °C	Rating factors applicable to the values in Table 1.2.6						
	Column 3	Column 9	Column 2	Column 6	Column 3	Column 2	Column 6
50	1.00		1.00		1.00	1.00	
55	0.94		1.00		1.00	1.00	
60	0.87		1.00		1.00	1.00	
65	0.79		1.00		1.00	1.00	
70	0.71		1.00		1.00	1.00	
75	0.61		1.00		1.00	1.00	
80	0.50		1.00		1.00	1.00	
85	0.35		1.00		1.00	1.00	
90	—		1.00		1.00	1.00	
95	—		0.91		1.00	1.00	
100	—		0.82		0.94	1.00	
105	—		0.71		0.87	1.00	
110	—		0.58		0.79	1.00	
115	—		0.41		0.71	1.00	
120	—		—		0.61	1.00	
125	—		—		0.50	1.00	
130	—		—		0.35	1.00	
135	—		—		—	1.00	
140	—		—		—	1.00	
145	—		—		—	1.00	
150	—		—		—	1.00	
155	—		—		—	0.91	
160	—		—		—	0.82	
165	—		—		—	0.71	
170	—		—		—	0.58	
175	—		—		—	0.41	

[1] Type designation codes and further details are given in Clause 1.1

[2] Not included in DIN VDE 0298 Part 4

Flexible Cables with Rated Voltages 0.6/1 kV and Above

The calculation method for the determination of the cross-sectional area of flexible cables with rated voltages 0.6/1 kV and above is given in Table 1.2.8 for installation on a wall. The rated currents I_r for the "reference operating conditions" listed in this table shall be taken from Table 1.2.9 as well as from the tables of Clause 2.

For "site operating conditions" the rated current I_r shall be multiplied by the product of the respective rating factors f given in Tables 1.2.10 to 1.2.14:

$$I_z = I_r \cdot \Pi f.$$

Instructions for this procedure shall be taken from the right column of Table 1.2.8.

The cables NSHTÖU with the trademarks CORDAFLEX, CORDAFLEX(K) and CORDA-FLEX(SM) as well as the cables NGFLGÖU with the trademark PLANOFLEX are designed for an operating temperature of 90 °C which deviates from DIN VDE 0298.

Therefore, the current-carrying capacities given in Table 1.2.9 apply to these types of cable. Up to now, however, the permissible operating temperature of 90 °C could not be considered for Tables 1.2.9 and 1.2.10.

Example for a Trailing Cable

A trailing cable NTSCGEWÖU $3 \times 50 + 3 \times 25/3$ 6/10 kV is used as connecting cable for a container crane. During operation the trailing cable is reeled with up to three layers on a drum. The ambient temperature is 40 °C. The current-carrying capacity is calculated as follows:

Rated current I_r (Table 2.8.11, Item 41):	202 A
Rating factor for	
– ambient temperature 40 °C (Table 1.2.10):	0.89
– reeled cable (Table 1.2.14):	0.49
Total rating factor Πf:	0.44
Current-carrying capacity I_z:	88.9 A

Table 1.2.8 Operating conditions for flexible cables with rated voltages 0.6/1 kV and above

Reference operating conditions for the determination of rated currents I_r	Site operating conditions and calculation of the current-carrying capacity $I_z = I_r \Pi f$
Type of operation	
Continuous operation with the current-carrying capacities according to Table 1.2.9	Current-carrying capacity at intermittent operation according to Figure 1.2.1 and Part 1, Clause 18.6
Cable types and installation conditions	
Multi-core rubber-insulated cables with three loaded cores and rated voltages up to 6/10 kV, on a surface	Rating factors for grouping according to Table 1.2.11,
Multi-core rubber-insulated cables with three loaded cores and a rated voltage above 6/10 kV, on a surface	for reeled cables according to Table 1.2.14, for multi-core cables according to Table 1.2.13
Ambient conditions (see Part 1, Clause 18.2.2)	
Ambient temperature 30 °C	Rating factors
Sufficiently large or ventilated rooms in which the ambient temperature is not noticeably increased by the heat losses from the cables	for deviating ambient temperatures Table 1.2.10
Protection from solar radiation etc.	see Part 1, Clause 18.4.2
System frequency 50 to 60 Hz	

Table 1.2.9 Current-carrying capacity, flexible cables with rated voltages 0.6/1 kV and above as well as HYDROFIRM cables TG ...

Insulation material	EPR		
Type designation codes[1]	NT... NSSH...ÖU (N)HXSHXÖ[2] NSHTÖU[2]	TGK[2][3] TGKT[2][3] TGW[2][3] TGFLW[2][3]	NT...
Rated voltage	up to 6/10 kV		above 6/10 kV
Permissible operating temperature	80 °C		
Number of loaded cores	3		
Installation method			
Copper conductor Nominal cross-sectional area mm²	Current-carrying capacity in A		
1.5[2]	23		—
2.5	30		—
4	41		—
6	53		—
10	74		—
16	99		105
25	131		139
35	162		172
50	202		215
70	250		265
95	301		319
120	352		371
150	404		428
185	461		488
240[2]	540		—

[1] Type designation codes and further details are given in Clause 1.1   [2] Not included in DIN VDE 0298 Part 4   [3] Rated voltage 450/750 V

Table 1.2.10 Rating factors for ambient temperatures deviating from 30 °C

Insulation material	NR/SR	PVC	EPR
Permissible operating temperature	60 °C	70 °C	80 °C
Ambient temperature °C	Rating factors		
10	1.29	1.22	1.18
15	1.22	1.17	1.14
20	1.15	1.12	1.10
25	1.08	1.06	1.05
30	1.00	1.00	1.00
35	0.91	0.94	0.95
40	0.82	0.87	0.89
45	0.71	0.79	0.84
50	0.58	0.71	0.77
55	0.41	0.61	0.71
60	—	0.50	0.63
65	—	0.35	0.55
70	—	—	0.45

Table 1.2.11 Rating factors for grouping, applicable to the values in Tables 1.2.3, 1.2.6 and 1.2.9

Arrangement		Number of multi-core cables or number of a.c. or three-phase circuits consisting of single-core cables (2 or 3 loaded cores)														
		1	2	3	4	5	6	7	8	9	10	12	14	16	18	20
Bunched directly on the wall, on the floor, in conduit or trunking, on or in the wall		1.00	0.80	0.70	0.65	0.60	0.57	0.54	0.52	0.50	0.48	0.45	0.43	0.41	0.39	0.38
One layer on the wall or on the floor, with contact		1.00	0.85	0.79	0.75	0.73	0.72	0.71	0.70	0.70	0.70	0.70	0.70	0.70	0.70	0.70
One layer on the wall or on the floor, with a clearance equal to the cable diameter d		1.00	0.94	0.90	0.90	0.90	0.90	0.90	0.90	0.90	0.90	0.90	0.90	0.90	0.90	0.90
One layer under the ceiling, with contact		0.95	0.81	0.72	0.68	0.66	0.64	0.63	0.62	0.61	0.61	0.61	0.61	0.61	0.61	0.61
One layer under the ceiling, with a clearance equal to the cable diameter d		0.95	0.85	0.85	0.85	0.85	0.85	0.85	0.85	0.85	0.85	0.85	0.85	0.85	0.85	0.85

○ Symbol for one single-core or one multi-core cable

Notes:
When applying these rating factors to the values in Tables 1.2.3, 1.2.6 and 1.2.9, the number of loaded cores, the type of cable and the installation method shall correspond.

For grouping of multi-core cables with two or three loaded cores, the rating factor is to be chosen for the total number of bunched cables and applied to the current-carrying capacity of cables with two or three loaded cores.

If the grouping of single-core cables consists of n loaded single-core cables, the rating factor shall be determined for $n/2$ or $n/3$ circuits and applied to the current-carrying capacity of two or three loaded cores.

Table 1.2.12 Rating factors for grouping, applicable to the values in Table 1.2.4

Arrangement		Number of trays or ladders	Number of cables					
			1	2	3	4	6	9
Non-perforated cable trays[1]		1	0.97	0.84	0.78	0.75	0.71	0.68
		2	0.97	0.83	0.76	0.72	0.68	0.63
		3	0.97	0.82	0.75	0.71	0.66	0.61
		6	0.97	0.81	0.73	0.69	0.63	0.58
Perforated cable trays[1]		1	1.0	0.87	0.81	0.78	0.75	0.73
		2	1.0	0.86	0.79	0.76	0.72	0.68
		3	1.0	0.85	0.78	0.75	0.70	0.66
		6	1.0	0.84	0.77	0.73	0.68	0.64
		1	1.0	0.88	0.82	0.77	0.73	0.72
		2	1.0	0.88	0.81	0.76	0.71	0.70
		1	1.0	0.91	0.89	0.88	0.87	–
		2	1.0	0.91	0.88	0.87	0.86	–
Cable ladders[2]		1	1.0	0.88	0.83	0.81	0.79	0.78
		2	1.0	0.86	0.81	0.78	0.75	0.73
		3	1.0	0.85	0.79	0.76	0.73	0.70
		6	1.0	0.83	0.76	0.73	0.69	0.66

⊙   Symbol for one multi-core cable with 2 or 3 loaded cores

[1] A cable tray is a continuous support plate with raised edges, but without covering.
   A cable tray is considered to be perforated when the perforation is at least 30 % of the plan area.
[2] A cable ladder is a support system where the supporting metal work covers less than 10 % of the plan area of this system.

Table 1.2.13
Rating factors for multi-core cables with nominal cross-sectional areas of conductors up to 10 mm^2

Number of loaded cores	Rating factor
5	0.75
7	0.65
10	0.55
14	0.50
19	0.45
24	0.40
40	0.35
61	0.30

Table 1.2.14
Rating factors for reeled cables[1]

Number of layers	1	2	3	4
Rating factors	0.80	0.61	0.49	0.42

[1] The characteristics of the design shall be checked in order to ensure suitability for this type of operation

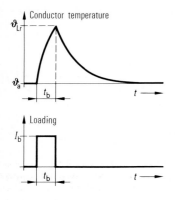

$$f_{KB}=\sqrt{\frac{1}{1-e^{-\frac{t_b}{\tau}}}}$$

without previous loading
$I_0=0$

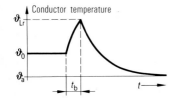

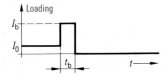

$$f_{KB}=\sqrt{\frac{1-\left(\frac{I_0}{I_b}\right)^2 e^{-\frac{t_b}{\tau}}}{1-e^{-\frac{t_b}{\tau}}}}$$

with previous loading
$I_0<I_b$

a) Short-time operation (KB[1])

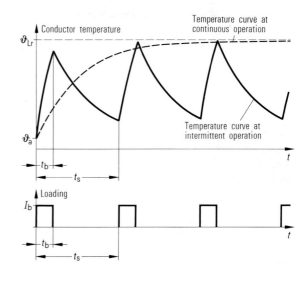

$$f_{AB}=\sqrt{\frac{1-e^{-\frac{t_s}{\tau}}}{1-e^{-\frac{t_b}{\tau}}}}=\sqrt{\frac{1-e^{-\frac{t_s}{\tau}}}{1-e^{-\frac{t_s}{\tau}\frac{ED}{100}}}}$$

$$ED=\frac{t_b}{t_s}100\%.$$

b) Intermittent operation (AB[2]) without previous loading

ϑ_{Lr} Permissible operating temperature
ϑ_a Initial temperature of the conductor
t_b Load period
t_s Duty cycle time
τ Minimum time value of the cable
   (see tables in Clause 2)
I_b Load current
I_0 Previous load current
ED Relative on-time period (%)

---

[1] The index KB is the abbreviation of the German term "Kurzzeitbetrieb"

[2] The index AB is the abbreviation of the German term "Aussetzbetrieb"

Fig. 1.2.1
Determination of the rating factors f_{KB} for short-time operation and f_{AB} for intermittent operation

Short-Time and Intermittent Operation

The current-carrying capacities determined for continuous operation may be exceeded for a short time provided it is ensured that a lower load current has flown or flows before and after continuous operation so that the permissible operating temperature is not exceeded.

In case of sufficiently short on-times t_n according to Table 1.2.15, e.g. for starting currents or re-acceleration currents of motors, the "average" heating effect of the current may be considered by means of the square mean value I_q:

$$I_q = \sqrt{\frac{I_{b1}^2 t_1 + I_{b2}^2 t_2 + ... + I_{bn}^2 t_n}{t_1 + t_2 + ... + t_n}}$$

with

$$t_1 + t_2 + ... + t_n = 24 \text{ h.}$$

The current I_q obtained from this equation shall be smaller than I_z or at least equal to I_z. The on-time of the occuring maximum current, however, may not exceed the values listed in Table 1.2.15.

If the permissible on-time is exceeded, then, for example, it shall be calculated with the minimum time value τ given in the tables of Clause 2, Item 42. The minimum time value is one fifth of the time taken from the temperature curve to almost reach the permissible final temperature at constant current.

This method can be applied to single current impulses with and without previous loading I_0 (short-time operation KB) and to periodic loadings (intermittent operation AB) according to Fig. 1.2.1. The permissible loading is:

at short-time operation $\quad I_{KB} = f_{KB} \cdot I_z$ or
at intermittent operation $\quad I_{AB} = f_{AB} \cdot I_z$.

Table 1.2.15
Permissible on-time when applying the square mean value

Nominal cross-sectional area mm^2	Permissible on-time s
≤ 6	4
10 – 25	8
35 – 50	15
70 – 150	30
185 and above	60

1.2.3 Determination of the Cross-Sectional Area for Short-Circuit

If the cross-sectional area is to be determined for short-circuit, it shall be ensured that

$$I_{thz} \geq I_{th}.$$

I_{thz} Thermal short-circuit capacity
I_{th} Thermal equivalent short-circuit current

The thermal equivalent short-circuit current I_{th} shall be determined by means of a suitable method, for example according to DIN VDE 0103, or obtained from the local utilities of installations or systems.

The short-circuit capacity I_{thz} results from the multiplication of the rated short-time current density J_{thr}, according to Table 1.2.16, by the cross-sectional area of conductor q_n at the short-circuit duration t_k where

$$I_{thz} = q_n J_{thr} \sqrt{\frac{t_{kr}}{t_k}}$$

with

$$t_{kr} = 1 \text{ s.}$$

The short-circuit duration t_k is given by the characteristic curve of the protective device used or the response time in view of the mechanical delay of the switch and the protective device.

If the cable is operated with a current below its current-carrying capacity before the beginning of the short-circuit, a lower conductor temperature ϑ_a may be assumed for the calculation (see Part 1, Clause 19.3).

Table 1.2.16 Permissible short-circuit temperatures ϑ_e and rated short-time current densities J_{thr} for cables

Insulation material[1]	Perm. operating temp. ϑ_{Lr}	Perm. short-circuit temp. ϑ_e	Conductor temperature at the beginning of short-circuit ϑ_a in °C												
	°C	°C	180	165	150	135	120	105	90	80	70	60	50	40	30
			Rated short-time current density J_{thr} in A/mm² for $t_{kr} = 1$ s												
Soft-soldered connections	–	160	–	–	36	58	74	87	100	107	115	122	129	136	143
Tinned conductors	–	200	49	65	79	91	102	112	122	128	135	141	147	153	159
NR, SR	60	200	–	–	–	–	–	–	–	–	–	141	147	153	159
PVC – flexible cables ≤ 300 mm²	70	150	–	–	–	–	–	–	–	–	109	117	124	131	138
– cables for fixed installation ≤ 300 mm²	70	160	–	–	–	–	–	–	–	–	115	122	129	136	143
> 300 mm²	70	140	–	–	–	–	–	–	–	–	103	111	118	126	133
EPR	90	250[2]	–	–	–	–	–	–	143	148	154	159	165	170	176
PVC (with increased thermal stability)	90	160	–	–	–	–	–	–	100	107	115	122	129	136	143
EVA	120	250	–	–	–	–	126	135	143	148	154	159	165	170	176
ETFE	135	250	–	–	–	117	126	135	143	148	154	159	165	170	176
SiR	180	350	132	139	146	153	160	166	173	178	182	187	191	196	201

[1] NR Natural rubber
SR Synthetic rubber
EPR Ethylene-propylene rubber (EPM) or
 Ethylene-propylene diene monomer rubber (EPDM)

PVC Polyvinyl chloride
EVA Ethylene vinyl-acetate copolymer
ETFE Ethylene-tetrafluoroethylene copolymer
SiR Silicone rubber

[2] In case of tinned conductors, the permissible short-circuit temperature is limited to 200 °C

Example for a Trailing Cable

A trailing cable
NTSCGEWÖU 3 x 50 + 3 x 25/3 3.6/6 kV used for a container crane is operated with a load current causing a conductor temperature of 60 °C at normal operation. It shall be checked whether the cross-sectional area of conductor selected is sufficient for the thermal equivalent short-circuit current $I_{th} = 15$ kA and the break-time of a series-connected power circuit breaker $t_k = 0.2$ s.

For this cable having tinned conductors ($\vartheta_e = 200$ °C!), the rated short-time current

$$I_{thr} = J_{thr} \cdot q_n = 141 \text{ A/mm}^2 \cdot 50 \text{ mm}^2$$
$$= 7050 \text{ A}$$

is obtained by means of Table 1.2.16.

For the given break time $t_k = 0.2$ s, the short-circuit capacity is

$$I_{thz} = I_{thr}\sqrt{\frac{t_{kr}}{t_k}} = 7050 \text{ A} \cdot \sqrt{\frac{1\text{s}}{0.2\text{s}}}$$
$$= 15.8 \text{ kA}.$$

With regard to the conditions given, it is thus ensured that the cross-sectional area is sufficient for the thermal equivalent short-circuit current required.

1.2.4 Determination of the Cross-Sectional Area for the Permitted Voltage Drop

Along a cable run the voltage drop ΔU is:

at direct current

$$\Delta U = 2I_b \, l \, R',$$

at single-phase alternating current

$$\Delta U = 2I_b \, l \, (R_w' \cos\varphi + X_L' \sin\varphi),$$

at three-phase current

$$\Delta U = \sqrt{3} I_b \, l \, (R_w' \cos\varphi + X_L' \sin\varphi).$$

ΔU Voltage drop
I_b Load current
l Length of cable run
R' d.c. resistance per unit length at operating temperature according to the tables in Clause 2, Item 30
R_w' Effective a.c. resistance per unit length at operating temperature
$X_L' = \omega L_m' = 2\pi f L_m'$
L_m' Mean inductance per unit length of the cable according to the tables in Clause 2, Item 32
$\cos\varphi$ Power factor of load

The voltage drop Δu in % related to the nominal voltage of the system U_n results in:

$$\Delta u = \frac{\Delta U}{U_n} \cdot 100\%.$$

For conductor cross-sectional areas $q_n \leq 16 \text{ mm}^2$, the inductance per unit length may be neglected and for ΔU it is only necessary to take into account the d.c. resistance per unit length at the operating temperature R'.

1.2.5 Protection against Overcurrents

It is possible to heat cables above the permissible operating temperatures by operational overloads as well as by short-circuits.

Thus cables shall be protected against overload and short-circuit by overcurrent protective devices, e.g. as given for low voltage equipment in DIN VDE 0636, DIN VDE 0641 and DIN VDE 0660.

Overcurrent protective devices are installed at the feeder point of each circuit as well as at all points where the current-carrying capacity or the short-circuit current-carrying capacity is reduced, e.g. because of a reduction in the cross-sectional area or other installation conditions.

For example, the allocation of fuses according to DIN VDE 0636 and circuit-breakers according to DIN VDE 0641 shall be made by means of the allocation rules given in DIN VDE 0100 Part 430. Thus the following equation applies:

$$I_b \leq I_n \leq I_z,$$

$$I_2 \leq 1.45 I_z.$$

I_b Load current of the circuit to be expected
I_z Current-carrying capacity of the cable
I_n Rated current of the protective device
I_2 Current causing a tripping or fusing of the protective device under the conditions specified in the installation specifications (triggering current)

The values for the tripping or fusing currents I_2 are to be taken from the installation specifications of the protective device, e.g. DIN VDE 0636 and 0641.

The application of the allocation rules given does not ensure complete protection in all cases. Owing to the characteristic curves of the protective devices moderate overcurrents may occur for a short time. These overcurrents can cause higher temperatures than permitted especially in the case of small conductor cross-sectional areas. Therefore, according to DIN VDE 0100 Part 430, the circuit shall basically be formed in such a way that small overloads lasting a long time do not occur. If necessary, a lower rating has to be chosen for the overload protection.

1.2.6 Maximum Permissible Cable Lengths in Power Cable Installations with Nominal Voltages up to 1000 V

For the erection of power cable installations with nominal voltages up to 1000 V not only the current-carrying capacity shall be considered, it shall also be ensured that the projected cable lengths neither endanger the protection at indirect contact and at short-circuit nor the conditions which are relevant for the permissible voltage drop.

The longer the cable the higher is the impedance of the cable. It can reach high values especially in the case of small cross-sectional areas. High values for the impedance, however, can lead to the following consequences:

▷ there are too high touch voltages on exposed conductive parts of electric equipment in case of a fault,

▷ the minimum current necessary for tripping or fusing the protective device is not sufficient in case of a short-circuit,

▷ the voltage drop along a cable reaches higher values than permitted.

More detailed information can be obtained from DIN VDE 0100 Supplement 5.

2 Project Planning Data of Cables

Summary of Tables

Design	Type designation code	Rated voltage									
		100 V	300 V	380 V	500 V	750 V	1 kV	3 kV	6 kV	10 kV	20 kV
		Page									
2.1 Building wires and cables	NYM 1×...				56						
	NYM 2×...				56						
	NYM 3×...				56						
	NYM 4×...				56						
	NYM 5×...				56						
	NYM ...×...				58						
	NHYRUZY ...×...				59						
	NYBUY ...×...				60						
	NYIF ...×...			60							
2.2 Single-core cables for internal wiring, general and special purposes	H05V-U 1×...				62						
	H05V-K 1×...				62						
	H07V-U 1×...					63					
	H07V-R 1×...					64					
	H07V-K 1×...					64					
	NSGAFÖU 1×...							66			
2.3 Heat-resistant cables and cords	N4GA 1×...					68					
	N4GAF 1×...					68					
	4GMH4G ...×...				70						
	SIA 1×...				70						
	SIAF 1×...				72		72				
	H05SJ-K 1×...				74						
	A05SJ-K 1×...				74						
	N2GMH2G ...×...				74						
2.4 Cables for measuring, controlling and regulating (MSR)	NYSLYÖ ...×0.75				76						
	NYSLYÖ ...×1				76						
	NYSLYÖ ...×1.5				78						
	NYSLYÖ ...×2.5				78						
	(N)YSLYÖ 2×...				80						
	YSLYÖ ...×...				80						
	NYSLYCYÖ ...×...				82						

Design	Type designation code	Rated voltage									
		100 V	300 V	380 V	500 V	750 V	1 kV	3 kV	6 kV	10 kV	20 kV
		Page									
2.4 Cables for measuring, controlling and regulating (MSR) (continued)	YSLYCYÖ ...×...				84						
	YSLYCY ...×...				84						
	SYSL ...×0.5				86						
	SYSL ...×0.75				86						
	SYSL ...×1				88						
	SYSL ...×1.5				88						
	SYSL ...×...				90						
2.5 Flexible PVC-sheathed cords	H03VH-H 2×0.75		92								
	H03VV-F 2×0.75		93								
	H03VV-F ...G0.75		93								
	H05VV-F 2×1.5				94						
	H05VV-F ...G...				94						
2.6 Flexible rubber-sheathed cords and cables	H05RR-F 2×...				96						
	H05RR-F ...G...				96						
	H05RR-F 3G0.75				96						
	H05RN-F 2×...				98						
	H05RN-F 3G...				98						
	A05RN-F 4G0.75				99						
	H05RN-F 3G0.75				99						
	H07RN-F ...×...					100					
	H07RN-F 3G...					100					
	H07RN-F 4G...					102					
	H07RN-F 5G...					102					
	A07RN-F ...×...					104					
	A07RN-F ...G...					104					
	H07RN-F ...G...					106					
	NSSHÖU 1×...						108				
	NSSHÖU ...×...						108				
	NSSHÖU 3×.../...						110				
	NSSHÖU ...×...						110				
	TGKT 4×...					112					
	TGK 3×...					114					
	TGK ...×...					114					
	TGW 1×...					116					
	TGFLW 3×...					118					
	TGFLW 4×...					118					
2.7 Welding cables	NSLFFÖU 1×...	120									

Table 2.1.1a
Light PVC-sheathed cables 1, 2 and 3 cores

Design				1	1	1	1
	1	Number of cores		1	1	1	1
	2	Nom. cross-sectional area of conductor	mm^2	1.5	2.5	4	6
	6	Type of conductor		E	E	E	E
	7	Coating of strands		plain	plain	plain	plain
	8	Number of strands		1	1	1	1
	10	Diameter of conductor	mm	1.5	1.9	2.4	2.9
	11	Thickness of insulation	mm	0.6	0.7	0.8	0.8
	12	Thickness of sheath	mm	1.4	1.4	1.4	1.4
	15	Overall diameter of cable (min. value)	mm	5.2	5.8	6.4	6.8
	16	Overall diameter of cable (max. value)	mm	6.2	6.8	7.6	8.2
	19	Weight of cable	kg/km	40	60	90	110
Mechanical properties	20	Minimum bending radius for fixed installation	mm	25	27	30	33
	21	Minimum bending radius, formed bend	mm	6	7	8	8
	28	Perm. pulling force for installation	N	75	125	200	300
Electrical properties	30	d.c. resistance/unit length at 70 °C	Ω/km	14.5	8.87	5.52	3.69
	31	d.c. resistance/unit length at 20 °C	Ω/km	12.1	7.41	4.61	3.08
Current-carrying capacity	37	Number of loaded cores		3	3	3	3
	39	Install. method C, on/in walls or under plaster	A	17.5	24	32	41
	42	Minimum time value	s	47	70	101	138
Short-circuit	43	Rated short-time current of conductor (1s)	kA	0.172	0.287	0.460	0.690

Table 2.1.1b
Light PVC-sheathed cables 4 and 5 cores

Design				4	4	4	4
	1	Number of cores		4	4	4	4
	2	Nom. cross-sectional area of conductor	mm^2	1.5	2.5	4	6
	6	Type of conductor		E	E	E	E
	7	Coating of strands		plain	plain	plain	plain
	8	Number of strands		1	1	1	1
	10	Diameter of conductor	mm	1.5	1.9	2.4	2.9
	11	Thickness of insulation	mm	0.6	0.7	0.8	0.8
	12	Thickness of sheath	mm	1.4	1.4	1.6	1.6
	15	Overall diameter of cable (min. value)	mm	9.5	11.0	12.5	14.5
	16	Overall diameter of cable (max. value)	mm	11.0	12.5	14.5	16.5
	19	Weight of cable	kg/km	160	230	350	460
Mechanical properties	20	Minimum bending radius for fixed installation	mm	44	50	58	66
	21	Minimum bending radius, formed bend	mm	22	25	29	33
	28	Perm. pulling force for installation	N	300	500	800	1200
Electrical properties	30	d.c. resistance/unit length at 70 °C	Ω/km	14.5	8.87	5.52	3.69
	31	d.c. resistance/unit length at 20 °C	Ω/km	12.1	7.41	4.61	3.08
Current-carrying capacity	37	Number of loaded cores		3	3	3	3
	39	Install. method C, on/in walls or under plaster	A	17.5	24	32	41
	42	Minimum time value	s	47	70	101	138
Short-circuit	43	Rated short-time current of conductor (1s)	kA	0.172	0.287	0.460	0.690

1 10	1 16	2 1.5	2 2.5	3 1.5	3 2.5	3 4	3 6	3 10
E plain	M plain	E plain	E plain	E plain	E plain	E plain	E plain	E plain
1 3.7	7 5.3	1 1.5	1 1.9	1 1.5	1 1.9	1 2.4	1 2.9	1 3.7
1.0 1.4	1.0 1.4	0.6 1.4	0.7 1.4	0.6 1.4	0.7 1.4	0.8 1.4	0.8 1.6	1.0 1.6
8.0 9.4	9.4 11.0	8.4 9.8	9.6 11.0	8.8 10.5	10.0 11.5	11.5 13.0	12.0 15.0	16.0 18.0
160	240	115	160	135	190	290	360	570
38 9	44 22	39 10	44 22	42 21	46 23	52 26	60 30	72 36
500	800	150	250	225	375	600	900	1500
2.19 1.83	1.38 1.15	14.5 12.1	8.87 7.41	14.5 12.1	8.87 7.41	5.52 4.61	3.69 3.08	2.19 1.83
3 57	3 76	2 19.5	2 26	2 19.5	2 26	2 35	2 46	2 63
199	286	38	60	38	60	84	110	163
1.15	1.84	0.172	0.287	0.172	0.287	0.460	0.690	1.15

4 10	4 16	4 25	4 35	5 1.5	5 2.5	5 4	5 6	5 10	5 16	5 25
E plain	M plain	M plain	M plain	E plain	E plain	E plain	E plain	E plain	M plain	M plain
1 3.7	7 5.3	7 6.6	7 7.9	1 1.5	1 1.9	1 2.4	1 2.9	1 3.7	7 5.3	7 6.6
1.0 1.6	1.0 1.6	1.2 1.8	1.2 1.8	0.6 1.4	0.7 1.4	0.8 1.6	0.8 1.6	1.0 1.6	1.0 1.8	1.2 1.8
17.0 19.5	20.5 23.5	25.0 28.5	27.5 32.0	9.9 12.0	11.5 13.5	14.0 16.5	15.5 18.0	18.5 21.5	22.5 26.0	27.5 31.5
700	1050	1650	2100	190	270	420	540	820	1300	2000
78 39	94 47	114 86	128 96	48 24	54 27	66 33	72 36	86 43	104 78	126 95
2000	3200	5000	7000	375	625	1000	1500	2500	4000	6250
2.19 1.83	1.38 1.15	0.870 0.727	0.627 0.524	14.5 12.1	8.87 7.41	5.52 4.61	3.69 3.08	2.19 1.83	1.38 1.15	0.870 0.727
3 57	3 76	3 96	3 119	3 17.5	3 24	3 32	3 41	3 57	3 76	3 96
199	286	438	559	47	70	101	138	199	286	438
1.15	1.84	2.88	4.02	0.172	0.287	0.460	0.690	1.15	1.84	2.88

Table 2.1.1c Light PVC-sheathed cables 7 and 10 cores				NYM ... ×... 300/500 V		
Design	1	Number of cores		7	7	10
	2	Nom. cross-sectional area of conductor	mm²	1.5	2.5	1.5
	6	Type of conductor		E	E	E
	7	Coating of strands		plain	plain	plain
	8	Number of strands		1	1	1
	10	Diameter of conductor	mm	1.5	1.9	1.5
	11	Thickness of insulation	mm	0.6	0.7	0.6
	12	Thickness of sheath	mm	1.4	1.6	1.6
	15	Overall diameter of cable (min. value)	mm	11.0	12.7	13.9
	16	Overall diameter of cable (max. value)	mm	13.0	14.3	15.6
	19	Weight of cable	kg/km	240	350	360
Mechanical properties	20	Minimum bending radius for fixed installation	mm	52	57	62
	21	Minimum bending radius, formed bend	mm	26	29	31
	28	Perm. pulling force for installation	N	525	875	750
Electrical properties	30	d.c. resistance/unit length at 70 °C	Ω/km	14.5	14.5	14.5
	31	d.c. resistance/unit length at 20 °C	Ω/km	12.1	12.1	12.1
Current-carrying capacity	37	Number of loaded cores		3	3	3
	39	Install. method C, on/in walls or under plaster	A	17.5	24	17.5
	42	Minimum time value	s	47	70	47
Short-circuit	43	Rated short-time current of conductor (1s)	kA	0.172	0.287	0.172

Table 2.1.2
Metalclad PVC-sheathed cables 3, 4 and 5 cores

NHYRUZY ...×... 300/500 V

				3	3	4	5	5
Design	1	Number of cores		3	3	4	5	5
	2	Nom. cross-sectional area of conductor	mm²	1.5	2.5	1.5	1.5	2.5
	6	Type of conductor		E	E	E	E	E
	7	Coating of strands		plain	plain	plain	plain	plain
	8	Number of strands		1	1	1	1	1
	10	Diameter of conductor	mm	1.5	1.9	1.5	1.5	1.9
	11	Thickness of insulation	mm	0.6	0.7	0.6	0.6	0.7
	12	Thickness of sheath	mm	1.0	1.0	1.0	1.0	1.0
	15	Overall diameter of cable (min. value)	mm	8.4	9.8	9.1	9.8	11.5
	16	Overall diameter of cable (max. value)	mm	10.6	12.0	11.5	12.5	14.0
	19	Weight of cable	kg/km	190	235	220	240	345
Mechanical properties	20	Minimum bending radius for fixed installation	mm	42	48	46	50	56
	21	Minimum bending radius, formed bend	mm	21	24	23	25	28
	28	Perm. pulling force for installation	N	225	375	300	375	625
Electrical properties	30	d.c. resistance/unit length at 70 °C	Ω/km	14.5	8.87	14.5	14.5	8.87
	31	d.c. resistance/unit length at 20 °C	Ω/km	12.1	7.41	12.1	12.1	7.41
Current-carrying capacity	37	Number of loaded cores		2	2	3	3	3
	39	Install. method C, on/in walls or under plaster	A	19.5	26	17.5	17.5	24
	42	Minimum time value	s	38	60	47	47	70
Short-circuit	43	Rated short-time current of conductor (1s)	kA	0.172	0.287	0.172	0.172	0.287

Table 2.1.3
Lead-covered PVC-sheathed cables 3, 4 and 5 cores

Design				3	4	4	4
	1	Number of cores		3	4	4	4
	2	Nom. cross-sectional area of conductor	mm^2	1.5	1.5	2.5	6
	6	Type of conductor		E	E	E	E
	7	Coating of strands		plain	plain	plain	plain
	8	Number of strands		1	1	1	1
	10	Diameter of conductor	mm	1.5	1.5	1.9	2.9
	11	Thickness of insulation	mm	0.6	0.6	0.7	0.8
	12	Thickness of sheath	mm	1.0	1.0	1.2	1.2
	15	Overall diameter of cable (min. value)	mm	9.6	10.0	12.0	15.0
	16	Overall diameter of cable (max. value)	mm	11.1	11.5	13.5	16.5
	19	Weight of cable	kg/km	335	380	495	840
Mechanical properties	20	Minimum bending radius for fixed installation	mm	44	46	54	66
	21	Minimum bending radius, formed bend	mm	22	23	27	33
	28	Perm. pulling force for installation	N	225	300	500	1200
Electrical properties	30	d.c. resistance/unit length at 70 °C	Ω/km	14.5	14.5	8.87	3.69
	31	d.c. resistance/unit length at 20 °C	Ω/km	12.1	12.1	7.41	3.08
Current-carrying capacity	37	Number of loaded cores		2	3	3	3
	39	Install. method C, on/in walls or under plaster	A	19.5	17.5	24	41
	42	Minimum time value	s	38	47	70	138
Short-circuit	43	Rated short-time current of conductor (1s)	kA	0.172	0.172	0.287	0.690

Table 2.1.4
SIFLA flat building wires 2, 3, 4 and 5 cores

Design				2	2	3
	1	Number of cores		2	2	3
	2	Nom. cross-sectional area of conductor	mm^2	1.5	2.5	1.5
	6	Type of conductor		E	E	E
	7	Coating of strands		plain	plain	plain
	8	Number of strands		1	1	1
	10	Diameter of conductor	mm	1.5	1.9	1.5
	11	Thickness of insulation	mm	0.4	0.5	0.4
	12	Thickness of sheath	mm	0.8	0.9	0.8
	17	Overall dimensions of cable (min. value)	mm × mm	3.7×10.5	4.5×12.0	3.7×17.5
	18	Overall dimensions of cable (max. value)	mm × mm	4.4×12.0	5.2×13.5	4.4×19.0
	19	Weight of cable	kg/km	75	105	115
Mechanical properties	20	Minimum bending radius for fixed installation	mm	18	21	18
	21	Minimum bending radius, formed bend	mm	4	5	4
	28	Perm. pulling force for installation	N	150	250	225
Electrical properties	30	d.c. resistance/unit length at 70 °C	Ω/km	14.5	8.87	14.5
	31	d.c. resistance/unit length at 20 °C	Ω/km	12.1	7.41	12.1
Current-carrying capacity	37	Number of loaded cores		2	2	2
	40	Installation method C, under plaster	A	19.5	26	19.5
	42	Minimum time value	s	38	60	38
Short-circuit	43	Rated short-time current of conductor (1s)	kA	0.172	0.287	0.172

NYBUY … × … 300/500 V

5	5
2.5	4
E	E
plain	plain
1	1
1.9	2.4
0.7	0.8
1.2	1.2
12.5	15.0
14.0	16.5
560	810
56	66
28	33
625	1000
8.87	5.52
7.41	4.61
3	3
24	32
70	101
0.287	0.460

NYIF … × … 220/380 V

3	3	4	4	4	5	5
2.5	4	1.5	2.5	4	1.5	2.5
E	E	E	E	E	E	E
plain	plain	plain	plain	plain	plain	plain
1	1	1	1	1	1	1
1.9	2.4	1.5	1.9	2.4	1.5	1.9
0.5	0.6	0.4	0.5	0.6	0.4	0.5
0.9	0.9	0.8	0.9	0.9	0.8	0.9
4.5×19.5	5.2×23.0	3.7×24.0	4.5×27.0	5.2×31.4	3.7×31.0	4.5×34.5
5.2×21.5	6.0×25.0	4.4×26.0	5.2×29.5	6.0×34.4	4.4×33.0	5.2×37.0
160	230	160	225	315	205	290
21	24	18	21	24	18	21
5	6	4	5	6	4	5
375	600	300	500	800	375	625
8.87	5.52	14.5	8.87	5.52	14.5	8.87
7.41	4.61	12.1	7.41	4.61	12.1	7.41
2	2	3	3	3	3	3
26	35	17.5	24	32	17.5	24
60	84	47	70	101	47	70
0.287	0.460	0.172	0.287	0.460	0.172	0.287

Table 2.2.1
PVC-insulated single-core non-sheathed cables for internal wiring solid conductor   H05V-U 1X... 300/500 V

Design	2	Nom. cross-sectional area of conductor	mm^2	0.5	0.75	1
	6	Type of conductor		E	E	E
	7	Coating of strands		plain	plain	plain
	8	Number of strands		1	1	1
	10	Diameter of conductor	mm	0.9	1.0	1.2
	11	Thickness of insulation	mm	0.6	0.6	0.6
	15	Overall diameter of cable (min. value)	mm	1.9	2.0	2.2
	16	Overall diameter of cable (max. value)	mm	2.4	2.6	2.8
	19	Weight of cable	kg/km	9	12	15
Mechanical properties	20	Minimum bending radius for fixed installation	mm	10	10	11
	21	Minimum bending radius, formed bend	mm	2	3	3
	28	Perm. pulling force for installation	N	25	38	50
Electrical properties	30	d.c. resistance/unit length at 70 °C	Ω/km	43.1	29.3	21.7
	31	d.c. resistance/unit length at 20 °C	Ω/km	36.0	24.5	18.1
Current-carrying capacity	37	Number of loaded cores		1	1	1
	38	Installation in free air	A	12.0	15.0	19.0
	42	Minimum time value	s	11	16	18
Short-circuit	43	Rated short-time current of conductor (1s)	kA	0.0575	0.0862	0.115

Table 2.2.2
PVC-insulated single-core non-sheathed cables for internal wiring finely stranded conductor   H05V-K 1X... 300/500 V

Design	2	Nom. cross-sectional area of conductor	mm^2	0.5	0.75	1
	6	Type of conductor		F	F	F
	7	Coating of strands		plain	plain	plain
	8	Number of strands		16	24	32
	9	Diameter of strand	mm	0.21	0.21	0.21
	10	Diameter of conductor	mm	1.1	1.3	1.5
	11	Thickness of insulation	mm	0.6	0.6	0.6
	15	Overall diameter of cable (min. value)	mm	1.9	2.1	2.3
	16	Overall diameter of cable (max. value)	mm	2.6	2.8	3.0
	19	Weight of cable	kg/km	10	13	16
Mechanical properties	20	Minimum bending radius for fixed installation	mm	10	11	12
	21	Minimum bending radius, formed bend	mm	3	3	3
	29	Perm. pulling force for install. and operation	N	8	11	15
Electrical properties	30	d.c. resistance/unit length at 70 °C	Ω/km	46.7	31.1	23.3
	31	d.c. resistance/unit length at 20 °C	Ω/km	39.0	26.0	19.5
Current-carrying capacity	37	Number of loaded cores		1	1	1
	38	Installation in free air	A	12.0	15.0	19.0
	42	Minimum time value	s	11	16	18
Short-circuit	43	Rated short-time current of conductor (1s)	kA	0.0575	0.0862	0.115

Table 2.2.3
PVC-insulated single-core non-sheathed cables for general purposes solid conductor | | | | | | | **H07V-U 1X... 450/750 V**

				1.5	2.5	4	6	10
esign	2	Nom. cross-sectional area of conductor	mm²	1.5	2.5	4	6	10
	6	Type of conductor		E	E	E	E	E
	7	Coating of strands		plain	plain	plain	plain	plain
	8	Number of strands		1	1	1	1	1
	10	Diameter of conductor	mm	1.5	1.9	2.4	2.9	3.7
	11	Thickness of insulation	mm	0.7	0.8	0.8	0.8	1.0
	15	Overall diameter of cable (min. value)	mm	2.6	3.2	3.6	4.1	5.3
	16	Overall diameter of cable (max. value)	mm	3.3	3.9	4.4	4.9	6.4
	19	Weight of cable	kg/km	20	35	50	70	120
echanical operties	20	Minimum bending radius for fixed installation	mm	13	16	18	20	26
	21	Minimum bending radius, formed bend	mm	3	4	4	5	6
	28	Perm. pulling force for installation	N	75	125	200	300	500
ectrical operties	30	d.c. resistance/unit length at 70 °C	Ω/km	14.5	8.87	5.52	3.69	2.19
	31	d.c. resistance/unit length at 20 °C	Ω/km	12.1	7.41	4.61	3.08	1.83
urrent-carrying pacity	37	Number of loaded cores		1	1	1	1	1
	38	Installation in free air	A	24	32	42	54	73
	42	Minimum time value	s	25	39	59	80	121
ort-circuit	43	Rated short-time current of conductor (1s)	kA	0.172	0.287	0.460	0.690	1.15

Table 2.2.4
PVC-insulated single-core non-sheathed cables for general purposes stranded conductor

Design	2	Nom. cross-sectional area of conductor	mm²	16	25	35	50
	6	Type of conductor		M	M	M	M
	7	Coating of strands		plain	plain	plain	plain
	8	Number of strands		7	7	7	19
	10	Diameter of conductor	mm	5.3	6.6	7.9	9.1
	11	Thickness of insulation	mm	1.0	1.2	1.2	1.4
	15	Overall diameter of cable (min. value)	mm	6.6	8.3	9.3	10.9
	16	Overall diameter of cable (max. value)	mm	8.0	9.8	11.0	13.0
	19	Weight of cable	kg/km	185	290	390	525
Mechanical properties	20	Minimum bending radius for fixed installation	mm	32	39	44	52
	21	Minimum bending radius, formed bend	mm	8	10	22	26
	28	Perm. pulling force for installation	N	800	1250	1750	2500
Electrical properties	30	d.c. resistance/unit length at 70 °C	Ω/km	1.38	0.870	0.627	0.463
	31	d.c. resistance/unit length at 20 °C	Ω/km	1.15	0.727	0.524	0.387
Current-carrying capacity	37	Number of loaded cores		1	1	1	1
	38	Installation in free air	A	98	129	158	198
	42	Minimum time value	s	172	243	317	412
Short-circuit	43	Rated short-time current of conductor (1s)	kA	1.84	2.88	4.02	5.75

Table 2.2.5
PVC-insulated single-core non-sheathed cables for general purposes finely stranded conductor

Design	2	Nom. cross-sectional area of conductor	mm²	1.5	2.5	4	6
	6	Type of conductor		F	F	F	F
	7	Coating of strands		plain	plain	plain	plain
	8	Number of strands		30	50	56	84
	9	Diameter of strand	mm	0.26	0.26	0.31	0.31
	10	Diameter of conductor	mm	1.8	2.6	3.2	3.9
	11	Thickness of insulation	mm	0.7	0.8	0.8	0.8
	15	Overall diameter of cable (min. value)	mm	2.7	3.3	3.8	4.4
	16	Overall diameter of cable (max. value)	mm	3.5	4.2	4.8	6.3
	19	Weight of cable	kg/km	23	35	51	71
Mechanical properties	20	Minimum bending radius for fixed installation	mm	14	17	19	25
	21	Minimum bending radius, formed bend	mm	4	4	5	6
	29	Perm. pulling force for install. and operation	N	23	38	60	90
Electrical properties	30	d.c. resistance/unit length at 70 °C	Ω/km	15.9	9.55	5.92	3.95
	31	d.c. resistance/unit length at 20 °C	Ω/km	13.3	7.98	4.95	3.30
Current-carrying capacity	37	Number of loaded cores		1	1	1	1
	38	Installation in free air	A	24	32	42	54
	42	Minimum time value	s	25	39	59	80
Short-circuit	43	Rated short-time current of conductor (1s)	kA	0.172	0.287	0.460	0.690

H07V-R 1X... 450/750 V

70	95	120	150	185	240	300
M plain	M plain	M plain	M plain	M plain	M plain	M plain
19	19	37	37	37	61	61
11.0	12.9	14.5	16.2	18.0	20.6	23.1
1.4	1.6	1.6	1.8	2.0	2.2	2.4
12.5	14.6	16.1	17.9	20.0	22.8	25.4
15.0	17.0	19.0	21.0	23.5	26.5	29.5
735	1010	1260	1540	1940	2550	3180
60	68	76	84	94	106	118
30	34	38	42	47	80	89
3500	4750	6000	7500	9250	12000	15000
0.321	0.231	0.183	0.148	0.119	0.0902	0.0719
0.268	0.193	0.153	0.124	0.0991	0.0754	0.0601
1	1	1	1	1	1	1
245	292	344	391	448	528	608
527	684	786	951	1101	1334	1572
8.05	10.9	13.8	17.3	21.3	27.6	34.5

H07V-K 1X... 450/750 V

10	16	25	35	50	70	95	120	150
F plain	F plain	F plain	F plain	F plain	F plain	F plain	F plain	F plain
80	126	196	276	396	360	475	605	756
0.41	0.41	0.41	0.41	0.41	0.51	0.51	0.51	0.51
5.1	6.3	7.8	9.2	11.0	13.1	15.1	17.0	19.0
1.0	1.0	1.2	1.2	1.4	1.4	1.6	1.6	1.8
5.6	6.8	8.3	9.6	11.4	13.1	15.0	16.6	18.5
7.6	8.8	11.0	12.5	14.5	17.0	19.0	21.0	23.5
125	195	300	410	585	810	1065	1335	1600
30	35	44	50	58	68	76	84	94
8	9	22	25	29	34	38	42	47
150	240	375	525	750	1050	1425	1800	2250
2.29	1.45	0.933	0.663	0.462	0.325	0.246	0.193	0.154
1.91	1.21	0.780	0.554	0.386	0.272	0.206	0.161	0.129
1	1	1	1	1	1	1	1	1
73	98	129	158	198	245	292	344	391
121	172	243	317	412	527	684	786	951
1.15	1.84	2.88	4.02	5.75	8.05	10.9	13.8	17.3

Table 2.2.6
Rubber-insulated single-core cables for special purposes finely stranded conductor

Design				1.5	2.5	4	6
Design	2	Nom. cross-sectional area of conductor	mm^2	1.5	2.5	4	6
	6	Type of conductor		F	F	F	F
	7	Coating of strands		tinned	tinned	tinned	tinned
	8	Number of strands		30	50	56	84
	9	Diameter of strand	mm	0.26	0.26	0.31	0.31
	10	Diameter of conductor	mm	1.8	2.6	3.2	3.9
	11	Thickness of insulation	mm	1.3	1.3	1.3	1.3
	12	Thickness of sheath	mm	0.8	0.8	0.8	0.8
	15	Overall diameter of cable (min. value)	mm	5.3	5.7	6.2	6.8
	16	Overall diameter of cable (max. value)	mm	7.0	7.5	9.0	9.5
	19	Weight of cable	kg/km	55	65	85	110
Mechanical properties	20	Minimum bending radius for fixed installation	mm	42	45	54	57
	21	Minimum bending radius, formed bend	mm	28	30	36	38
	29	Perm. pulling force for install. and operation	N	23	38	60	90
Electrical properties	30	d.c. resistance/unit length at 90 °C	Ω/km	17.5	10.5	6.49	4.32
	31	d.c. resistance/unit length at 20 °C	Ω/km	13.7	8.21	5.09	3.39
Current-carrying capacity	37	Number of loaded cores		1	1	1	1
	38	Installation in free air	A	35	45	62	80
	42	Minimum time value	s	17	28	38	51
Short-circuit	43	Rated short-time current of conductor (1s)	kA	0.183	0.305	0.488	0.732

NSGAFÖU 1×... 1.8/3 kV

10	16	25	35	50	70	95	120	150	185	240
F	F	F	F	F	F	F	F	F	F	F
tinned	tinned	tinned	tinned	tinned	tinned	tinned	tinned	tinned	tinned	tinned
80	126	196	276	396	360	475	608	756	925	1221
0.41	0.41	0.41	0.41	0.41	0.51	0.51	0.51	0.51	0.51	0.51
5.1	6.3	7.8	9.2	11.0	13.1	15.1	17.0	19.0	21.0	24.0
1.5	1.5	1.8	1.8	1.8	1.8	2.2	2.2	2.2	2.4	2.6
0.8	0.8	1.0	1.0	1.0	1.0	1.0	1.0	1.2	1.2	1.2
8.2	9.2	11.3	12.6	14.0	15.7	17.9	19.6	21.5	23.5	26.4
11.0	13.0	15.0	16.5	18.0	20.5	24.0	26.0	28.0	31.0	34.5
160	240	365	475	640	850	1110	1350	1650	2000	2600
66	78	90	99	108	123	144	156	168	186	207
44	52	60	66	72	82	96	104	112	124	138
150	240	375	525	750	1050	1425	1800	2250	2775	3600
2.49	1.58	1.01	0.720	0.501	0.353	0.268	0.209	0.168	0.138	0.104
1.95	1.24	0.795	0.565	0.393	0.277	0.210	0.164	0.132	0.108	0.0817
1	1	1	1	1	1	1	1	1	1	1
111	149	197	244	304	376	453	529	608	693	823
74	105	146	187	246	315	400	468	553	648	773
1.22	1.95	3.05	4.27	6.10	8.54	11.6	14.6	18.3	22.6	29.3

Table 2.3.1
Heat-resistant rubber-insulated single-core non-sheathed cables solid conductor

N4GA 1×... **450/750 V**

Design					
	2	Nom. cross-sectional area of conductor	mm²	1.5	2.5
	6	Type of conductor		E	E
	7	Coating of strands		tinned	tinned
	8	Number of strands		1	1
	10	Diameter of conductor	mm	1.5	1.9
	11	Thickness of insulation	mm	0.8	0.9
	15	Overall diameter of cable (min. value)	mm	2.8	3.4
	16	Overall diameter of cable (max. value)	mm	3.4	4.1
	19	Weight of cable	kg/km	25	35
Mechanical properties	20	Minimum bending radius for fixed installation	mm	14	16
	21	Minimum bending radius, formed bend	mm	3	4
	28	Perm. pulling force for installation	N	75	125
Electrical properties	30	d.c. resistance/unit length at 120 °C	Ω/km	19.1	11.4
	31	d.c. resistance/unit length at 20 °C	Ω/km	13.7	8.21
Current-carrying capacity	37	Number of loaded cores		1	1
	38	Installation in free air	A	24	32
	42	Minimum time value	s	49	76
Short-circuit	43	Rated short-time current of conductor (1s)	kA	0.153	0.255

Table 2.3.2
Heat-resistant rubber-insulated single-core non-sheathed cables finely stranded conductor

Design							
	2	Nom. cross-sectional area of conductor	mm²	0.75	1	1.5	2.5
	6	Type of conductor		F	F	F	F
	7	Coating of strands		tinned	tinned	tinned	tinned
	8	Number of strands		24	32	30	50
	9	Diameter of strand	mm	0.21	0.21	0.26	0.26
	10	Diameter of conductor	mm	1.3	1.5	1.8	2.6
	11	Thickness of insulation	mm	0.8	0.8	0.8	0.9
	15	Overall diameter of cable (min. value)	mm	2.5	2.6	2.9	3.5
	16	Overall diameter of cable (max. value)	mm	3.2	3.4	3.7	4.4
	19	Weight of cable	kg/km	15	20	25	40
Mechanical properties	20	Minimum bending radius for fixed installation	mm	13	14	15	18
	21	Minimum bending radius, formed bend	mm	3	3	4	4
	29	Perm. pulling force for install. and operation	N	11	15	23	38
Electrical properties	30	d.c. resistance/unit length at 120 °C	Ω/km	37.2	27.9	19.1	11.4
	31	d.c. resistance/unit length at 20 °C	Ω/km	26.7	20.0	13.7	8.21
Current-carrying capacity	37	Number of loaded cores		1	1	1	1
	38	Installation in free air	A	15.0	19.0	24	32
	42	Minimum time value	s	31	35	49	76
Short-circuit	43	Rated short-time current of conductor (1s)	kA	0.0765	0.102	0.153	0.255

N4GAF 1×... 450/750 V

4	6	10	16	25	35
F	F	F	F	F	F
tinned	tinned	tinned	tinned	tinned	tinned
56	84	80	126	196	276
0.31	0.31	0.41	0.41	0.41	0.41
3.2	3.9	5.1	6.3	7.8	9.2
1.0	1.0	1.2	1.2	1.4	1.4
4.2	4.8	6.0	7.2	8.7	10.0
5.5	6.3	7.9	9.0	11.0	12.4
60	80	130	210	320	420
22	25	32	36	44	50
6	6	8	9	22	25
60	90	150	240	375	525
7.09	4.72	2.72	1.73	1.11	0.787
5.09	3.39	1.95	1.24	0.795	0.565
1	1	1	1	1	1
42	54	73	98	129	158
113	154	234	333	469	613
0.408	0.612	1.02	1.63	2.55	3.57

Table 2.3.3
Heat-resistant rubber-sheathed flexible cords 3, 4, 5 and 7 cores

Design							
	1	Number of cores		3	5	7	3
	2	Nom. cross-sectional area of conductor	mm²	0.75	0.75	0.75	1.5
	6	Type of conductor		F	F	F	F
	7	Coating of strands		tinned	tinned	tinned	tinned
	8	Number of strands		24	24	24	30
	9	Diameter of strand	mm	0.21	0.21	0.21	0.26
	10	Diameter of conductor	mm	1.3	1.3	1.3	1.8
	11	Thickness of insulation	mm	0.8	0.8	0.8	0.8
	12	Thickness of sheath	mm	0.8	1.0	1.2	1.0
	15	Overall diameter of cable (min. value)	mm	6.9	9.1	9.6	8.2
	16	Overall diameter of cable (max. value)	mm	8.6	11.0	12.0	10.0
	19	Weight of cable	kg/km	80	135	175	120
Mechanical properties	20	Minimum bending radius for fixed installation	mm	26	33	36	30
	22	Minimum bending radius for free flexing	mm	34	44	48	40
	23	Minimum bending radius for cable entry	mm	34	44	48	40
	29	Perm. pulling force for install. and operation	N	34	56	79	68
Electrical properties	30	d.c. resistance/unit length at 120 °C	Ω/km	37.2	37.2	37.2	19.1
	31	d.c. resistance/unit length at 20 °C	Ω/km	26.7	26.7	26.7	13.7
Current-carrying capacity	37	Number of loaded cores		2	3	3	2
	41	Installation on walls	A	12.0	12.0	12.0	18.0
	42	Minimum time value	s	49	49	49	87
Short-circuit	43	Rated short-time current of conductor (1s)	kA	0.0765	0.0765	0.0765	0.153

Table 2.3.4
SINOTHERM heat-resistant silicone rubber-insulated single-core non-sheathed cables solid conductor   **SIA 1×... 300/500 V**

Design								
	2	Nom. cross-sectional area of conductor	mm²	1.5	2.5	4	6	10
	6	Type of conductor		E	E	E	E	E
	7	Coating of strands		plain	plain	plain	plain	plain
	8	Number of strands		1	1	1	1	1
	10	Diameter of conductor	mm	1.5	1.9	2.4	2.9	3.7
	11	Thickness of insulation	mm	0.6	0.7	0.8	0.8	1.0
	15	Overall diameter of cable (min. value)	mm	2.4	3.0	3.6	4.1	5.3
	16	Overall diameter of cable (max. value)	mm	3.3	3.9	4.4	4.9	6.4
	19	Weight of cable	kg/km	20	30	50	65	115
Mechanical properties	20	Minimum bending radius for fixed installation	mm	13	16	18	20	26
	21	Minimum bending radius, formed bend	mm	3	4	4	5	6
	28	Perm. pulling force for installation	N	75	125	200	300	500
Electrical properties	30	d.c. resistance/unit length at 180 °C	Ω/km	19.7	12.1	7.51	5.02	2.98
	31	d.c. resistance/unit length at 20 °C	Ω/km	12.1	7.41	4.61	3.08	1.83
Current-carrying capacity	37	Number of loaded cores		1	1	1	1	1
	38	Installation in free air	A	24	32	42	54	73
	42	Minimum time value	s	70	109	161	220	334
Short-circuit	43	Rated short-time current of conductor (1s)	kA	0.198	0.330	0.528	0.792	1.32

4	5	7	5
1.5	1.5	1.5	2.5
F	F	F	F
tinned	tinned	tinned	tinned
30	30	30	50
0.26	0.26	0.26	0.26
1.8	1.8	1.8	2.6
0.8	0.8	0.8	0.9
1.2	1.2	1.2	1.2
9.3	9.9	10.6	11.6
11.0	12.0	13.5	14.5
140	185	220	250
33	36	54	58
44	48	68	73
44	48	68	73
90	113	158	188
19.1	19.1	19.1	11.4
13.7	13.7	13.7	8.21
3	3	3	3
18.0	18.0	18.0	26
87	87	87	115
0.153	0.153	0.153	0.255

Table 2.3.5
SINOTHERM heat-resistant silicone rubber-insulated single-core non-sheathed cables finely stranded conductor

Design	2	Nom. cross-sectional area of conductor	mm²	0.75	1	1.5	2.5
	6	Type of conductor		F	F	F	F
	7	Coating of strands		plain	plain	plain	plain
	8	Number of strands		24	32	30	50
	9	Diameter of strand	mm	0.21	0.21	0.26	0.26
	10	Diameter of conductor	mm	1.3	1.5	1.8	2.6
	11	Thickness of insulation	mm	0.6	0.6	0.6	0.7
	15	Overall diameter of cable (min. value)	mm	2.1	2.3	2.5	3.1
	16	Overall diameter of cable (max. value)	mm	2.8	3.0	3.5	4.2
	19	Weight of cable	kg/km	10	15	20	30
Mechanical properties	20	Minimum bending radius for fixed installation	mm	11	12	14	17
	21	Minimum bending radius, formed bend	mm	3	3	4	4
	29	Perm. pulling force for install. and operation	N	11	15	23	38
Electrical properties	30	d.c. resistance/unit length at 180 °C	Ω/km	42.3	31.8	21.7	13.0
	31	d.c. resistance/unit length at 20 °C	Ω/km	26.0	19.5	13.3	7.98
Current-carrying capacity	37	Number of loaded cores		1	1	1	1
	38	Installation in free air	A	15.0	19.0	24	32
	42	Minimum time value	s	44	49	70	109
Short-circuit	43	Rated short-time current of conductor (1s)	kA	0.0990	0.132	0.198	0.330

Table 2.3.6
SINOTHERM heat-resistant silicone rubber-insulated single-core non-sheathed cables finely stranded conductor

Design	2	Nom. cross-sectional area of conductor	mm²	1.5	2.5	4	6
	6	Type of conductor		F	F	F	F
	7	Coating of strands		plain	plain	plain	plain
	8	Number of strands		30	50	56	84
	9	Diameter of strand	mm	0.26	0.26	0.31	0.31
	10	Diameter of conductor	mm	1.8	2.6	3.2	3.9
	11	Thickness of insulation	mm	0.8	0.9	1.0	1.0
	15	Overall diameter of cable (min. value)	mm	2.9	3.5	4.2	4.8
	16	Overall diameter of cable (max. value)	mm	3.9	4.6	5.2	6.7
	19	Weight of cable	kg/km	20	30	50	65
Mechanical properties	20	Minimum bending radius for fixed installation	mm	23	28	31	40
	21	Minimum bending radius, formed bend	mm	16	18	21	27
	29	Perm. pulling force for install. and operation	N	23	38	60	90
Electrical properties	30	d.c. resistance/unit length at 180 °C	Ω/km	21.7	13.0	8.06	5.38
	31	d.c. resistance/unit length at 20 °C	Ω/km	13.3	7.98	4.95	3.30
Current-carrying capacity	37	Number of loaded cores		1	1	1	1
	38	Installation in free air	A	24	32	42	54
	42	Minimum time value	s	70	109	161	220
Short-circuit	43	Rated short-time current of conductor (1s)	kA	0.198	0.330	0.528	0.792

SIAF 1×... 300/500 V

4	6	10	16	25	35	50	70	95	120
F	F	F	F	F	F	F	F	F	F
plain	plain	plain	plain	plain	plain	plain	plain	plain	plain
56	84	80	126	196	276	396	360	475	608
0.31	0.31	0.41	0.41	0.41	0.41	0.41	0.51	0.51	0.51
3.2	3.9	5.1	6.3	7.8	9.2	11.0	13.1	15.1	17.0
0.8	0.8	1.0	1.0	1.2	1.2	1.4	1.4	1.6	1.6
3.8	4.4	5.8	6.8	8.3	9.6	11.4	13.1	15.0	16.6
4.8	6.3	7.6	8.9	11.0	13.0	15.0	17.5	19.5	21.5
45	65	110	175	280	375	530	750	975	1230
19	25	30	36	44	52	60	70	78	86
5	6	8	9	22	26	30	35	39	43
60	90	150	240	375	525	750	1050	1425	1800
8.06	5.38	3.11	1.97	1.27	0.902	0.634	0.443	0.336	0.262
4.95	3.30	1.91	1.21	0.780	0.554	0.389	0.272	0.206	0.161
1	1	1	1	1	1	1	1	1	1
42	54	73	98	129	158	198	245	292	344
161	220	334	474	668	873	1135	1452	1883	2165
0.528	0.792	1.32	2.11	3.30	4.62	6.60	9.24	12.5	15.8

SIAF 1×... 0.6/1 kV

10	16	25	35	50	70	95
F	F	F	F	F	F	F
plain	plain	plain	plain	plain	plain	plain
80	126	196	276	396	360	475
0.41	0.41	0.41	0.41	0.41	0.51	0.51
5.1	6.3	7.8	9.2	11.0	13.1	15.1
1.2	1.2	1.4	1.4	1.6	1.6	1.8
6.2	7.2	8.7	10.0	11.8	13.5	15.4
8.0	9.3	11.4	13.4	15.4	17.9	20.0
115	180	275	365	520	710	930
48	56	68	80	92	107	120
32	37	46	54	62	72	80
150	240	375	525	750	1050	1425
3.11	1.97	1.27	0.902	0.629	0.443	0.336
1.91	1.21	0.780	0.554	0.386	0.272	0.206
1	1	1	1	1	1	1
73	98	129	158	198	245	292
334	474	668	873	1135	1452	1883
1.32	2.11	3.30	4.62	6.60	9.24	12.5

Table 2.3.7
SINOTHERM heat-resistant silicone rubber-insulated single-core non-sheathed cables glass fibre braid, finely stranded conductor

Design	2	Nom. cross-sectional area of conductor	mm^2	0.75	1	1.5	2.5
	6	Type of conductor		F	F	F	F
	7	Coating of strands		plain	plain	plain	plain
	8	Number of strands		24	32	30	50
	9	Diameter of strand	mm	0.21	0.21	0.26	0.26
	10	Diameter of conductor	mm	1.3	1.5	1.8	2.6
	11	Thickness of insulation	mm	0.6	0.6	0.7	0.8
	15	Overall diameter of cable (min. value)	mm	2.6	2.7	3.2	3.8
	16	Overall diameter of cable (max. value)	mm	3.6	3.8	4.3	5.0
	19	Weight of cable	kg/km	15	20	25	40
Mechanical properties	20	Minimum bending radius for fixed installation	mm	14	15	17	20
	21	Minimum bending radius, formed bend	mm	4	4	4	5
	29	Perm. pulling force for install. and operation	N	11	15	23	38
Electrical properties	30	d.c. resistance/unit length at 180 °C	Ω/km	42.4	31.8	21.7	13.0
	31	d.c. resistance/unit length at 20 °C	Ω/km	26.0	19.5	13.3	7.98
Current-carrying capacity	37	Number of loaded cores		1	1	1	1
	38	Installation in free air	A	15.0	19.0	24	32
	42	Minimum time value	s	44	49	70	109
Short-circuit	43	Rated short-time current of conductor (1s)	kA	0.0990	0.132	0.198	0.330

Table 2.3.8
SINOTHERM heat-resistant silicone rubber-insulated and -sheathed cords 2, 3, 4 and 5 cores

Design	1	Number of cores		2	2	2	2	2
	2	Nom. cross-sectional area of conductor	mm^2	0.75 ·	1	1.5	2.5	4
	6	Type of conductor		F	F	F	F	F
	7	Coating of strands		plain	plain	plain	plain	plain
	8	Number of strands		24	32	30	50	56
	9	Diameter of strand	mm	0.21	0.21	0.26	0.26	0.31
	10	Diameter of conductor	mm	1.3	1.5	1.8	2.6	3.2
	11	Thickness of insulation	mm	0.6	0.6	0.7	0.8	0.8
	12	Thickness of sheath	mm	0.8	0.9	1.0	1.1	1.2
	15	Overall diameter of cable (min. value)	mm	6.0	6.4	7.6	9.0	10.0
	16	Overall diameter of cable (max. value)	mm	7.2	7.8	9.0	10.5	12.0
	19	Weight of cable	kg/km	50	60	80	125	180
Mechanical properties	20	Minimum bending radius for fixed installation	mm	22	23	27	32	36
	22	Minimum bending radius for free flexing	mm	22	23	36	42	48
	23	Minimum bending radius for cable entry	mm	22	23	36	42	48
	29	Perm. pulling force for install. and operation	N	23	30	45	75	120
Electrical properties	30	d.c. resistance/unit length at 180 °C	Ω/km	42.3	31.8	21.7	13.0	8.06
	31	d.c. resistance/unit length at 20 °C	Ω/km	26.0	19.5	13.3	7.98	4.95
Current-carrying capacity	37	Number of loaded cores		2	2	2	2	2
	41	Installation on walls	A	12.0	15.0	18.0	26	34
	42	Minimum time value	s	70	79	124	164	246
Short-circuit	43	Rated short-time current of conductor (1s)	kA	0.0990	0.132	0.198	0.330	0.528

H05SJ-K 1X... 300/500 V | A05SJ-K 1X... 300/500 V

4	6	10	16	25	35	50	70	95
F	F	F	F	F	F	F	F	F
plain	plain	plain	plain	plain	plain	plain	plain	plain
56	84	80	126	196	276	396	360	475
0.31	0.31	0.41	0.41	0.41	0.41	0.41	0.51	0.51
3.2	3.9	5.1	6.3	7.8	9.2	11.0	13.1	15.1
0.8	0.8	1.0	1.0	1.2	1.2	1.4	1.4	1.6
4.3	4.8	6.0	8.1	9.7	11.6	13.5	15.6	17.5
5.6	6.2	8.2	9.6	12.0	13.5	15.5	18.0	20.0
55	75	130	200	305	420	595	825	1075
22	25	33	38	48	54	62	72	80
6	6	8	10	24	27	31	36	40
60	90	150	240	375	525	750	1050	1425
8.06	8.06	3.11	1.97	1.97	0.902	0.629	0.443	0.336
4.95	4.95	1.91	1.21	1.21	0.554	0.386	0.272	0.206
1	1	1	1	1	1	1	1	1
42	54	73	98	129	158	198	245	292
161	220	334	474	668	873	1135	1452	1883
0.528	0.792	1.32	2.11	3.30	4.62	6.60	9.24	12.5

N2GMH2G ... ×... 300/500 V

| 3 | 3 | 3 | 3 | 3 | 4 | 4 | 4 | 4 | 4 | 5 |
0.75	1	1.5	2.5	4	0.75	1	1.5	2.5	4	1.5
F	F	F	F	F	F	F	F	F	F	F
plain	plain	plain	plain	plain	plain	plain	plain	plain	plain	plain
24	32	30	50	56	24	32	30	50	56	30
0.21	0.21	0.26	0.26	0.31	0.21	0.21	0.26	0.26	0.31	0.26
1.3	1.5	1.8	2.6	3.2	1.3	1.5	1.8	2.6	3.2	1.8
0.6	0.6	0.7	0.8	0.8	0.6	0.6	0.7	0.8	0.8	0.7
0.9	0.9	1.0	1.1	1.2	0.9	0.9	1.1	1.2	1.3	1.1
6.6	6.8	8.0	9.6	11.0	7.2	7.6	9.0	10.5	12.0	9.8
7.8	8.2	9.6	11.5	13.0	8.4	8.8	10.5	12.5	14.0	11.5
60	75	100	150	225	80	95	125	185	290	155
23	25	29	35	52	25	26	32	50	56	35
23	33	38	46	65	34	35	42	63	70	46
23	33	38	46	65	34	35	42	63	70	46
34	45	68	113	180	45	60	90	150	240	113
42.3	31.8	21.7	13.0	8.06	42.3	31.8	21.7	13.0	8.06	21.7
26.0	19.5	13.3	7.98	4.95	26.0	19.5	13.3	7.98	4.95	13.3
2	2	2	2	2	3	3	3	3	3	3
12.0	15.0	18.0	26	34	12.0	15.0	18.0	26	34	18.0
70	79	124	164	246	70	79	124	164	246	124
0.0990	0.132	0.198	0.330	0.528	0.0990	0.132	0.198	0.330	0.528	0.198

Table 2.4.1a
PROTOFLEX PVC control cables nominal cross-sectional area of conductor = 0.75 mm^2

Design					3	4	5	7
	6	Type of conductor			F	F	F	F
	7	Coating of strands			plain	plain	plain	plain
	8	Number of strands			24	24	24	24
	9	Diameter of strand		mm	0.21	0.21	0.21	0.21
	10	Diameter of conductor		mm	1.3	1.3	1.3	1.3
	11	Thickness of insulation		mm	0.6	0.6	0.6	0.6
	12	Thickness of sheath		mm	0.8	0.8	1.0	1.0
	15	Overall diameter of cable (min. value)		mm	6.2	6.8	7.8	9.1
	16	Overall diameter of cable (max. value)		mm	7.9	8.2	9.6	11.2
	19	Weight of cable		kg/km	70	80	110	150
Mechanical properties	22	Minimum bending radius for free flexing		mm	24	33	38	45
	23	Minimum bending radius for cable entry		mm	24	33	38	45
	29	Perm. pulling force for install. and operation		N	34	45	56	79
Electrical properties	30	d.c. resistance/unit length at 70 °C		Ω/km	31.1	31.1	31.1	31.1
	31	d.c. resistance/unit length at 20 °C		Ω/km	26.0	26.0	26.0	26.0
Current-carrying capacity	37	Number of loaded cores			2	3	3	3
	41	Installation on walls		A	12.0	12.0	12.0	12.0
	42	Minimum time value		s	25	25	25	25
Short-circuit	43	Rated short-time current of conductor (1s)		kA	0.0817	0.0817	0.0817	0.0817

Table 2.4.1b
PROTOFLEX PVC control cables nominal cross-sectional area of conductor = 1 mm^2

Design					3	4	5	7
	6	Type of conductor			F	F	F	F
	7	Coating of strands			plain	plain	plain	plain
	8	Number of strands			32	32	32	32
	9	Diameter of strand		mm	0.21	0.21	0.21	0.21
	10	Diameter of conductor		mm	1.5	1.5	1.5	1.5
	11	Thickness of insulation		mm	0.6	0.6	0.6	0.6
	12	Thickness of sheath		mm	0.8	0.8	1.0	1.0
	15	Overall diameter of cable (min. value)		mm	6.5	7.1	8.1	9.5
	16	Overall diameter of cable (max. value)		mm	8.3	8.5	10.2	11.6
	19	Weight of cable		kg/km	75	95	125	170
Mechanical properties	22	Minimum bending radius for free flexing		mm	33	34	41	46
	23	Minimum bending radius for cable entry		mm	33	34	41	46
	29	Perm. pulling force for install. and operation		N	45	60	75	105
Electrical properties	30	d.c. resistance/unit length at 70 °C		Ω/km	23.3	23.3	23.3	23.3
	31	d.c. resistance/unit length at 20 °C		Ω/km	19.5	19.5	19.5	19.5
Current-carrying capacity	37	Number of loaded cores			2	3	3	3
	41	Installation on walls		A	15.0	15.0	15.0	15.0
	42	Minimum time value		s	29	29	29	29
Short-circuit	43	Rated short-time current of conductor (1s)		kA	0.109	0.109	0.109	0.109

NYSLYÖ ... ×0.75 300/500 V

12	18	25	34	50	60
F	F	F	F	F	F
plain	plain	plain	plain	plain	plain
24	24	24	24	24	24
0.21	0.21	0.21	0.21	0.21	0.21
1.3	1.3	1.3	1.3	1.3	1.3
0.6	0.6	0.6	0.6	0.6	0.6
1.2	1.2	1.5	1.8	1.8	2.1
11.7	13.5	16.7	19.1	22.2	24.3
14.2	15.8	19.6	22.1	25.4	28.6
215	300	430	585	790	950
71	79	98	110	127	143
71	79	98	110	127	143
135	203	281	383	563	675
31.1	31.1	31.1	31.1	31.1	31.1
26.0	26.0	26.0	26.0	26.0	26.0
3	3	3	3	3	3
12.0	12.0	12.0	12.0	12.0	12.0
25	25	25	25	25	25
0.0817	0.0817	0.0817	0.0817	0.0817	0.0817

NYSLYÖ ... ×1 300/500 V

12	18	25	34	50	60
F	F	F	F	F	F
plain	plain	plain	plain	plain	plain
32	32	32	32	32	32
0.21	0.21	0.21	0.21	0.21	0.21
1.5	1.5	1.5	1.5	1.5	1.5
0.6	0.6	0.6	0.6	0.6	0.6
1.2	1.5	1.8	1.8	2.1	2.1
12.3	14.8	18.1	20.0	23.9	25.5
14.3	17.3	21.3	23.9	28.9	29.9
250	375	520	670	980	1120
71	86	106	119	144	149
71	86	106	119	144	149
180	270	375	510	750	900
23.3	23.3	23.3	23.3	23.3	23.3
19.5	19.5	19.5	19.5	19.5	19.5
3	3	3	3	3	3
15.0	15.0	15.0	15.0	15.0	15.0
29	29	29	29	29	29
0.109	0.109	0.109	0.109	0.109	0.109

Table 2.4.1c
PROTOFLEX PVC control cables nominal cross-sectional area of conductor = 1.5 mm^2

Design					3	4	5	7
Design	1	Number of cores			3	4	5	7
	6	Type of conductor			F	F	F	F
	7	Coating of strands			plain	plain	plain	plain
	8	Number of strands			30	30	30	30
	9	Diameter of strand	mm		0.26	0.26	0.26	0.26
	10	Diameter of conductor	mm		1.8	1.8	1.8	1.8
	11	Thickness of insulation	mm		0.6	0.6	0.6	0.6
	12	Thickness of sheath	mm		0.8	1.0	1.0	1.2
	15	Overall diameter of cable (min. value)	mm		7.1	8.0	8.8	10.7
	16	Overall diameter of cable (max. value)	mm		8.9	9.6	11.0	12.9
	19	Weight of cable	kg/km		95	125	155	225
Mechanical properties	22	Minimum bending radius for free flexing	mm		36	38	44	64
	23	Minimum bending radius for cable entry	mm		36	38	44	64
	29	Perm. pulling force for install. and operation	N		68	90	113	158
Electrical properties	30	d.c. resistance/unit length at 70 °C	Ω/km		15.9	15.9	15.9	15.9
	31	d.c. resistance/unit length at 20 °C	Ω/km		13.3	13.3	13.3	13.3
Current-carrying capacity	37	Number of loaded cores			2	3	3	3
	41	Installation on walls	A		18.0	18.0	18.0	18.0
	42	Minimum time value	s		45	45	45	45
Short-circuit	43	Rated short-time current of conductor (1s)	kA		0.163	0.163	0.163	0.163

Table 2.4.1d
PROTOFLEX PVC control cables nominal cross-sectional area of conductor = 2.5 mm^2

Design					3	4	5	7
Design	1	Number of cores			3	4	5	7
	6	Type of conductor			F	F	F	F
	7	Coating of strands			plain	plain	plain	plain
	8	Number of strands			50	50	50	50
	9	Diameter of strand	mm		0.26	0.26	0.26	0.26
	10	Diameter of conductor	mm		2.6	2.6	2.6	2.6
	11	Thickness of insulation	mm		0.7	0.7	0.7	0.7
	12	Thickness of sheath	mm		1.0	1.2	1.2	1.2
	15	Overall diameter of cable (min. value)	mm		8.7	9.1	10.8	12.6
	16	Overall diameter of cable (max. value)	mm		10.7	12.0	12.1	14.9
	19	Weight of cable	kg/km		150	195	240	330
Mechanical properties	22	Minimum bending radius for free flexing	mm		43	48	60	74
	23	Minimum bending radius for cable entry	mm		43	48	60	74
	29	Perm. pulling force for install. and operation	N		113	150	188	263
Electrical properties	30	d.c. resistance/unit length at 70 °C	Ω/km		9.55	9.55	9.55	9.55
	31	d.c. resistance/unit length at 20 °C	Ω/km		7.98	7.98	7.98	7.98
Current-carrying capacity	37	Number of loaded cores			2	3	3	3
	41	Installation on walls	A		26	26	26	26
	42	Minimum time value	s		60	60	60	60
Short-circuit	43	Rated short-time current of conductor (1s)	kA		0.272	0.272	0.272	0.272

NYSLYÖ ... ×1.5 300/500 V

12	18	25	34	41	50	60
F	F	F	F	F	F	F
plain	plain	plain	plain	plain	plain	plain
30	30	30	30	30	30	30
0.26	0.26	0.26	0.26	0.26	0.26	0.26
1.8	1.8	1.8	1.8	1.8	1.8	1.8
0.6	0.6	0.6	0.6	0.6	0.6	0.6
1.2	1.5	1.8	1.8	2.1	2.1	2.1
13.3	16.0	19.6	21.7	23.1	25.9	27.8
16.1	18.8	23.2	26.1	27.1	31.5	32.3
315	470	670	850	1100	1240	1440
80	94	116	130	135	158	161
80	94	116	130	135	158	161
270	405	563	765	923	1125	1350
15.9	15.9	15.9	15.9	15.9	15.9	15.9
13.3	13.3	13.3	13.3	13.3	13.3	13.3
3	3	3	3	3	3	3
18.0	18.0	18.0	18.0	18.0	18.0	18.0
45	45	45	45	45	45	45
0.163	0.163	0.163	0.163	0.163	0.163	0.163

NYSLYÖ ... ×2.5 300/500 V

12	14	18
F	F	F
plain	plain	plain
50	50	50
0.26	0.26	0.26
2.6	2.6	2.6
0.7	0.7	0.7
1.5	1.8	1.8
16.4	17.7	19.6
19.3	20.8	22.6
500	590	735
96	104	113
96	104	113
450	525	675
9.55	9.55	9.55
7.98	7.98	7.98
3	3	3
26	26	26
60	60	60
0.272	0.272	0.272

Table 2.4.2
PROTOFLEX PVC control cables 2 cores

(N)YSLYÖ 2×... 300/500 V

Design	2	Nom. cross-sectional area of conductor	mm²	1	1.5
	6	Type of conductor		F	F
	7	Coating of strands		plain	plain
	8	Number of strands		32	30
	9	Diameter of strand	mm	0.21	0.26
	10	Diameter of conductor	mm	1.5	1.8
	11	Thickness of insulation	mm	0.6	0.6
	12	Thickness of sheath	mm	0.7	0.7
	15	Overall diameter of cable (min. value)	mm	5.8	6.3
	16	Overall diameter of cable (max. value)	mm	7.2	7.7
	19	Weight of cable	kg/km	65	80
Mechanical properties	22	Minimum bending radius for free flexing	mm	22	23
	23	Minimum bending radius for cable entry	mm	22	23
	29	Perm. pulling force for install. and operation	N	30	45
Electrical properties	30	d.c. resistance/unit length at 70 °C	Ω/km	23.3	15.9
	31	d.c. resistance/unit length at 20 °C	Ω/km	19.5	13.3
Current-carrying capacity	37	Number of loaded cores		2	2
	41	Installation on walls	A	15.0	18.0
	42	Minimum time value	s	29	45
Short-circuit	43	Rated short-time current of conductor (1s)	kA	0.109	0.163

Table 2.4.3
PROTOFLEX PVC control cables nominal cross-sectional area of conductor = 4 and 6 mm²

Design	1	Number of cores		4	5	7	4
	2	Nom. cross-sectional area of conductor	mm²	4	4	4	6
	6	Type of conductor		F	F	F	F
	7	Coating of strands		plain	plain	plain	plain
	8	Number of strands		56	56	56	84
	9	Diameter of strand	mm	0.31	0.31	0.31	0.31
	10	Diameter of conductor	mm	3.2	3.2	3.2	3.9
	11	Thickness of insulation	mm	0.8	0.8	0.8	0.8
	12	Thickness of sheath	mm	1.2	1.2	1.2	1.5
	15	Overall diameter of cable (min. value)	mm	11.5	12.8	13.9	12.6
	16	Overall diameter of cable (max. value)	mm	13.8	15.1	16.3	15.4
	19	Weight of cable	kg/km	280	340	460	400
Mechanical properties	22	Minimum bending radius for free flexing	mm	69	75	81	77
	23	Minimum bending radius for cable entry	mm	69	75	81	77
	29	Perm. pulling force for install. and operation	N	240	300	420	360
Electrical properties	30	d.c. resistance/unit length at 70 °C	Ω/km	5.92	5.92	5.92	3.95
	31	d.c. resistance/unit length at 20 °C	Ω/km	4.95	4.95	4.95	3.30
Current-carrying capacity	37	Number of loaded cores		3	3	3	3
	41	Installation on walls	A	34	34	34	44
	42	Minimum time value	s	89	89	89	120
Short-circuit	43	Rated short-time current of conductor (1s)	kA	0.436	0.436	0.436	0.654

YSLYÖ ... × ... 300/500 V

5	7
6	6
F	F
plain	plain
84	84
0.31	0.31
3.9	3.9
0.8	0.8
1.5	1.8
14.4	16.0
17.6	19.0
500	650
88	95
88	95
450	630
3.95	3.95
3.30	3.30
3	3
44	44
120	120
0.654	0.654

Table 2.4.4 a
Screened PROTOFLEX PVC control cables nominal cross-sectional area of conductor = 0.75 and 1.5 mm^2

Design				3	4	5	7
	1	Number of cores		3	4	5	7
	2	Nom. cross-sectional area of conductor	mm^2	0.75	0.75	0.75	0.75
	6	Type of conductor		F	F	F	F
	7	Coating of strands		plain	plain	plain	plain
	8	Number of strands		24	24	24	24
	9	Diameter of strand	mm	0.21	0.21	0.21	0.21
	10	Diameter of conductor	mm	1.3	1.3	1.3	1.3
	11	Thickness of insulation	mm	0.6	0.6	0.6	0.6
	13	Thickness of inner sheath	mm	0.6	0.7	0.7	0.7
	14	Thickness of outer sheath	mm	1.0	1.2	1.2	1.2
	15	Overall diameter of cable (min. value)	mm	8.3	9.5	10.1	11.4
	16	Overall diameter of cable (max. value)	mm	9.8	11.0	12.0	13.6
	19	Weight of cable	kg/km	125	150	180	230
Mechanical properties	22	Minimum bending radius for free flexing	mm	39	44	48	68
	23	Minimum bending radius for cable entry	mm	39	44	48	68
	29	Perm. pulling force for install. and operation	N	34	45	56	79
Electrical properties	30	d.c. resistance/unit length at 70 °C	Ω/km	31.1	31.1	31.1	31.1
	31	d.c. resistance/unit length at 20 °C	Ω/km	26.0	26.0	26.0	26.0
Current-carrying capacity	37	Number of loaded cores		2	3	3	3
	41	Installation on walls	A	12.0	12.0	12.0	12.0
	42	Minimum time value	s	25	25	25	25
Short-circuit	43	Rated short-time current of conductor (1s)	kA	0.0817	0.0817	0.0817	0.0817

Table 2.4.4 b
Screened PROTOFLEX PVC control cables nominal cross-sectional area of conductor = 1.5 and 2.5 mm^2

Design				3	4	5	7
	1	Number of cores		3	4	5	7
	2	Nom. cross-sectional area of conductor	mm^2	1.5	1.5	1.5	1.5
	6	Type of conductor		F	F	F	F
	7	Coating of strands		plain	plain	plain	plain
	8	Number of strands		30	30	30	30
	9	Diameter of strand	mm	0.26	0.26	0.26	0.26
	10	Diameter of conductor	mm	1.8	1.8	1.8	1.8
	11	Thickness of insulation	mm	0.6	0.6	0.6	0.6
	13	Thickness of inner sheath	mm	0.7	0.7	0.7	0.8
	14	Thickness of outer sheath	mm	1.2	1.2	1.2	1.2
	15	Overall diameter of cable (min. value)	mm	9.7	10.3	11.1	12.8
	16	Overall diameter of cable (max. value)	mm	11.9	12.5	13.3	15.1
	19	Weight of cable	kg/km	170	200	235	330
Mechanical properties	22	Minimum bending radius for free flexing	mm	48	63	66	75
	23	Minimum bending radius for cable entry	mm	48	63	66	75
	29	Perm. pulling force for install. and operation	N	68	90	113	158
Electrical properties	30	d.c. resistance/unit length at 70 °C	Ω/km	15.9	15.9	15.9	15.9
	31	d.c. resistance/unit length at 20 °C	Ω/km	13.3	13.3	13.3	13.3
Current-carrying capacity	37	Number of loaded cores		2	3	3	3
	41	Installation on walls	A	18.0	18.0	18.0	18.0
	42	Minimum time value	s	45	45	45	45
Short-circuit	43	Rated short-time current of conductor (1s)	kA	0.163	0.163	0.163	0.163

NYSLYCYÖ ... × ... 300/500 V

12	18	25	3	4	5	7	12	18	25
0.75	0.75	0.75	1	1	1	1	1	1	1
F	F	F	F	F	F	F	F	F	F
plain	plain	plain	plain	plain	plain	plain	plain	plain	plain
24	24	24	32	32	32	32	32	32	32
0.21	0.21	0.21	0.21	0.21	0.21	0.21	0.21	0.21	0.21
1.3	1.3	1.3	1.5	1.5	1.5	1.5	1.5	1.5	1.5
0.6	0.6	0.6	0.6	0.6	0.6	0.6	0.6	0.6	0.6
0.8	1.0	1.0	0.7	0.7	0.7	0.7	1.0	1.0	1.2
1.2	1.5	1.8	1.0	1.2	1.2	1.2	1.5	1.8	1.8
14.0	16.8	19.9	8.8	9.7	10.4	11.8	15.5	18.0	21.1
16.4	19.7	23.0	10.3	11.9	12.6	14.1	18.4	21.0	24.3
345	470	640	140	170	200	255	410	550	770
82	98	115	41	48	63	70	92	105	121
82	98	115	41	48	63	70	92	105	121
135	203	281	45	60	75	105	180	270	375
31.1	31.1	31.1	23.3	23.3	23.3	23.3	23.3	23.3	23.3
26.0	26.0	26.0	19.5	19.5	19.5	19.5	19.5	19.5	19.5
3	3	3	2	3	3	3	3	3	3
12.0	12.0	12.0	15.0	15.0	15.0	15.0	15.0	15.0	15.0
25	25	25	29	29	29	29	29	29	29
0.0817	0.0817	0.0817	0.109	0.109	0.109	0.109	0.109	0.109	0.109

NYSLYCYÖ ... × ... 300/500 V

12	18	25	3	4	5	7	12	18	25
1.5	1.5	1.5	2.5	2.5	2.5	2.5	2.5	2.5	2.5
F	F	F	F	F	F	F	F	F	F
plain	plain	plain	plain	plain	plain	plain	plain	plain	plain
30	30	30	50	50	50	50	50	50	50
0.26	0.26	0.26	0.26	0.26	0.26	0.26	0.26	0.26	0.26
1.8	1.8	1.8	2.6	2.6	2.6	2.6	2.6	2.6	2.6
0.6	0.6	0.6	0.7	0.7	0.7	0.7	0.7	0.7	0.7
1.0	1.0	1.2	0.7	0.8	0.8	0.8	1.0	1.2	1.4
1.5	1.8	2.1	1.2	1.2	1.2	1.2	1.8	2.1	2.1
16.6	19.2	23.2	11.0	12.0	13.1	14.8	19.6	23.2	27.5
19.5	22.3	26.5	13.2	14.3	15.4	17.7	22.7	26.5	31.0
500	680	930	230	270	340	450	690	990	1390
98	111	133	66	71	77	88	113	133	155
98	111	133	66	71	77	88	113	133	155
270	405	563	113	150	188	263	450	675	938
15.9	15.9	15.9	9.55	9.55	9.55	9.55	9.55	9.55	9.55
13.3	13.3	13.3	7.98	7.98	7.98	7.98	7.98	7.98	7.98
3	3	3	2	3	3	3	3	3	3
18.0	18.0	18.0	26	26	26	26	26	26	26
45	45	45	60	60	60	60	60	60	60
0.163	0.163	0.163	0.272	0.272	0.272	0.272	0.272	0.272	0.272

Table 2.4.5
Screened PROTOFLEX PVC control cables nominal cross-sectional area of conductor = 4 and 6 mm^2

Design							
	1	Number of cores		4	5	7	4
	2	Nom. cross-sectional area of conductor	mm^2	4	4	4	6
	6	Type of conductor		F	F	F	F
	7	Coating of strands		plain	plain	plain	plain
	8	Number of strands		56	56	56	84
	9	Diameter of strand	mm	0.31	0.31	0.31	0.31
	10	Diameter of conductor	mm	3.2	3.2	3.2	3.9
	11	Thickness of insulation	mm	0.8	0.8	0.8	0.8
	13	Thickness of inner sheath	mm	0.8	1.0	1.0	1.0
	14	Thickness of outer sheath	mm	1.2	1.5	1.8	1.8
	15	Overall diameter of cable (min. value)	mm	13.9	16.0	17.7	16.4
	16	Overall diameter of cable (max. value)	mm	16.3	18.9	20.7	19.4
	19	Weight of cable	kg/km	405	520	645	560
Mechanical properties	22	Minimum bending radius for free flexing	mm	81	94	103	97
	23	Minimum bending radius for cable entry	mm	81	94	103	97
	29	Perm. pulling force for install. and operation	N	240	300	420	360
Electrical properties	30	d.c. resistance/unit length at 70 °C	Ω/km	5.92	5.92	5.92	3.95
	31	d.c. resistance/unit length at 20 °C	Ω/km	4.95	4.95	4.95	3.30
Current-carrying capacity	37	Number of loaded cores		3	3	3	3
	41	Installation on walls	A	34	34	34	44
	42	Minimum time value	s	89	89	89	120
Short-circuit	43	Rated short-time current of conductor (1s)	kA	0.436	0.436	0.436	0.654

Table 2.4.6
Screened PROTOFLEX PVC control cables coloured cores

Design							
	1	Number of cores		3	4	3	5
	2	Nom. cross-sectional area of conductor	mm^2	0.75	0.75	1.5	1.5
	6	Type of conductor		F	F	F	F
	7	Coating of strands		plain	plain	plain	plain
	8	Number of strands		24	24	30	30
	9	Diameter of strand	mm	0.21	0.21	0.26	0.26
	10	Diameter of conductor	mm	1.3	1.3	1.8	1.8
	11	Thickness of insulation	mm	0.6	0.6	0.6	0.6
	13	Thickness of inner sheath	mm	0.6	0.7	0.7	0.7
	14	Thickness of outer sheath	mm	1.0	1.2	1.2	1.2
	15	Overall diameter of cable (min. value)	mm	7.9	9.5	9.0	10.5
	16	Overall diameter of cable (max. value)	mm	9.4	11.0	10.5	12.9
	19	Weight of cable	kg/km	130	150	165	230
Mechanical properties	22	Minimum bending radius for free flexing	mm	38	44	42	64
	23	Minimum bending radius for cable entry	mm	38	44	42	64
	29	Perm. pulling force for install. and operation	N	34	45	68	113
Electrical properties	30	d.c. resistance/unit length at 70 °C	Ω/km	31.1	31.1	15.9	15.9
	31	d.c. resistance/unit length at 20 °C	Ω/km	26.0	26.0	13.3	13.3
Current-carrying capacity	37	Number of loaded cores		2	3	2	3
	41	Installation on walls	A	12.0	12.0	18.0	18.0
	42	Minimum time value	s	25	25	45	45
Short-circuit	43	Rated short-time current of conductor (1s)	kA	0.0817	0.0817	0.163	0.163

YSLYCYÖ ... × ... 300/500 V

5	7
6	6
F	F
plain	plain
84	84
0.31	0.31
3.9	3.9
0.8	0.8
1.0	1.2
1.8	1.8
18.4	20.9
21.4	23.9
690	850
107	119
107	119
450	630
3.95	3.95
3.30	3.30
3	3
44	44
120	120
0.654	0.654

YSLYCY ... × ... 300/500 V

3	5	5	5
2.5	2.5	4	6
F	F	F	F
plain	plain	plain	plain
50	50	56	84
0.26	0.26	0.31	0.31
2.6	2.6	3.2	3.9
0.7	0.7	0.8	0.8
0.7	0.8	1.0	1.0
1.2	1.2	1.5	1.8
10.4	12.8	15.2	16.8
12.8	15.2	18.2	19.5
225	330	505	640
64	76	91	98
64	76	91	98
113	188	300	450
9.55	9.55	5.92	3.95
7.98	7.98	4.95	3.30
2	3	3	3
26	26	34	44
60	60	89	120
0.272	0.272	0.436	0.654

Table 2.4.7 a
PVC control cables nominal cross-sectional area of conductor = 0.5 mm²

Design	1	Number of cores		3	4	5	7
	6	Type of conductor		F	F	F	F
	7	Coating of strands		plain	plain	plain	plain
	8	Number of strands		16	16	16	16
	9	Diameter of strand	mm	0.21	0.21	0.21	0.21
	10	Diameter of conductor	mm	1.1	1.1	1.1	1.1
	11	Thickness of insulation	mm	0.6	0.6	0.6	0.6
	12	Thickness of sheath	mm	0.8	0.8	0.8	1.0
	15	Overall diameter of cable (min. value)	mm	5.6	6.1	6.6	7.3
	16	Overall diameter of cable (max. value)	mm	6.9	7.3	8.0	9.3
	19	Weight of cable	kg/km	55	65	75	105
Mechanical properties	20	Minimum bending radius for fixed installation	mm	28	29	32	37
	21	Minimum bending radius, formed bend	mm	7	7	8	9
	29	Perm. pulling force for install. and operation	N	23	30	38	53
Electrical properties	30	d.c. resistance/unit length at 70 °C	Ω/km	46.7	46.7	46.7	46.7
	31	d.c. resistance/unit length at 20 °C	Ω/km	39.0	39.0	39.0	39.0
Current-carrying capacity	37	Number of loaded cores		2	3	3	3
	41	Installation on walls	A	9.5	9.5	9.5	9.5
	42	Minimum time value	s	18	18	18	18
Short-circuit	43	Rated short-time current of conductor (1s)	kA	0.0545	0.0545	0.0545	0.0545

Table 2.4.7 b
PVC control cables nominal cross-sectional area of conductor = 0.75 mm²

Design	1	Number of cores		3	4	5	7
	6	Type of conductor		F	F	F	F
	7	Coating of strands		plain	plain	plain	plain
	8	Number of strands		24	24	24	24
	9	Diameter of strand	mm	0.21	0.21	0.21	0.21
	10	Diameter of conductor	mm	1.3	1.3	1.3	1.3
	11	Thickness of insulation	mm	0.6	0.6	0.6	0.6
	12	Thickness of sheath	mm	0.8	0.8	1.0	1.0
	15	Overall diameter of cable (min. value)	mm	5.9	6.5	7.3	7.9
	16	Overall diameter of cable (max. value)	mm	7.3	7.9	9.1	9.9
	19	Weight of cable	kg/km	65	80	105	125
Mechanical properties	20	Minimum bending radius for fixed installation	mm	29	32	36	40
	21	Minimum bending radius, formed bend	mm	7	8	9	10
	29	Perm. pulling force for install. and operation	N	34	45	56	79
Electrical properties	30	d.c. resistance/unit length at 70 °C	Ω/km	31.1	31.1	31.1	31.1
	31	d.c. resistance/unit length at 20 °C	Ω/km	26.0	26.0	26.0	26.0
Current-carrying capacity	37	Number of loaded cores		2	3	3	3
	41	Installation on walls	A	12.0	12.0	12.0	12.0
	42	Minimum time value	s	25	25	25	25
Short-circuit	43	Rated short-time current of conductor (1s)	kA	0.0817	0.0817	0.0817	0.0817

12	18	25	34	50	60
F	F	F	F	F	F
plain	plain	plain	plain	plain	plain
16	16	16	16	16	16
0.21	0.21	0.21	0.21	0.21	0.21
1.1	1.1	1.1	1.1	1.1	1.1
0.6	0.6	0.6	0.6	0.6	0.6
1.2	1.2	1.5	1.8	1.8	1.8
10.1	11.9	14.9	16.4	19.9	20.6
12.3	14.1	17.7	19.2	22.9	24.6
165	235	350	435	615	705
49	56	71	77	92	98
25	28	35	38	46	49
90	135	188	255	375	450
46.7	46.7	46.7	46.7	46.7	47.3
39.0	39.0	39.0	39.0	39.0	39.5
3	3	3	3	3	3
9.5	9.5	9.5	9.5	9.5	9.5
18	18	18	18	18	18
0.0545	0.0545	0.0545	0.0545	0.0545	0.0545

12	18	25	34	50	60
F	F	F	F	F	F
plain	plain	plain	plain	plain	plain
24	24	24	24	24	24
0.21	0.21	0.21	0.21	0.21	0.21
1.3	1.3	1.3	1.3	1.3	1.3
0.6	0.6	0.6	0.6	0.6	0.6
1.2	1.2	1.5	1.8	1.8	2.1
11.0	12.7	16.4	18.3	21.5	22.9
13.0	15.1	19.2	21.3	24.7	27.1
210	300	445	575	780	935
52	60	77	85	99	108
26	30	38	43	49	81
135	203	281	383	563	675
31.1	31.1	31.1	31.1	31.1	31.1
26.0	26.0	26.0	26.0	26.0	26.0
3	3	3	3	3	3
12.0	12.0	12.0	12.0	12.0	12.0
25	25	25	25	25	25
0.0817	0.0817	0.0817	0.0817	0.0817	0.0817

Table 2.4.7 c
PVC control cables nominal cross-sectional area of conductor $= 1$ mm^2

Design					2	3	4	5
	1	Number of cores			2	3	4	5
	6	Type of conductor			F	F	F	F
	7	Coating of strands			plain	plain	plain	plain
	8	Number of strands			32	32	32	32
	9	Diameter of strand	mm		0.21	0.21	0.21	0.21
	10	Diameter of conductor	mm		1.5	1.5	1.5	1.5
	11	Thickness of insulation	mm		0.6	0.6	0.6	0.6
	12	Thickness of sheath	mm		0.7	0.8	0.8	1.0
	15	Overall diameter of cable (min. value)	mm		5.6	6.0	6.6	7.6
	16	Overall diameter of cable (max. value)	mm		6.8	7.8	8.4	9.6
	19	Weight of cable	kg/km		60	75	90	115
Mechanical properties	20	Minimum bending radius for fixed installation	mm		27	31	34	38
	21	Minimum bending radius, formed bend	mm		7	8	8	10
	29	Perm. pulling force for install. and operation	N		30	45	60	75
Electrical properties	30	d.c. resistance/unit length at 70 °C	Ω/km		23.3	23.3	23.3	23.3
	31	d.c. resistance/unit length at 20 °C	Ω/km		19.5	19.5	19.5	19.5
Current-carrying capacity	37	Number of loaded cores			2	2	3	3
	41	Installation on walls	A		15.0	15.0	15.0	15.0
	42	Minimum time value	s		29	29	29	29
Short-circuit	43	Rated short-time current of conductor (1s)	kA		0.109	0.109	0.109	0.109

Table 2.4.7 d
PVC control cables nominal cross-sectional area of conductor $= 1.5$ mm^2

Design					2	3	4	5
	1	Number of cores			2	3	4	5
	6	Type of conductor			F	F	F	F
	7	Coating of strands			plain	plain	plain	plain
	8	Number of strands			30	30	30	30
	9	Diameter of strand	mm		0.26	0.26	0.26	0.26
	10	Diameter of conductor	mm		1.8	1.8	1.8	1.8
	11	Thickness of insulation	mm		0.6	0.6	0.6	0.6
	12	Thickness of sheath	mm		0.7	0.8	1.0	1.0
	15	Overall diameter of cable (min. value)	mm		6.1	6.4	7.9	8.2
	16	Overall diameter of cable (max. value)	mm		7.3	8.4	9.9	10.4
	19	Weight of cable	kg/km		75	95	130	150
Mechanical properties	20	Minimum bending radius for fixed installation	mm		29	34	40	42
	21	Minimum bending radius, formed bend	mm		7	8	10	21
	29	Perm. pulling force for install. and operation	N		45	68	90	113
Electrical properties	30	d.c. resistance/unit length at 70 °C	Ω/km		15.9	15.9	15.9	15.9
	31	d.c. resistance/unit length at 20 °C	Ω/km		13.3	13.3	13.3	13.3
Current-carrying capacity	37	Number of loaded cores			2	2	3	3
	41	Installation on walls	A		18.0	18.0	18.0	18.0
	42	Minimum time value	s		45	45	45	45
Short-circuit	43	Rated short-time current of conductor (1s)	kA		0.163	0.163	0.163	0.163

SYSL ... ×1 300/500 V

7	12	18	25	34	50	60
F	F	F	F	F	F	F
plain	plain	plain	plain	plain	plain	plain
32	32	32	32	32	32	32
0.21	0.21	0.21	0.21	0.21	0.21	0.21
1.5	1.5	1.5	1.5	1.5	1.5	1.5
0.6	0.6	0.6	0.6	0.6	0.6	0.6
1.0	1.2	1.5	1.8	1.8	2.1	2.1
8.3	11.5	14.0	17.6	19.2	20.9	23.8
10.3	13.7	16.4	20.8	22.9	25.3	28.8
145	245	370	530	665	950	1100
41	55	66	83	92	101	115
21	27	33	42	46	76	86
105	180	270	375	510	750	900
23.3	23.3	23.3	23.3	23.3	23.3	23.3
19.5	19.5	19.5	19.5	19.5	19.5	19.5
3	3	3	3	3	3	3
15.0	15.0	15.0	15.0	15.0	15.0	15.0
29	29	29	29	29	29	29
0.109	0.109	0.109	0.109	0.109	0.109	0.109

SYSL ... ×1.5 300/500 V

7	12	18	25	34	41	50	60
F	F	F	F	F	F	F	F
plain	plain	plain	plain	plain	plain	plain	plain
30	30	30	30	30	30	30	30
0.26	0.26	0.26	0.26	0.26	0.26	0.26	0.26
1.8	1.8	1.8	1.8	1.8	1.8	1.8	1.8
0.6	0.6	0.6	0.6	0.6	0.6	0.6	0.6
1.2	1.2	1.5	1.8	1.8	2.1	2.1	2.1
9.4	12.3	15.1	19.0	20.3	22.8	24.1	25.8
11.6	15.1	17.9	22.6	24.7	26.8	29.7	31.4
195	315	465	680	855	1050	1220	1410
46	60	72	90	99	107	119	126
23	30	36	45	49	80	89	94
158	270	405	563	765	923	1125	1350
15.9	15.9	15.9	15.9	15.9	15.9	15.9	15.9
13.3	13.3	13.3	13.3	13.3	13.3	13.3	13.3
3	3	3	3	3	3	3	3
18.0	18.0	18.0	18.0	18.0	18.0	18.0	18.0
45	45	45	45	45	45	45	45
0.163	0.163	0.163	0.163	0.163	0.163	0.163	0.163

Table 2.4.7 e
PVC control cables nominal cross-sectional area of conductor = 2.5, 4 and 6 mm²

Design				3	4	5	7
	1	Number of cores		3	4	5	7
	2	Nom. cross-sectional area of conductor	mm²	2.5	2.5	2.5	2.5
	6	Type of conductor		F	F	F	F
	7	Coating of strands		plain	plain	plain	plain
	8	Number of strands		50	50	50	50
	9	Diameter of strand	mm	0.26	0.26	0.26	0.26
	10	Diameter of conductor	mm	2.6	2.6	2.6	2.6
	11	Thickness of insulation	mm	0.7	0.7	0.7	0.7
	12	Thickness of sheath	mm	1.0	1.2	1.2	1.2
	15	Overall diameter of cable (min. value)	mm	8.2	9.3	10.3	11.3
	16	Overall diameter of cable (max. value)	mm	10.2	11.5	12.5	13.5
	19	Weight of cable	kg/km	150	190	230	285
Mechanical properties	20	Minimum bending radius for fixed installation	mm	41	46	50	54
	21	Minimum bending radius, formed bend	mm	20	23	25	27
	29	Perm. pulling force for install. and operation	N	113	150	188	263
Electrical properties	30	d.c. resistance/unit length at 70 °C	Ω/km	9.55	9.55	9.55	9.55
	31	d.c. resistance/unit length at 20 °C	Ω/km	7.98	7.98	7.98	7.98
Current-carrying capacity	37	Number of loaded cores		2	3	3	3
	41	Installation on walls	A	26	26	26	26
	42	Minimum time value	s	60	60	60	60
Short-circuit	43	Rated short-time current of conductor (1s)	kA	0.272	0.272	0.272	0.272

12	14	18	4	5	7	4	5	7
2.5	2.5	2.5	4	4	4	6	6	6
F	F	F	F	F	F	F	F	F
plain	plain	plain	plain	plain	plain	plain	plain	plain
50	50	50	56	56	56	84	84	84
0.26	0.26	0.26	0.31	0.31	0.31	0.31	0.31	0.31
2.6	2.6	2.6	3.2	3.2	3.2	3.9	3.9	3.9
0.7	0.7	0.7	0.8	0.8	0.8	0.8	0.8	0.8
1.5	1.8	1.8	1.2	1.2	1.2	1.5	1.5	1.8
15.5	16.0	16.5	11.1	12.3	13.4	12.6	14.4	15.9
18.3	19.0	19.5	13.3	14.5	15.8	15.4	17.6	19.1
490	540	710	275	335	425	385	495	625
73	76	78	53	58	63	62	70	76
37	38	39	27	29	32	31	35	38
450	525	675	240	300	420	360	450	630
9.55	9.55	9.55	5.92	5.92	5.92	3.95	3.95	3.95
7.98	7.98	7.98	4.95	4.95	4.95	3.30	3.30	3.30
3	3	3	3	3	3	3	3	3
26	26	26	34	34	34	44	44	44
60	60	60	89	89	89	120	120	120
0.272	0.272	0.272	0.436	0.436	0.436	0.654	0.654	0.654

Table 2.5.1
Flat non-sheathed cord extra finely stranded conductor   H03VH-H 2X0.75 300/300 V

Design				FF
	6	Type of conductor		FF
	7	Coating of strands		plain
	8	Number of strands		42
	9	Diameter of strand	mm	0.16
	10	Diameter of conductor	mm	1.3
	11	Thickness of insulation	mm	0.8
	17	Overall dimensions of cable (min. value)	mm × mm	2.7×5.4
	18	Overall dimensions of cable (max. value)	mm × mm	3.2×6.4
	19	Weight of cable	kg/km	27
Mechanical properties	22	Minimum bending radius for free flexing	mm	10
	23	Minimum bending radius for cable entry	mm	10
	29	Perm. pulling force for install. and operation	N	23
Electrical properties	30	d.c. resistance/unit length at 70 °C	Ω/km	31.1
	31	d.c. resistance/unit length at 20 °C	Ω/km	26.0
Current-carrying capacity	37	Number of loaded cores		2
	41	Installation on walls	A	6.0
	42	Minimum time value	s	101
Short-circuit	43	Rated short-time current of conductor (1s)	kA	0.0817

Design					
	6	Type of conductor		F	
	7	Coating of strands		plain	
	8	Number of strands		24	
	9	Diameter of strand	mm	0.21	
	10	Diameter of conductor	mm	1.3	
	11	Thickness of insulation	mm	0.5	
	12	Thickness of sheath	mm	0.6	
	15	Overall diameter of cable (min. value)	mm	5.2	
	16	Overall diameter of cable (max. value)	mm	6.4	
	19	Weight of cable	kg/km	45	
Mechanical properties	22	Minimum bending radius for free flexing	mm	19	
	23	Minimum bending radius for cable entry	mm	19	
	29	Perm. pulling force for install. and operation	N	23	
Electrical properties	30	d.c. resistance/unit length at 70 °C	Ω/km	31.1	
	31	d.c. resistance/unit length at 20 °C	Ω/km	26.0	
Current-carrying capacity	37	Number of loaded cores		2	
	41	Installation on walls	A	6.0	
	42	Minimum time value	s	101	
Short-circuit	43	Rated short-time current of conductor (1s)	kA	0.0817	

Design					
	1	Number of cores		3	4
	6	Type of conductor		F	F
	7	Coating of strands		plain	plain
	8	Number of strands		24	24
	9	Diameter of strand	mm	0.21	0.21
	10	Diameter of conductor	mm	1.3	1.3
	11	Thickness of insulation	mm	0.5	0.5
	12	Thickness of sheath	mm	0.6	0.6
	15	Overall diameter of cable (min. value)	mm	5.4	6.0
	16	Overall diameter of cable (max. value)	mm	6.8	7.4
	19	Weight of cable	kg/km	55	70
Mechanical properties	22	Minimum bending radius for free flexing	mm	20	22
	23	Minimum bending radius for cable entry	mm	20	22
	29	Perm. pulling force for install. and operation	N	34	45
Electrical properties	30	d.c. resistance/unit length at 70 °C	Ω/km	31.1	31.1
	31	d.c. resistance/unit length at 20 °C	Ω/km	26.0	26.0
Current-carrying capacity	37	Number of loaded cores		2	3
	41	Installation on walls	A	6.0	6.0
	42	Minimum time value	s	101	101
Short-circuit	43	Rated short-time current of conductor (1s)	kA	0.0817	0.0817

Table 2.5.3 a Light PVC-sheathed cord 2 cores, without protective conductor				H05VV-F 2X1.5 300/500 V
Design	6	Type of conductor		F
	7	Coating of strands		plain
	8	Number of strands		30
	9	Diameter of strand	mm	0.26
	10	Diameter of conductor	mm	1.8
	11	Thickness of insulation	mm	0.7
	12	Thickness of sheath	mm	0.8
	15	Overall diameter of cable (min. value)	mm	7.4
	16	Overall diameter of cable (max. value)	mm	9.0
	19	Weight of cable	kg/km	85
Mechanical properties	20	Minimum bending radius for fixed installation	mm	27
	22	Minimum bending radius for free flexing	mm	36
	23	Minimum bending radius for cable entry	mm	36
	29	Perm. pulling force for install. and operation	N	45
Electrical properties	30	d.c. resistance/unit length at 70 °C	Ω/km	15.9
	31	d.c. resistance/unit length at 20 °C	Ω/km	13.3
Current-carrying capacity	37	Number of loaded cores		2
	41	Installation on walls	A	16.0
	42	Minimum time value	s	57
Short-circuit	43	Rated short-time current of conductor (1s)	kA	0.163

Table 2.5.3 b Light PVC-sheathed cords 3, 4 and 5 cores, protective conductor							
Design	1	Number of cores		3	3	3	3
	2	Nom. cross-sectional area of conductor	mm^2	0.75	1	1.5	2.5
	6	Type of conductor		F	F	F	F
	7	Coating of strands		plain	plain	plain	plain
	8	Number of strands		24	32	30	50
	9	Diameter of strand	mm	0.21	0.21	0.26	0.26
	10	Diameter of conductor	mm	1.3	1.5	1.8	2.6
	11	Thickness of insulation	mm	0.6	0.6	0.7	0.8
	12	Thickness of sheath	mm	0.8	0.8	0.9	1.1
	15	Overall diameter of cable (min. value)	mm	6.4	6.8	8.0	9.6
	16	Overall diameter of cable (max. value)	mm	8.0	8.4	9.8	12.0
	19	Weight of cable	kg/km	70	75	105	165
Mechanical properties	20	Minimum bending radius for fixed installation	mm	24	25	29	36
	22	Minimum bending radius for free flexing	mm	24	34	39	48
	23	Minimum bending radius for cable entry	mm	24	34	39	48
	29	Perm. pulling force for install. and operation	N	34	45	68	113
Electrical properties	30	d.c. resistance/unit length at 70 °C	Ω/km	31.1	23.3	15.9	9.55
	31	d.c. resistance/unit length at 20 °C	Ω/km	26.0	19.5	13.3	7.98
Current-carrying capacity	37	Number of loaded cores		2	2	2	2
	41	Installation on walls	A	6.0	10.0	16.0	25
	42	Minimum time value	s	101	65	57	65
Short-circuit	43	Rated short-time current of conductor (1s)	kA	0.0817	0.109	0.163	0.272

4	4	5	5
1.5	2.5	1.5	2.5
F	F	F	F
plain	plain	plain	plain
30	50	30	50
0.26	0.26	0.26	0.26
1.8	2.6	1.8	2.6
0.7	0.8	0.7	0.8
1.0	1.1	1.1	1.2
9.0	10.5	10.0	11.5
11.0	13.0	12.0	14.0
140	200	170	250
33	52	36	56
44	65	48	70
44	65	48	70
90	150	113	188
15.9	9.55	15.9	9.55
13.3	7.98	13.3	7.98
3	3	3	3
16.0	20	16.0	20
57	101	57	101
0.163	0.272	0.163	0.272

Table 2.6.1a
Ordinary rubber-sheathed cords 2 cores, without protective conductor

H05RR-F 2X... 300/500 V

Design	2	Nom. cross-sectional area of conductor	mm^2	0.75	1	1.5	2.5
	6	Type of conductor		F	F	F	F
	7	Coating of strands		tinned	tinned	tinned	tinned
	8	Number of strands		24	32	30	50
	9	Diameter of strand	mm	0.21	0.21	0.26	0.26
	10	Diameter of conductor	mm	1.3	1.5	1.8	2.6
	11	Thickness of insulation	mm	0.6	0.6	0.8	0.9
	12	Thickness of sheath	mm	0.8	0.9	1.0	1.1
	15	Overall diameter of cable (min. value)	mm	6.0	6.6	8.0	9.5
	16	Overall diameter of cable (max. value)	mm	8.2	8.8	10.5	12.5
	19	Weight of cable	kg/km	60	75	115	160
Mechanical properties	20	Minimum bending radius for fixed installation	mm	25	26	32	50
	22	Minimum bending radius for free flexing	mm	33	35	42	63
	23	Minimum bending radius for cable entry	mm	33	35	42	63
	29	Perm. pulling force for install. and operation	N	23	30	45	75
Electrical properties	30	d.c. resistance/unit length at 60 °C	Ω/km	30.9	23.1	15.9	9.50
	31	d.c. resistance/unit length at 20 °C	Ω/km	26.7	20.0	13.7	8.21
Current-carrying capacity	37	Number of loaded cores		2	2	2	2
	41	Installation on walls	A	6.0	10.0	16.0	25
	42	Minimum time value	s	78	50	44	50
Short-circuit	43	Rated short-time current of conductor (1s)	kA	0.106	0.141	0.211	0.352

Table 2.6.1b
Ordinary rubber-sheathed cords 3, 4 and 5 cores, protective conductor

Design	1	Number of cores		3	3	3	3
	2	Nom. cross-sectional area of conductor	mm^2	0.75[1]	1	1.5	2.5
	6	Type of conductor		F	F	F	F
	7	Coating of strands		tinned	tinned	tinned	tinned
	8	Number of strands		24	32	30	50
	9	Diameter of strand	mm	0.21	0.21	0.26	0.26
	10	Diameter of conductor	mm	1.3	1.5	1.8	2.6
	11	Thickness of insulation	mm	0.6	0.6	0.8	0.9
	12	Thickness of sheath	mm	0.9	0.9	1.0	1.1
	15	Overall diameter of cable (min. value)	mm	6.5	7.0	8.6	10.0
	16	Overall diameter of cable (max. value)	mm	8.8	9.2	11.0	13.0
	19	Weight of cable	kg/km	75	85	135	190
Mechanical properties	20	Minimum bending radius for fixed installation	mm	26	28	33	52
	22	Minimum bending radius for free flexing	mm	35	37	44	65
	23	Minimum bending radius for cable entry	mm	35	37	44	65
	29	Perm. pulling force for install. and operation	N	34	45	68	113
Electrical properties	30	d.c. resistance/unit length at 60 °C	Ω/km	30.9	23.1	15.9	9.50
	31	d.c. resistance/unit length at 20 °C	Ω/km	26.7	20.0	13.7	8.21
Current-carrying capacity	37	Number of loaded cores		2	2	2	2
	41	Installation on walls	A	6.0	10.0	16.0	25
	42	Minimum time value	s	78	50	44	50
Short-circuit	43	Rated short-time current of conductor (1s)	kA	0.106	0.141	0.211	0.352

[1] Diese Angaben gelten auch für die THERMOSTABIL-Ordinary rubber-sheathed cords H05RR-F 3G 0.75

4	4	4	4	5	5
0.75	1	1.5	2.5	1.5	2.5
F	F	F	F	F	F
tinned	tinned	tinned	tinned	tinned	tinned
24	32	30	50	30	50
0.21	0.21	0.26	0.26	0.26	0.26
1.3	1.5	1.8	2.6	1.8	2.6
0.6	0.6	0.8	0.9	0.8	0.9
0.9	0.9	1.1	1.2	1.1	1.3
7.1	7.6	9.6	11.0	10.5	12.5
9.6	10.0	12.5	14.0	13.5	15.5
90	105	165	235	190	285
29	30	50	56	54	62
38	40	63	70	68	78
38	40	63	70	68	78
45	60	90	150	113	188
30.9	23.1	15.9	9.50	15.9	9.50
26.7	20.0	13.7	8.21	13.7	8.21
3	3	3	3	3	3
6.0	10.0	16.0	20	16.0	20
78	50	44	78	44	78
0.106	0.141	0.211	0.352	0.211	0.352

Table 2.6.2a **Ordinary rubber-sheathed cords** 2 cores, without protective conductor				**H05RN-F 2X... 300/500 V**
Design	2	Nom. cross-sectional area of conductor	mm^2	0.75	1
	6	Type of conductor		F	F
	7	Coating of strands		tinned	tinned
	8	Number of strands		24	32
	9	Diameter of strand	mm	0.21	0.21
	10	Diameter of conductor	mm	1.3	1.5
	11	Thickness of insulation	mm	0.6	0.6
	12	Thickness of sheath	mm	0.8	0.9
	15	Overall diameter of cable (min. value)	mm	6.0	6.6
	16	Overall diameter of cable (max. value)	mm	8.2	8.8
	19	Weight of cable	kg/km	65	75
Mechanical properties	20	Minimum bending radius for fixed installation	mm	25	26
	22	Minimum bending radius for free flexing	mm	33	35
	23	Minimum bending radius for cable entry	mm	33	35
	29	Perm. pulling force for install. and operation	N	23	30
Electrical properties	30	d.c. resistance/unit length at 60 °C	Ω/km	30.9	23.1
	31	d.c. resistance/unit length at 20 °C	Ω/km	26.7	20.0
Current-carrying capacity	37	Number of loaded cores		2	2
	41	Installation on walls	A	6.0	10.0
	42	Minimum time value	s	78	50
Short-circuit	43	Rated short-time current of conductor (1s)	kA	0.106	0.141

Table 2.6.2b **Ordinary rubber-sheathed cords** 3 cores, protective conductor				**H05RN-F 3G... 300/500 V**
Design	2	Nom. cross-sectional area of conductor	mm^2	0.75	1
	6	Type of conductor		F	F
	7	Coating of strands		tinned	tinned
	8	Number of strands		24	32
	9	Diameter of strand	mm	0.21	0.21
	10	Diameter of conductor	mm	1.3	1.5
	11	Thickness of insulation	mm	0.6	0.6
	12	Thickness of sheath	mm	0.9	0.9
	15	Overall diameter of cable (min. value)	mm	6.5	7.0
	16	Overall diameter of cable (max. value)	mm	8.8	9.2
	19	Weight of cable	kg/km	80	90
Mechanical properties	20	Minimum bending radius for fixed installation	mm	26	28
	22	Minimum bending radius for free flexing	mm	35	37
	23	Minimum bending radius for cable entry	mm	35	37
	29	Perm. pulling force for install. and operation	N	34	45
Electrical properties	30	d.c. resistance/unit length at 60 °C	Ω/km	30.9	23.1
	31	d.c. resistance/unit length at 20 °C	Ω/km	26.7	20.0
Current-carrying capacity	37	Number of loaded cores		2	2
	41	Installation on walls	A	6.0	10.0
	42	Minimum time value	s	78	50
Short-circuit	43	Rated short-time current of conductor (1s)	kA	0.106	0.141

Table 2.6.3						A05RN-F 4G0.75 300/500 V
Ordinary rubber-sheathed cord 4 cores, protective conductor						
...ign	6	Type of conductor			F	
	7	Coating of strands			tinned	
	8	Number of strands			24	
	9	Diameter of strand		mm	0.21	
	10	Diameter of conductor		mm	1.3	
	11	Thickness of insulation		mm	0.6	
	12	Thickness of sheath		mm	0.6	
	15	Overall diameter of cable (min. value)		mm	7.1	
	16	Overall diameter of cable (max. value)		mm	9.6	
	19	Weight of cable		kg/km	95	
...chanical	20	Minimum bending radius for fixed installation		mm	29	
...perties	22	Minimum bending radius for free flexing		mm	38	
	23	Minimum bending radius for cable entry		mm	38	
	29	Perm. pulling force for install. and operation		N	45	
...ctrical	30	d.c. resistance/unit length at 60 °C		Ω/km	30.9	
...perties	31	d.c. resistance/unit length at 20 °C		Ω/km	26.7	
...rrent-carrying	37	Number of loaded cores			3	
...acity	41	Installation on walls		A	6.0	
	42	Minimum time value		s	78	
...ort-circuit	43	Rated short-time current of conductor (1s)		kA	0.106	

Table 2.6.4						H05RN-F 3G0.75 300/500 V
Highly flexible rubber-sheathed cord 3 cores, protective conductor						
...sign	6	Type of conductor			FS	
	7	Coating of strands			tinned	
	8	Number of strands			42	
	9	Diameter of strand		mm	0.16	
	10	Diameter of conductor		mm	1.3	
	11	Thickness of insulation		mm	0.6	
	12	Thickness of sheath		mm	0.9	
	15	Overall diameter of cable (min. value)		mm	6.5	
	16	Overall diameter of cable (max. value)		mm	8.8	
	19	Weight of cable		kg/km	65	
...echanical	20	Minimum bending radius for fixed installation		mm	26	
...operties	22	Minimum bending radius for free flexing		mm	35	
	23	Minimum bending radius for cable entry		mm	35	
	29	Perm. pulling force for install. and operation		N	34	
...ectrical	30	d.c. resistance/unit length at 60 °C		Ω/km	30.9	
...operties	31	d.c. resistance/unit length at 20 °C		Ω/km	26.7	
...rrent-carrying	37	Number of loaded cores			2	
...pacity	41	Installation on walls		A	6.0	
	42	Minimum time value		s	78	
...ort-circuit	43	Rated short-time current of conductor (1s)		kA	0.106	

Table 2.6.5a
OZOFLEX rubber-sheathed flexible cables 1 and 2 cores, without protective conductor

Design							
	1	Number of cores		1	1	1	1
	2	Nom. cross-sectional area of conductor	mm^2	10	16	25	35
	6	Type of conductor		F	F	F	F
	7	Coating of strands		plain	plain	plain	plain
	8	Number of strands		80	126	196	276
	9	Diameter of strand	mm	0.41	0.41	0.41	0.41
	10	Diameter of conductor	mm	5.1	6.3	7.8	9.2
	11	Thickness of insulation	mm	1.2	1.2	1.4	1.4
	12	Thickness of sheath	mm	1.8	1.9	2.0	2.2
	15	Overall diameter of cable (min. value)	mm	9.8	11.0	12.5	14.0
	16	Overall diameter of cable (max. value)	mm	12.5	14.5	16.5	18.5
	19	Weight of cable	kg/km	210	300	420	550
Mechanical properties	20	Minimum bending radius for fixed installation	mm	50	58	66	74
	22	Minimum bending radius for free flexing	mm	63	73	83	93
	23	Minimum bending radius for cable entry	mm	63	73	83	93
	29	Perm. pulling force for install. and operation	N	150	240	375	525
Electrical properties	30	d.c. resistance/unit length at 60 °C	Ω/km	2.21	1.40	0.903	0.641
	31	d.c. resistance/unit length at 20 °C	Ω/km	1.91	1.21	0.780	0.554
Current-carrying capacity	37	Number of loaded cores		1	1	1	1
	38	Installation in free air	A	73	98	129	158
	41	Installation on walls	A	–	–	–	–
	42	Minimum time value	s	94	134	188	246
Short-circuit	43	Rated short-time current of conductor (1s)	kA	1.59	2.54	3.97	5.56

Table 2.6.5b
OZOFLEX rubber-sheathed flexible cables 3 cores, protective conductor

Design							
	2	Nom. cross-sectional area of conductor	mm^2	1	1.5	2.5	4
	6	Type of conductor		F	F	F	F
	7	Coating of strands		tinned	tinned	tinned	tinned
	8	Number of strands		32	30	50	56
	9	Diameter of strand	mm	0.21	0.26	0.26	0.31
	10	Diameter of conductor	mm	1.5	1.8	2.6	3.2
	11	Thickness of insulation	mm	0.8	0.8	0.9	1.0
	13	Thickness of inner sheath	mm	–	-	–	-
	14	Thickness of outer sheath	mm	1.4	1.6	1.8	1.9
	15	Overall diameter of cable (min. value)	mm	8.6	9.6	11.5	13.0
	16	Overall diameter of cable (max. value)	mm	11.5	12.5	14.5	16.0
	19	Weight of cable	kg/km	120	160	230	320
Mechanical properties	20	Minimum bending radius for fixed installation	mm	35	50	58	64
	22	Minimum bending radius for free flexing	mm	46	63	73	80
	23	Minimum bending radius for cable entry	mm	46	63	73	80
	29	Perm. pulling force for install. and operation	N	45	68	113	180
Electrical properties	30	d.c. resistance/unit length at 60 °C	Ω/km	23.1	15.9	9.50	5.89
	31	d.c. resistance/unit length at 20 °C	Ω/km	20.0	13.7	8.21	5.09
Current-carrying capacity	37	Number of loaded cores		2	2	2	2
	41	Installation on walls	A	15.0	18.0	26	34
	42	Minimum time value	s	22	35	46	69
Short-circuit	43	Rated short-time current of conductor (1s)	kA	0.141	0.211	0.352	0.564

H07RN-F ...X... 450/750 V

1 50	1 70	1 95	1 120	1 150	1 185	1 240	1 300	2 1.5	2 2.5	2 4
F plain	F plain	F plain	F plain	F plain	F plain	F plain	F plain	F tinned	F tinned	F tinned
396	360	475	608	756	925	1221	1525	30	50	56
0.41	0.51	0.51	0.51	0.51	0.51	0.51	0.51	0.26	0.26	0.31
11.0	13.0	15.1	17.0	19.0	21.0	24.0	27.0	1.8	2.6	3.2
1.6	1.6	1.8	1.8	2.0	2.2	2.4	2.6	0.8	0.9	1.0
2.4	2.6	2.8	3.0	3.2	3.4	3.5	3.6	1.5	1.7	1.8
16.5	18.5	21.0	23.5	26.0	26.0	30.5	33.5	9.0	10.5	12.0
21.0	23.5	26.0	28.5	31.5	31.5	38.0	41.5	11.5	13.5	15.0
760	990	1280	1590	1950	2300	3100	3800	130	190	270
84	94	104	114	126	126	152	166	35	54	60
105	118	130	143	158	158	190	208	46	68	75
105	118	130	143	158	158	190	208	46	68	75
750	1050	1425	1800	2250	2775	3600	4500	45	75	120
0.447	0.315	0.238	0.186	0.149	0.123	0.0927	0.0742	15.9	9.50	5.89
0.386	0.272	0.206	0.161	0.129	0.106	0.0801	0.0641	13.7	8.21	5.09
1	1	1	1	1	1	1	1	2	2	2
198	245	292	344	391	448	528	608	–	–	–
–	–	–	–	–	–	–	–	18.0	26	34
319	409	530	609	737	854	1035	1219	35	46	69
7.95	11.1	15.1	19.1	23.8	29.4	38.2	47.7	0.211	0.352	0.564

H07RN-F 3G... 450/750 V

6	10	16
F tinned	F plain	F plain
84	80	126
0.31	0.41	0.41
3.9	5.1	6.3
1.0	1.2	1.2
–	-	1.4
2.1	3.3	2.1
14.5	20.0	22.5
20.0	25.5	29.5
420	800	1130
80	102	118
100	128	148
100	128	148
270	450	720
3.92	2.21	1.40
3.39	1.91	1.21
2	2	2
44	61	82
93	135	191
0.846	1.59	2.54

Table 2.6.5c
OZOFLEX rubber-sheathed flexible cables 4 cores, protective conductor

Design							
	2	Nom. cross-sectional area of conductor	mm²	1.5	2.5	4	6
	6	Type of conductor		F	F	F	F
	7	Coating of strands		tinned	tinned	tinned	tinned
	8	Number of strands		30	50	56	84
	9	Diameter of strand	mm	0.26	0.26	0.31	0.31
	10	Diameter of conductor	mm	1.8	2.6	3.2	3.9
	11	Thickness of insulation	mm	0.8	0.9	1.0	1.0
	13	Thickness of inner sheath	mm	–	-	–	-
	14	Thickness of outer sheath	mm	1.7	1.9	2.0	2.3
	15	Overall diameter of cable (min. value)	mm	10.5	12.5	14.5	16.5
	16	Overall diameter of cable (max. value)	mm	13.5	15.5	18.0	22.0
	19	Weight of cable	kg/km	200	280	390	530
Mechanical properties	20	Minimum bending radius for fixed installation	mm	54	62	72	88
	22	Minimum bending radius for free flexing	mm	68	78	90	110
	23	Minimum bending radius for cable entry	mm	68	78	90	110
	29	Perm. pulling force for install. and operation	N	90	150	240	360
Electrical properties	30	d.c. resistance/unit length at 60 °C	Ω/km	15.9	9.50	5.89	3.92
	31	d.c. resistance/unit length at 20 °C	Ω/km	13.7	8.21	5.09	3.39
	32	Inductance/unit length per conductor	mH/km	0.340	0.320	0.310	0.290
Current-carrying capacity	37	Number of loaded cores		3	3	3	3
	41	Installation on walls	A	18.0	26	34	44
	42	Minimum time value	s	35	46	69	93
Short-circuit	43	Rated short-time current of conductor (1s)	kA	0.211	0.352	0.564	0.846

Table 2.6.5d
OZOFLEX rubber-sheathed flexible cables 5 cores, protective conductor

Design							
	2	Nom. cross-sectional area of conductor	mm²	1	1.5	2.5	4
	6	Type of conductor		F	F	F	F
	7	Coating of strands		tinned	tinned	tinned	tinned
	8	Number of strands		32	30	50	56
	9	Diameter of strand	mm	0.21	0.26	0.26	0.31
	10	Diameter of conductor	mm	1.5	1.8	2.6	3.2
	11	Thickness of insulation	mm	0.8	0.8	0.9	1.0
	13	Thickness of inner sheath	mm	–	-	–	-
	14	Thickness of outer sheath	mm	1.6	1.8	2.0	2.2
	15	Overall diameter of cable (min. value)	mm	10.5	11.5	13.5	16.0
	16	Overall diameter of cable (max. value)	mm	13.5	15.0	17.0	19.5
	19	Weight of cable	kg/km	180	240	330	480
Mechanical properties	20	Minimum bending radius for fixed installation	mm	54	60	68	78
	22	Minimum bending radius for free flexing	mm	68	75	85	98
	23	Minimum bending radius for cable entry	mm	68	75	85	98
	29	Perm. pulling force for install. and operation	N	75	113	188	300
Electrical properties	30	d.c. resistance/unit length at 60 °C	Ω/km	23.1	15.9	9.50	5.89
	31	d.c. resistance/unit length at 20 °C	Ω/km	20.0	13.7	8.21	5.09
Current-carrying capacity	37	Number of loaded cores		3	3	3	3
	41	Installation on walls	A	15.0	18.0	26	34
	42	Minimum time value	s	22	35	46	69
Short-circuit	43	Rated short-time current of conductor (1s)	kA	0.141	0.211	0.352	0.564

H07RN-F 4G... 450/750 V

10	16	25	35	50	70	95	120
F	F	F	F	F	F	F	F
plain	plain	plain	plain	plain	plain	plain	plain
80	126	196	276	396	360	475	608
0.41	0.41	0.41	0.41	0.41	0.51	0.51	0.51
5.1	6.3	7.8	9.2	11.0	13.1	15.1	17.0
1.2	1.2	1.4	1.4	1.6	1.6	1.8	1.8
–	1.4	1.6	1.7	1.9	2.0	2.3	2.4
3.4	2.2	2.5	2.7	2.9	3.2	3.6	3.6
21.5	24.5	29.5	33.0	38.0	43.0	49.0	53.0
28.0	32.0	37.5	42.0	48.5	54.5	60.5	65.5
960	1370	1950	2570	3490	4660	6110	7680
112	128	150	168	194	218	242	262
140	160	188	210	243	273	303	328
140	160	188	210	243	273	303	328
600	960	1500	2100	3000	4200	5700	7200
2.21	1.40	0.903	0.641	0.447	0.315	0.238	0.186
1.91	1.21	0.780	0.554	0.386	0.272	0.206	0.161
0.290	0.280	0.270	0.260	0.260	0.260	0.250	0.250
3	3	3	3	3	3	3	3
61	82	108	135	168	207	250	292
135	191	268	337	444	573	723	846
1.59	2.54	3.97	5.56	7.95	11.1	15.1	19.1

H07RN-F 5G... 450/750 V

6	10	16	25
F	F	F	F
tinned	plain	plain	plain
84	80	126	196
0.31	0.41	0.41	0.41
3.9	5.1	6.3	7.8
1.0	1.2	1.2	1.4
–	-	1.5	1.7
2.5	3.6	2.4	2.7
18.0	24.0	27.0	32.5
24.5	30.5	35.5	41.5
650	1160	1680	2400
98	122	142	166
123	153	178	208
123	153	178	208
450	750	1200	1875
3.92	2.21	1.40	0.903
3.39	1.91	1.21	0.780
3	3	3	3
44	61	82	108
93	135	191	268
0.846	1.59	2.54	3.97

Table 2.6.6 a
OZOFLEX rubber-sheathed flexible cables 3 and 4 cores, without protective conductor

Design							
	1	Number of cores		3	3	3	3
	2	Nom. cross-sectional area of conductor	mm²	1.5	2.5	4	6
	6	Type of conductor		F	F	F	F
	7	Coating of strands		tinned	tinned	tinned	tinned
	8	Number of strands		30	50	56	84
	9	Diameter of strand	mm	0.26	0.26	0.31	0.31
	10	Diameter of conductor	mm	1.8	2.6	3.2	3.9
	11	Thickness of insulation	mm	0.8	0.9	1.0	1.0
	13	Thickness of inner sheath	mm	–	–	–	–
	14	Thickness of outer sheath	mm	1.6	1.8	1.9	2.1
	15	Overall diameter of cable (min. value)	mm	9.6	11.5	13.0	14.5
	16	Overall diameter of cable (max. value)	mm	12.5	14.5	16.0	20.0
	19	Weight of cable	kg/km	160	230	320	420
Mechanical properties	20	Minimum bending radius for fixed installation	mm	50	58	64	80
	22	Minimum bending radius for free flexing	mm	63	73	80	100
	23	Minimum bending radius for cable entry	mm	63	73	80	100
	29	Perm. pulling force for install. and operation	N	68	113	180	270
Electrical properties	30	d.c. resistance/unit length at 60 °C	Ω/km	15.9	9.50	5.89	3.92
	31	d.c. resistance/unit length at 20 °C	Ω/km	13.7	8.21	5.09	3.39
	32	Inductance/unit length per conductor	mH/km	0.320	0.290	0.290	0.270
Current-carrying capacity	37	Number of loaded cores		3	3	3	3
	41	Installation on walls	A	18.0	26	34	44
	42	Minimum time value	s	35	46	69	93
Short-circuit	43	Rated short-time current of conductor (1s)	kA	0.211	0.352	0.564	0.846

Table 2.6.6 b
OZOFLEX rubber-sheathed flexible cables more than 5 cores, protective conductor

Design							
	1	Number of cores		7	7	8	12
	2	Nom. cross-sectional area of conductor	mm²	1.5	2.5	2.5	2.5
	6	Type of conductor		F	F	F	F
	7	Coating of strands		tinned	tinned	tinned	tinned
	8	Number of strands		30	50	50	50
	9	Diameter of strand	mm	0.26	0.26	0.26	0.26
	10	Diameter of conductor	mm	1.8	2.6	2.6	2.6
	11	Thickness of insulation	mm	0.8	0.9	0.9	0.9
	13	Thickness of inner sheath	mm	1.0	1.2	1.2	1.3
	14	Thickness of outer sheath	mm	1.6	1.7	1.7	1.9
	15	Overall diameter of cable (min. value)	mm	14.0	16.5	18.0	18.5
	16	Overall diameter of cable (max. value)	mm	17.5	20.0	21.5	26.5
	19	Weight of cable	kg/km	380	520	590	800
Mechanical properties	20	Minimum bending radius for fixed installation	mm	70	80	86	106
	22	Minimum bending radius for free flexing	mm	88	100	108	133
	23	Minimum bending radius for cable entry	mm	88	100	108	133
	29	Perm. pulling force for install. and operation	N	158	263	300	450
Electrical properties	30	d.c. resistance/unit length at 60 °C	Ω/km	15.9	9.50	9.50	9.50
	31	d.c. resistance/unit length at 20 °C	Ω/km	13.7	8.21	8.21	8.21
Current-carrying capacity	37	Number of loaded cores		3	3	3	3
	41	Installation on walls	A	18.0	26	26	26
	42	Minimum time value	s	35	46	46	46
Short-circuit	43	Rated short-time current of conductor (1s)	kA	0.211	0.352	0.352	0.352

A07RN-F ...X... 450/750 V

3	3	3	3	3	3	3	4	4	4	4
10	16	25	35	50	70	95	10	16	25	35
F	F	F	F	F	F	F	F	F	F	F
plain	plain	plain	plain	plain	plain	plain	plain	plain	plain	plain
80	126	196	276	396	360	475	80	126	196	276
0.41	0.41	0.41	0.41	0.41	0.51	0.51	0.41	0.41	0.41	0.41
5.1	6.3	7.8	9.2	11.0	13.1	15.1	5.1	6.3	7.8	9.2
1.2	1.2	1.4	1.4	1.6	1.6	1.8	1.2	1.2	1.4	1.4
–	1.4	1.5	1.6	1.8	1.9	2.1	–	1.4	1.6	1.7
3.3	2.1	2.3	2.5	2.7	2.9	3.2	3.4	2.2	2.5	2.7
20.0	22.5	26.5	29.5	34.5	39.0	44.0	21.5	24.5	29.5	33.0
25.5	29.5	34.0	38.0	44.0	49.5	54.0	28.0	32.0	37.5	42.0
800	1130	1150	2050	2780	3690	4770	960	1370	1950	2570
102	118	136	152	176	198	216	112	128	150	168
128	148	170	190	220	248	270	140	160	188	210
128	148	170	190	220	248	270	140	160	188	210
450	720	1125	1575	2250	3150	4275	600	960	1500	2100
2.21	1.40	0.903	0.641	0.447	0.315	0.238	2.21	1.40	0.903	0.641
1.91	1.21	0.780	0.554	0.386	0.272	0.206	1.91	1.21	0.780	0.554
0.270	0.250	0.250	0.240	0.240	0.230	0.230	0.290	0.280	0.270	0.260
3	3	3	3	3	3	3	3	3	3	3
61	82	108	135	168	207	250	61	82	108	135
135	191	268	337	444	573	723	135	191	268	337
1.59	2.54	3.97	5.56	7.95	11.1	15.1	1.59	2.54	3.97	5.56

A07RN-F ...G... 450/750 V

18	24
2.5	2.5
F	F
tinned	tinned
50	50
0.26	0.26
2.6	2.6
0.9	0.9
1.5	1.5
2.3	2.3
25.0	28.5
31.5	35.0
1160	1470
126	140
158	175
158	175
675	900
9.50	9.50
8.21	8.21
3	3
26	26
46	46
0.352	0.352

Table 2.6.7
Highly flexible OZOFLEX rubber-sheathed cords 3, 4 and 5 cores, protective conductor

				3	3	4	4
Design	1	Number of cores					
	2	Nom. cross-sectional area of conductor	mm^2	1	1.5	1	1.5
	6	Type of conductor		FS	FS	FS	FS
	7	Coating of strands		tinned	tinned	tinned	tinned
	8	Number of strands		57	48	57	48
	9	Diameter of strand	mm	0.16	0.21	0.16	0.21
	10	Diameter of conductor	mm	1.5	1.8	1.5	1.8
	11	Thickness of insulation	mm	0.8	0.8	0.8	0.8
	12	Thickness of sheath	mm	1.4	1.6	1.5	1.7
	15	Overall diameter of cable (min. value)	mm	8.6	9.6	9.6	10.5
	16	Overall diameter of cable (max. value)	mm	11.5	12.5	12.5	13.5
	19	Weight of cable	kg/km	115	140	140	180
Mechanical properties	20	Minimum bending radius for fixed installation	mm	35	50	50	54
	22	Minimum bending radius for free flexing	mm	46	63	63	68
	23	Minimum bending radius for cable entry	mm	46	63	63	68
	29	Perm. pulling force for install. and operation	N	45	68	60	90
Electrical properties	30	d.c. resistance/unit length at 60 °C	Ω/km	23.1	15.9	23.1	15.9
	31	d.c. resistance/unit length at 20 °C	Ω/km	20.0	13.7	20.0	13.7
Current-carrying capacity	37	Number of loaded cores		2	2	3	3
	41	Installation on walls	A	15.0	18.0	15.0	18.0
	42	Minimum time value	s	22	35	22	35
Short-circuit	43	Rated short-time current of conductor (1s)	kA	0.141	0.211	0.141	0.211

5	5
1	1.5
FS	FS
tinned	tinned
57	48
0.16	0.21
1.5	1.8
0.8	0.8
1.6	1.8
10.5	11.5
13.5	15.0
180	220
54	60
68	75
68	75
75	113
23.1	15.9
20.0	13.7
3	3
15.0	18.0
22	35
0.141	0.211

Table 2.6.8 a
Heavy duty PROTOMONT rubber-sheathed flexible cables 1 core

Design								
Design	2	Nom. cross-sectional area of conductor	mm²	16	25	35	50
	6	Type of conductor		F	F	F	F
	7	Coating of strands		tinned	tinned	tinned	tinned
	8	Number of strands		126	196	276	396
	9	Diameter of strand	mm	0.41	0.41	0.41	0.41
	10	Diameter of conductor	mm	6.3	7.8	9.2	11.0
	11	Thickness of insulation	mm	1.2	1.4	1.4	1.6
	12	Thickness of sheath	mm	1.6	2.0	2.0	2.0
	15	Overall diameter of cable (min. value)	mm	10.0	12.0	13.5	15.5
	16	Overall diameter of cable (max. value)	mm	13.5	16.5	18.0	20.0
	19	Weight of cable	kg/km	260	400	500	680
Mechanical properties	20	Minimum bending radius for fixed installation	mm	81	99	108	120
	22	Minimum bending radius for free flexing	mm	135	165	180	200
	23	Minimum bending radius for cable entry	mm	135	165	180	200
	25	Minimum bending radius for festoon system	mm	135	165	180	200
	26	Minimum bending radius for power track system	mm	135	165	180	200
	27	Minimum bending radius for guide rollers	mm	203	248	270	300
	29	Perm. pulling force for install. and operation	N	240	375	525	750
Electrical properties	30	d.c. resistance/unit length at 90 °C	Ω/km	1.58	1.01	0.720	0.501
	31	d.c. resistance/unit length at 20 °C	Ω/km	1.24	0.795	0.565	0.393
Current-carrying capacity	37	Number of loaded cores		3	3	3	3
	41	Installation on walls	A	99	131	162	202
	42	Minimum time value	s	237	331	424	557
Short-circuit	43	Rated short-time current of conductor (1s)	kA	1.95	3.05	4.27	6.10

Table 2.6.8 b
Heavy duty PROTOMONT rubber-sheathed flexible cables 2, 3 and 4 cores

Design								
Design	1	Number of cores		2	3	3	3
	2	Nom. cross-sectional area of conductor	mm²	1.5	1.5	2.5	4
	6	Type of conductor		F	F	F	F
	7	Coating of strands		tinned	tinned	tinned	tinned
	8	Number of strands		30	30	50	56
	9	Diameter of strand	mm	0.26	0.26	0.26	0.31
	10	Diameter of conductor	mm	1.8	1.8	2.6	3.2
	11	Thickness of insulation	mm	0.8	0.8	0.9	1.0
	13	Thickness of inner sheath	mm	1.0	1.0	1.0	1.2
	14	Thickness of outer sheath	mm	1.6	1.6	1.6	2.0
	15	Overall diameter of cable (min. value)	mm	10.0	10.5	12.0	14.5
	16	Overall diameter of cable (max. value)	mm	14.5	15.0	16.5	20.0
	19	Weight of cable	kg/km	190	220	280	420
Mechanical properties	20	Minimum bending radius for fixed installation	mm	87	90	99	120
	22	Minimum bending radius for free flexing	mm	145	150	165	200
	23	Minimum bending radius for cable entry	mm	145	150	165	200
	29	Perm. pulling force for install. and operation	N	45	68	113	180
Electrical properties	30	d.c. resistance/unit length at 90 °C	Ω/km	17.5	17.5	10.5	6.49
	31	d.c. resistance/unit length at 20 °C	Ω/km	13.7	13.7	8.21	5.09
	32	Inductance/unit length per conductor	mH/km	0.320	0.320	0.290	0.290
Current-carrying capacity	37	Number of loaded cores		2	2	2	2
	41	Installation on walls	A	23	23	30	41
	42	Minimum time value	s	39	39	63	87
Short-circuit	43	Rated short-time current of conductor (1s)	kA	0.183	0.183	0.305	0.488

NSSHÖU 1×... 0.6/1 kV

70	95	120	150
F	F	F	F
tinned	tinned	tinned	tinned
360	475	608	756
0.51	0.51	0.51	0.51
13.1	15.1	17.0	19.0
1.6	1.8	1.8	2.0
2.2	2.2	2.5	2.5
17.5	19.5	21.5	24.5
22.0	25.0	27.5	30.0
910	1170	1470	1770
132	150	165	180
220	250	275	300
220	250	275	300
220	250	275	300
220	250	275	300
330	375	413	450
1050	1425	1800	2250
0.353	0.268	0.209	0.168
0.277	0.210	0.164	0.132
3	3	3	3
250	301	352	404
713	906	1057	1253
8.54	11.6	14.6	18.3

NSSHÖU ...×... 0.6/1 kV

3	4	4	4	4	4	4	4	4	4
6	1.5	2.5	4	6	10	16	25	35	50
F	F	F	F	F	F	F	F	F	F
tinned	tinned	tinned	tinned	tinned	tinned	tinned	tinned	tinned	tinned
84	30	50	56	84	80	126	196	276	396
0.31	0.26	0.26	0.31	0.31	0.41	0.41	0.41	0.41	0.41
3.9	1.8	2.6	3.2	3.9	5.1	6.3	7.8	9.2	11.0
1.0	0.8	0.9	1.0	1.0	1.2	1.2	1.4	1.4	1.6
1.2	1.0	1.2	1.2	1.2	1.4	1.6	1.8	1.8	2.0
2.0	1.6	2.0	2.0	2.0	2.2	2.5	3.0	3.0	3.5
15.5	11.0	13.5	15.5	17.0	20.5	25.0	30.0	33.5	39.0
22.0	16.0	19.0	21.5	23.0	27.5	32.0	39.0	42.5	49.0
520	250	370	490	620	940	1420	2120	2750	3780
132	96	114	129	138	165	192	234	255	294
220	160	190	215	230	275	320	390	425	490
220	160	190	215	230	275	320	390	425	490
270	90	150	240	360	600	960	1500	2100	3000
4.32	17.5	10.5	6.49	4.32	2.49	1.58	1.01	0.720	0.501
3.39	13.7	8.21	5.09	3.39	1.95	1.24	0.795	0.565	0.393
0.270	0.340	0.320	0.310	0.290	0.290	0.280	0.270	0.260	0.260
2	3	3	3	3	3	3	3	3	3
53	23	30	41	53	74	99	131	162	202
117	39	63	87	117	166	237	331	424	557
0.732	0.183	0.305	0.488	0.732	1.22	1.95	3.05	4.27	6.10

Table 2.6.8 c
Heavy duty PROTOMONT rubber-sheathed flexible cables 3 cores, reduced protective conductor   NSSHÖU 3×.../... 0.6/1 kV

Design					
Design	3	Nom. cross-sectional area of phase conductor	mm²	70	95
	4	Nom. cross-sectional area of PE conductor	mm²	35	50
	6	Type of conductor		F	F
	7	Coating of strands		tinned	tinned
	8	Number of strands		360	475
	9	Diameter of strand	mm	0.51	0.51
	10	Diameter of conductor	mm	13.1	15.1
	11	Thickness of insulation	mm	1.6	1.8
	13	Thickness of inner sheath	mm	2.0	2.4
	14	Thickness of outer sheath	mm	3.5	4.0
	15	Overall diameter of cable (min. value)	mm	44.0	50.0
	16	Overall diameter of cable (max. value)	mm	53.5	61.5
	19	Weight of cable	kg/km	4710	6190
Mechanical properties	20	Minimum bending radius for fixed installation	mm	321	369
	22	Minimum bending radius for free flexing	mm	535	615
	23	Minimum bending radius for cable entry	mm	535	615
	29	Perm. pulling force for install. and operation	N	3150	4275
Electrical properties	30	d.c. resistance/unit length at 90 °C	Ω/km	0.353	0.268
	31	d.c. resistance/unit length at 20 °C	Ω/km	0.277	0.210
	32	Inductance/unit length per conductor	mH/km	0.230	0.230
Current-carrying capacity	37	Number of loaded cores		3	3
	41	Installation on walls	A	250	301
	42	Minimum time value	s	713	906
Short-circuit	43	Rated short-time current of conductor (1s)	kA	8.54	11.6

Table 2.6.8 d
Heavy duty PROTOMONT rubber-sheathed flexible cables 5 cores and more

Design							
Design	1	Number of cores		5	5	5	5
	2	Nom. cross-sectional area of conductor	mm²	1.5	2.5	4	6
	6	Type of conductor		F	F	F	F
	7	Coating of strands		tinned	tinned	tinned	tinned
	8	Number of strands		30	50	56	84
	9	Diameter of strand	mm	0.26	0.26	0.31	0.31
	10	Diameter of conductor	mm	1.8	2.6	3.2	3.9
	11	Thickness of insulation	mm	0.8	0.9	1.0	1.0
	13	Thickness of inner sheath	mm	1.0	1.2	1.2	1.4
	14	Thickness of outer sheath	mm	1.6	2.0	2.0	2.2
	15	Overall diameter of cable (min. value)	mm	12.0	14.5	16.0	18.5
	16	Overall diameter of cable (max. value)	mm	17.0	20.0	23.0	26.5
	19	Weight of cable	kg/km	280	410	560	740
Mechanical properties	20	Minimum bending radius for fixed installation	mm	102	120	138	159
	22	Minimum bending radius for free flexing	mm	170	200	230	265
	23	Minimum bending radius for cable entry	mm	170	200	230	265
	29	Perm. pulling force for install. and operation	N	113	188	300	450
Electrical properties	30	d.c. resistance/unit length at 90 °C	Ω/km	17.5	10.5	6.49	4.32
	31	d.c. resistance/unit length at 20 °C	Ω/km	13.7	8.21	5.09	3.39
Current-carrying capacity	37	Number of loaded cores		3	3	3	3
	41	Installation on walls	A	23	30	41	53
	42	Minimum time value	s	39	63	87	117
Short-circuit	43	Rated short-time current of conductor (1s)	kA	0.183	0.305	0.488	0.732

NSSHÖU …×… 0.6/1 kV

5	5	5	7	12	18
10	16	25	2.5	2.5	2.5
F	F	F	F	F	F
tinned	tinned	tinned	tinned	tinned	tinned
80	126	196	50	50	50
0.41	0.41	0.41	0.26	0.26	0.26
5.1	6.3	7.8	2.6	2.6	2.6
1.2	1.2	1.4	0.9	0.9	0.9
1.4	1.6	1.8	1.2	1.4	1.6
2.2	2.5	3.0	2.0	2.0	2.0
22.0	27.0	32.5	16.5	20.5	26.0
30.0	34.0	42.0	21.5	28.0	33.0
1070	1670	2460	540	820	1180
180	204	252	129	168	198
300	340	420	215	280	330
300	340	420	215	280	330
750	1200	1875	263	450	675
2.49	1.58	1.01	10.5	10.5	10.5
1.95	1.24	0.795	8.21	8.21	8.21
3	3	3	3	3	3
74	99	131	30	30	30
166	237	331	63	63	63
1.22	1.95	3.05	0.305	0.305	0.305

Table 2.6.9
HYDROFIRM(T) rubber-sheathed flexible cables 4 cores

Design	2	Nom. cross-sectional area of conductor	mm²	1.5	2.5	4	6
	6	Type of conductor		F	F	F	F
	7	Coating of strands		plain	plain	plain	plain
	8	Number of strands		28	46	52	80
	9	Diameter of strand	mm	0.26	0.26	0.31	0.31
	10	Diameter of conductor	mm	1.8	2.6	3.2	3.9
	11	Thickness of insulation	mm	0.8	0.9	1.0	1.0
	12	Thickness of sheath	mm	1.7	1.9	2.0	2.3
	15	Overall diameter of cable (min. value)	mm	10.4	12.3	14.3	15.8
	16	Overall diameter of cable (max. value)	mm	12.0	13.9	15.9	17.8
	19	Weight of cable	kg/km	175	250	355	470
Mechanical properties	20	Minimum bending radius for fixed installation	mm	36	56	64	71
	22	Minimum bending radius for free flexing	mm	48	69	79	89
	23	Minimum bending radius for cable entry	mm	48	69	79	89
	29	Perm. pulling force for install. and operation	N	90	150	240	360
Electrical properties	30	d.c. resistance/unit length at 90 °C	Ω/km	17.0	10.2	6.31	4.21
	31	d.c. resistance/unit length at 20 °C	Ω/km	13.3	7.98	4.95	3.30
Current-carrying capacity	37	Number of loaded cores		3	3	3	3
	41	Installation on walls	A	23	30	41	53
	42	Minimum time value	s	39	63	87	117
Short-circuit	43	Rated short-time current of conductor (1s)	kA	0.214	0.357	0.572	0.858

10	16
F	F
plain	plain
80	130
0.41	0.41
5.1	6.3
1.2	1.2
3.4	3.6
21.2	25.8
23.2	28.8
810	1260
93	115
116	144
116	144
600	960
2.44	1.54
1.91	1.21
3	3
74	99
166	237
1.43	2.29

Table 2.6.10 a
HYDROFIRM rubber-sheathed flexible cables 3 cores

				1	1.5	2.5	4
Design	2	Nom. cross-sectional area of conductor	mm^2				
	6	Type of conductor		F	F	F	F
	7	Coating of strands		plain	plain	plain	plain
	8	Number of strands		32	28	46	52
	9	Diameter of strand	mm	0.21	0.26	0.26	0.31
	10	Diameter of conductor	mm	1.5	1.8	2.6	3.2
	11	Thickness of insulation	mm	0.8	0.8	0.9	1.0
	12	Thickness of sheath	mm	1.4	1.6	1.8	1.9
	15	Overall diameter of cable (min. value)	mm	8.6	9.6	11.2	13.1
	16	Overall diameter of cable (max. value)	mm	9.7	10.6	12.8	14.7
	19	Weight of cable	kg/km	110	140	200	285
Mechanical properties	20	Minimum bending radius for fixed installation	mm	29	32	51	59
	22	Minimum bending radius for free flexing	mm	39	42	64	73
	23	Minimum bending radius for cable entry	mm	39	42	64	73
	29	Perm. pulling force for install. and operation	N	45	68	113	180
Electrical properties	30	d.c. resistance/unit length at 90 °C	Ω/km	24.9	17.0	10.2	6.31
	31	d.c. resistance/unit length at 20 °C	Ω/km	19.5	13.3	7.98	4.95
	32	Inductance/unit length per conductor	mH/km	0.330	0.320	0.290	0.290
Current-carrying capacity	37	Number of loaded cores		2	2	2	2
	41	Installation on walls	A	18.0	23	30	41
	42	Minimum time value	s	28	39	63	87
Short-circuit	43	Rated short-time current of conductor (1s)	kA	0.143	0.214	0.357	0.572

Table 2.6.10 b
HYDROFIRM rubber-sheathed flexible cables 4 and 5 cores

Design	1	Number of cores		4	4	4	4
	2	Nom. cross-sectional area of conductor	mm^2	1.5	2.5	4	6
	6	Type of conductor		F	F	F	F
	7	Coating of strands		plain	plain	plain	plain
	8	Number of strands		28	46	52	80
	9	Diameter of strand	mm	0.26	0.26	0.31	0.31
	10	Diameter of conductor	mm	1.8	2.6	3.2	3.9
	11	Thickness of insulation	mm	0.8	0.9	1.0	1.0
	12	Thickness of sheath	mm	1.7	1.9	2.0	2.3
	15	Overall diameter of cable (min. value)	mm	10.4	12.3	14.5	16.4
	16	Overall diameter of cable (max. value)	mm	12.0	13.9	16.1	18.4
	19	Weight of cable	kg/km	175	250	355	470
Mechanical properties	20	Minimum bending radius for fixed installation	mm	36	56	64	74
	22	Minimum bending radius for free flexing	mm	48	69	80	92
	23	Minimum bending radius for cable entry	mm	48	69	80	92
	29	Perm. pulling force for install. and operation	N	90	150	240	360
Electrical properties	30	d.c. resistance/unit length at 90 °C	Ω/km	17.0	10.2	6.31	4.21
	31	d.c. resistance/unit length at 20 °C	Ω/km	13.3	7.98	4.95	3.30
	32	Inductance/unit length per conductor	mH/km	0.340	0.320	0.310	0.290
Current-carrying capacity	37	Number of loaded cores		3	3	3	3
	41	Installation on walls	A	23	30	41	53
	42	Minimum time value	s	39	63	87	117
Short-circuit	43	Rated short-time current of conductor (1s)	kA	0.214	0.357	0.572	0.858

TGK 3×... 450/750 V

6	10	16	25	35	50	70
F	F	F	F	F	F	F
plain	plain	plain	plain	plain	plain	plain
80	80	130	200	270	400	350
0.31	0.41	0.41	0.41	0.41	0.41	0.51
3.9	5.1	6.3	7.8	9.2	11.0	13.1
1.0	1.2	1.2	1.4	1.4	1.6	1.6
2.1	3.3	3.5	3.8	4.1	4.5	4.8
14.5	19.4	23.9	28.0	32.3	37.6	42.1
16.0	21.4	25.9	31.0	35.3	40.6	45.1
370	655	1020	1460	1990	2720	3470
64	86	104	124	141	162	180
80	107	129	155	176	203	225
80	107	129	155	176	203	225
270	450	720	1125	1575	2250	3150
4.21	2.44	1.54	0.995	0.706	0.492	0.347
3.30	1.91	1.21	0.780	0.554	0.386	0.272
0.270	0.270	0.250	0.250	0.240	0.240	0.230
2	2	3	3	3	3	3
53	74	99	131	162	202	250
117	166	237	331	424	557	713
0.858	1.43	2.29	3.57	5.00	7.15	10.0

TGK ... ×... 450/750 V

4	4	4	4	4	4	5	5
10	16	25	35	50	70	1.5	2.5
F	F	F	F	F	F	F	F
plain	plain	plain	plain	plain	plain	plain	plain
80	130	200	270	400	350	28	46
0.41	0.41	0.41	0.41	0.41	0.51	0.26	0.26
5.1	6.3	7.8	9.2	11.0	13.1	1.8	2.6
1.2	1.2	1.4	1.4	1.6	1.6	0.8	0.9
3.4	3.6	4.1	4.4	4.8	5.2	1.8	2.0
21.9	25.8	31.0	35.8	41.6	46.8	11.5	13.6
23.9	28.8	34.0	38.8	44.6	49.8	13.1	15.2
810	1260	1870	2500	3420	4560	215	310
96	115	136	155	178	199	52	61
119	144	170	194	223	249	65	76
119	144	170	194	223	249	65	76
600	960	1500	2100	3000	4200	113	188
2.44	1.54	0.995	0.706	0.492	0.347	17.0	10.2
1.91	1.21	0.780	0.554	0.386	0.272	13.3	7.98
0.290	0.280	0.270	0.260	0.260	0.260	0.350	0.330
3	3	3	3	3	3	3	3
74	99	131	162	202	250	23	30
166	237	331	424	557	713	39	63
1.43	2.29	3.57	5.00	7.15	10.0	0.214	0.357

Table 2.6.11
HYDROFIRM rubber-sheathed flexible cables 1 core

Design	2	Nom. cross-sectional area of conductor	mm^2	1.5	2.5	4	6
	6	Type of conductor		F	F	F	F
	7	Coating of strands		plain	plain	plain	plain
	8	Number of strands		28	46	52	80
	9	Diameter of strand	mm	0.26	0.26	0.31	0.31
	10	Diameter of conductor	mm	1.8	2.6	3.2	3.9
	11	Thickness of insulation	mm	0.8	0.9	1.0	1.0
	12	Thickness of sheath	mm	1.4	1.4	1.5	1.6
	15	Overall diameter of cable (min. value)	mm	5.7	6.3	7.3	7.8
	16	Overall diameter of cable (max. value)	mm	6.7	7.4	8.4	8.8
	19	Weight of cable	kg/km	47	62	86	120
Mechanical properties	20	Minimum bending radius for fixed installation	mm	20	22	25	26
	22	Minimum bending radius for free flexing	mm	20	22	34	35
	23	Minimum bending radius for cable entry	mm	20	22	34	35
	29	Perm. pulling force for install. and operation	N	23	38	60	90
Electrical properties	30	d.c. resistance/unit length at 90 °C	Ω/km	17.0	10.2	6.31	4.21
	31	d.c. resistance/unit length at 20 °C	Ω/km	13.3	7.98	4.95	3.30
Current-carrying capacity	37	Number of loaded cores		3	3	3	3
	41	Installation on walls	A	23	30	41	53
	42	Minimum time value	s	39	63	87	117
Short-circuit	43	Rated short-time current of conductor (1s)	kA	0.214	0.357	0.572	0.858

10	16	25	35	50	70
F	F	F	F	F	F
plain	plain	plain	plain	plain	plain
80	130	200	270	400	350
0.41	0.41	0.41	0.41	0.41	0.51
5.1	6.3	7.8	9.2	11.0	13.1
1.2	1.2	1.4	1.4	1.6	1.6
1.8	1.9	2.0	2.2	2.4	2.6
9.6	11.5	13.6	15.4	17.8	20.1
10.6	13.1	15.2	17.4	19.8	22.1
160	255	360	485	670	890
32	52	61	70	79	88
42	65	76	87	99	110
42	65	76	87	99	110
150	240	375	525	750	1050
2.44	1.54	0.995	0.706	0.492	0.347
1.91	1.21	0.780	0.554	0.386	0.272
3	3	3	3	3	3
74	99	131	162	202	250
166	237	331	424	557	713
1.43	2.29	3.57	5.00	7.15	10.0

Table 2.6.12 a
HYDROFIRM rubber-sheathed flexible cables flat design, 3 cores

Design	2	Nom. cross-sectional area of conductor	mm^2	1.5	2.5	4
	6	Type of conductor		F	F	F
	7	Coating of strands		plain	plain	plain
	8	Number of strands		28	46	52
	9	Diameter of strand	mm	0.26	0.26	0.31
	10	Diameter of conductor	mm	1.8	2.6	3.2
	11	Thickness of insulation	mm	0.8	0.9	1.0
	12	Thickness of sheath	mm	1.6	1.8	1.9
	17	Overall dimensions of cable (min. value)	mm × mm	6.1×12.3	7.1×14.4	8.1×17.0
	18	Overall dimensions of cable (max. value)	mm × mm	7.1×13.9	8.1×16.4	9.1×19.0
	19	Weight of cable	kg/km	140	180	250
Mechanical properties	20	Minimum bending radius for fixed installation	mm	21	24	27
	22	Minimum bending radius for free flexing	mm	21	32	36
	23	Minimum bending radius for cable entry	mm	21	32	36
	29	Perm. pulling force for install. and operation	N	68	113	180
Electrical properties	30	d.c. resistance/unit length at 90 °C	Ω/km	17.0	10.2	6.31
	31	d.c. resistance/unit length at 20 °C	Ω/km	13.3	7.98	4.95
Current-carrying capacity	37	Number of loaded cores		3	3	3
	41	Installation on walls	A	23	30	41
	42	Minimum time value	s	39	63	87
Short-circuit	43	Rated short-time current of conductor (1s)	kA	0.214	0.357	0.572

Table 2.6.12 b
HYDROFIRM rubber-sheathed flexible cables flat design, 4 cores

Design	2	Nom. cross-sectional area of conductor	mm^2	1.5	2.5	4
	6	Type of conductor		F	F	F
	7	Coating of strands		plain	plain	plain
	8	Number of strands		28	46	52
	9	Diameter of strand	mm	0.26	0.26	0.31
	10	Diameter of conductor	mm	1.8	2.6	3.2
	11	Thickness of insulation	mm	0.8	0.9	1.0
	12	Thickness of sheath	mm	1.7	1.9	2.0
	17	Overall dimensions of cable (min. value)	mm × mm	6.3×16.4	7.3×19.4	8.3×22.5
	18	Overall dimensions of cable (max. value)	mm × mm	7.3×18.4	8.3×21.4	9.3×25.5
	19	Weight of cable	kg/km	175	250	345
Mechanical properties	20	Minimum bending radius for fixed installation	mm	22	25	28
	22	Minimum bending radius for free flexing	mm	22	33	37
	23	Minimum bending radius for cable entry	mm	22	33	37
	29	Perm. pulling force for install. and operation	N	90	150	240
Electrical properties	30	d.c. resistance/unit length at 90 °C	Ω/km	17.0	10.2	6.31
	31	d.c. resistance/unit length at 20 °C	Ω/km	13.3	7.98	4.95
Current-carrying capacity	37	Number of loaded cores		3	3	3
	41	Installation on walls	A	23	30	41
	42	Minimum time value	s	39	63	87
Short-circuit	43	Rated short-time current of conductor (1s)	kA	0.214	0.357	0.572

TGFLW 3×... 450/750 V

6	10	16	25	35	50	70
F	F	F	F	F	F	F
plain	plain	plain	plain	plain	plain	plain
80	80	130	200	270	400	350
0.31	0.41	0.41	0.41	0.41	0.41	0.51
3.9	5.1	6.3	7.8	9.2	11.0	13.1
1.0	1.2	1.2	1.4	1.4	1.6	1.6
2.1	3.3	3.5	3.8	4.1	4.5	4.8
9.1×19.1	12.6×25.0	14.6×31.0	17.0×36.8	19.2×42.3	22.0×48.8	24.2×55.1
10.1×21.1	14.2×28.0	16.6×34.0	19.0×39.8	21.2×45.3	24.0×52.8	27.2×59.1
340	600	890	1160	1800	2300	3030
30	57	66	76	85	96	109
40	71	83	95	106	120	136
40	71	83	95	106	120	136
270	450	720	1125	1575	2250	3150
4.21	2.44	1.54	0.995	0.706	0.492	0.347
3.30	1.91	1.21	0.780	0.554	0.386	0.272
3	3	3	3	3	3	3
53	74	99	131	162	202	250
117	166	237	331	424	557	713
0.858	1.43	2.29	3.57	5.00	7.15	10.0

TGFLW 4×... 450/750 V

6	10	16	25	35	50	70
F	F	F	F	F	F	F
plain	plain	plain	plain	plain	plain	plain
80	80	130	200	270	400	350
0.31	0.41	0.41	0.41	0.41	0.41	0.51
3.9	5.1	6.3	7.8	9.2	11.0	13.1
1.0	1.2	1.2	1.4	1.4	1.6	1.6
2.3	3.4	3.6	4.1	4.4	4.8	5.2
9.5×25.6	12.8×33.4	14.9×41.3	17.6×49.1	19.8×56.5	22.6×65.5	25.3×74.1
10.5×28.6	14.5×36.4	16.9×44.3	19.6×53.1	21.8×60.5	24.6×69.5	27.3×78.1
470	810	1210	1750	2320	3180	4180
32	58	68	78	87	98	109
42	73	84	98	109	123	136
42	73	84	98	109	123	136
360	600	960	1500	2100	3000	4200
4.21	2.44	1.54	0.995	0.706	0.492	0.347
3.30	1.91	1.21	0.780	0.554	0.386	0.272
3	3	3	3	3	3	3
53	74	99	131	162	202	250
117	166	237	331	424	557	713
0.858	1.43	2.29	3.57	5.00	7.15	10.0

Table 2.7.1
ARCOFLEX welding cables extra finely stranded conductor

Design	2	Nom. cross-sectional area of conductor	mm^2	16	25	35	50
	6	Type of conductor		FF	FF	FF	FF
	7	Coating of strands		plain	plain	plain	plain
	8	Number of strands		500	760	1080	1530
	9	Diameter of strand	mm	0.21	0.21	0.21	0.21
	10	Diameter of conductor	mm	6.3	7.8	9.2	11.0
	12	Thickness of sheath	mm	2.0	2.0	2.0	2.2
	15	Overall diameter of cable (min. value)	mm	8.5	10.0	11.5	13.0
	16	Overall diameter of cable (max. value)	mm	11.5	13.0	14.5	17.0
	19	Weight of cable	kg/km	220	310	410	570
Mechanical properties	22	Minimum bending radius for free flexing	mm	46	65	73	85
	23	Minimum bending radius for cable entry	mm	46	65	73	85
	29	Perm. pulling force for install. and operation	N	240	375	525	750
Electrical properties	30	d.c. resistance/unit length at 80 °C	Ω/km	1.43	0.937	0.662	0.468
	31	d.c. resistance/unit length at 20 °C	Ω/km	1.16	0.758	0.536	0.379
Current-carrying capacity	37	Number of loaded cores		1	1	1	1
	41	Installation on walls	A	117	153	188	236
	42	Minimum time value	s	146	209	271	351
Short-circuit	43	Rated short-time current of conductor (1s)	kA	2.37	3.70	5.18	7.40

Table 2.7.2
FLEXIPREN welding cables extremely finely stranded conductor   **NSLFFÖU 1×...**

Design	2	Nom. cross-sectional area of conductor	mm^2	25	35	50	70
	6	Type of conductor		FF	FF	FF	FF
	7	Coating of strands		plain	plain	plain	plain
	8	Number of strands		1350	1920	2720	3860
	9	Diameter of strand	mm	0.16	0.16	0.16	0.16
	10	Diameter of conductor	mm	7.8	9.2	11.0	13.1
	12	Thickness of sheath	mm	2.0	2.0	2.2	2.4
	15	Overall diameter of cable (min. value)	mm	10.0	11.5	13.0	15.5
	16	Overall diameter of cable (max. value)	mm	13.0	14.5	17.0	19.5
	19	Weight of cable	kg/km	310	415	570	790
Mechanical properties	22	Minimum bending radius for free flexing	mm	65	73	85	98
	23	Minimum bending radius for cable entry	mm	65	73	85	98
	29	Perm. pulling force for install. and operation	N	375	525	750	1050
Electrical properties	30	d.c. resistance/unit length at 80 °C	Ω/km	0.937	0.662	0.468	0.331
	31	d.c. resistance/unit length at 20 °C	Ω/km	0.758	0.536	0.379	0.268
Current-carrying capacity	37	Number of loaded cores		1	1	1	1
	41	Installation on walls	A	153	188	236	291
	42	Minimum time value	s	209	271	351	452
Short-circuit	43	Rated short-time current of conductor (1s)	kA	3.70	5.18	7.40	10.4

70	95	120
FF	FF	FF
plain	plain	plain
2160	2930	1660
0.21	0.21	0.31
13.1	15.1	19.0
2.4	2.6	2.8
15.0	17.0	19.5
19.5	22.0	24.0
790	1050	1330
98	110	120
98	110	120
1050	1425	1800
0.331	0.245	0.192
0.268	0.198	0.155
1	1	1
291	347	409
452	586	673
10.4	14.1	17.8

Table 2.8.1
CORDAFLEX rubber-sheathed flexible cables for travel speeds up to 60 m/min

Design				5	7	12	24
	1	Number of cores		5	7	12	24
	2	Nom. cross-sectional area of conductor	mm²	1.5	1.5	1.5	1.5
	6	Type of conductor		F	F	F	F
	7	Coating of strands		tinned	tinned	tinned	tinned
	8	Number of strands		30	30	30	30
	9	Diameter of strand	mm	0.26	0.26	0.26	0.26
	10	Diameter of conductor	mm	1.8	1.8	1.8	1.8
	11	Thickness of insulation	mm	0.8	0.8	0.8	0.8
	13	Thickness of inner sheath	mm	1.0	1.2	1.4	1.4
	14	Thickness of outer sheath	mm	1.6	2.0	2.2	2.2
	15	Overall diameter of cable (min. value)	mm	13.0	16.0	19.0	26.0
	16	Overall diameter of cable (max. value)	mm	16.0	19.5	23.0	31.0
	19	Weight of cable	kg/km	290	440	600	1150
Mechanical properties	20	Minimum bending radius for fixed installation	mm	96	117	138	186
	22	Minimum bending radius for free flexing	mm	160	195	230	310
	23	Minimum bending radius for cable entry	mm	160	195	230	310
	24	Minimum bending radius for reeling operation	mm	192	234	276	372
	26	Minimum bending radius for power track system	mm	160	195	230	310
	27	Minimum bending radius for guide rollers	mm	240	293	345	465
	29	Perm. pulling force for install. and operation	N	113	158	270	540
Electrical properties	30	d.c. resistance/unit length at 90 °C	Ω/km	17.5	17.5	17.5	17.5
	31	d.c. resistance/unit length at 20 °C	Ω/km	13.7	13.7	13.7	13.7
Current-carrying capacity[1]	37	Number of loaded cores		3	3	3	3
	41	Installation on walls	A	23	23	23	23
	42	Minimum time value	s	39	39	39	39
Short-circuit	43	Rated short-time current of conductor (1s)	kA	0.183	0.183	0.183	0.183

[1] Current-carrying capacity not according to DIN VDE 0298 Part 4, see Clause 1.2

| 7 | 12 | 24 | 4 | 4 | 4 | 4 | 4 | 4 | 4 | 4 |
2.5	2.5	2.5	2.5	4	6	10	16	25	35	50
F	F	F	F	F	F	F	F	F	F	F
tinned	tinned	tinned	tinned	tinned	tinned	tinned	tinned	tinned	tinned	tinned
50	50	50	50	56	84	80	126	196	276	396
0.26	0.26	0.26	0.26	0.31	0.31	0.41	0.41	0.41	0.41	0.41
2.6	2.6	2.6	2.6	3.2	3.9	5.1	6.3	7.8	9.2	11.0
0.9	0.9	0.9	0.9	1.0	1.0	1.2	1.2	1.4	1.4	1.6
1.2	1.4	1.6	1.2	1.2	1.2	1.4	1.6	1.8	1.8	2.0
2.0	2.2	2.5	2.0	2.0	2.0	2.2	2.5	3.0	3.0	3.5
18.0	21.5	29.5	14.5	16.5	18.0	23.0	26.5	31.5	35.5	41.0
22.0	26.0	35.5	18.0	20.0	22.0	28.0	32.0	38.0	41.5	48.5
590	810	1580	340	450	570	960	1300	1940	2460	3450
132	156	213	108	120	132	168	192	228	249	291
220	260	355	180	200	220	280	320	380	415	485
220	260	355	180	200	220	280	320	380	415	485
264	312	426	216	240	264	336	384	456	498	582
220	260	355	180	200	220	280	320	380	415	485
330	390	533	270	300	330	420	480	570	623	728
263	450	900	150	240	360	600	960	1500	2100	3000
10.5	10.5	10.5	10.5	6.49	4.32	2.49	1.58	1.01	0.720	0.501
8.21	8.21	8.21	8.21	5.09	3.39	1.95	1.24	0.795	0.565	0.393
3	3	3	3	3	3	3	3	3	3	3
30	30	30	30	41	53	74	99	131	162	202
63	63	63	63	87	117	166	237	331	424	557
0.305	0.305	0.305	0.305	0.488	0.732	1.22	1.95	3.05	4.27	6.10

Table 2.8.2 a
CORDAFLEX(K) rubber-sheathed flexible cables (1.5 and 2.5 mm^2) travel speeds up to 120 m/min

Design							
Design	1	Number of cores		5	7	12	24
	2	Nom. cross-sectional area of conductor	mm^2	1.5	1.5	1.5	1.5
	6	Type of conductor		FS	FS	FS	FS
	7	Coating of strands		tinned	tinned	tinned	tinned
	8	Number of strands		50	50	50	50
	9	Diameter of strand	mm	0.21	0.21	0.21	0.21
	10	Diameter of conductor	mm	1.8	1.8	1.8	1.8
	11	Thickness of insulation	mm	0.8	0.8	0.8	0.8
	13	Thickness of inner sheath	mm	1.0	1.2	1.4	1.4
	14	Thickness of outer sheath	mm	1.6	2.0	2.2	2.2
	15	Overall diameter of cable (min. value)	mm	13.1	16.0	19.0	25.5
	16	Overall diameter of cable (max. value)	mm	16.0	20.0	23.0	31.0
	19	Weight of cable	kg/km	270	450	640	1180
Mechanical properties	20	Minimum bending radius for fixed installation	mm	96	120	138	186
	22	Minimum bending radius for free flexing	mm	160	200	230	310
	23	Minimum bending radius for cable entry	mm	160	200	230	310
	24	Minimum bending radius for reeling operation	mm	192	240	276	372
	26	Minimum bending radius for power track system	mm	160	200	230	310
	27	Minimum bending radius for guide rollers	mm	240	300	345	465
	29	Perm. pulling force for install. and operation	N	150	210	360	720
Electrical properties	30	d.c. resistance/unit length at 90 °C	Ω/km	17.5	17.5	17.5	17.5
	31	d.c. resistance/unit length at 20 °C	Ω/km	13.7	13.7	13.7	13.7
Current-carrying capacity[1]	37	Number of loaded cores		3	3	3	3
	41	Installation on walls	A	23	23	23	23
	42	Minimum time value	s	39	39	39	39
Short-circuit	43	Rated short-time current of conductor (1s)	kA	0.183	0.183	0.183	0.183

[1] Current-carrying capacity not according to DIN VDE 0298 Part 4, see Clause 1.2

Table 2.8.2 b
CORDAFLEX(K) rubber-sheathed flexible cables travel speeds up to 120 m/min, screened cores

Design					
Design	1	Number of cores		19	25
	6	Type of conductor		FS	FS
	7	Coating of strands		tinned	tinned
	8	Number of strands		80	80
	9	Diameter of strand	mm	0.21	0.21
	10	Diameter of conductor	mm	2.6	2.6
	11	Thickness of insulation	mm	0.9	0.9
	13	Thickness of inner sheath	mm	1.6	1.8
	14	Thickness of outer sheath	mm	2.5	3.0
	15	Overall diameter of cable (min. value)	mm	29.0	31.5
	16	Overall diameter of cable (max. value)	mm	35.0	37.5
	19	Weight of cable	kg/km	1550	1840

NSHTÖU ... × ... 0.6/1 kV

4	7	12	18	24	30
2.5	2.5	2.5	2.5	2.5	2.5
FS tinned	FS tinned	FS tinned	FS tinned	FS tinned	FS tinned
80	80	80	80	80	80
0.21	0.21	0.21	0.21	0.21	0.21
2.6	2.6	2.6	2.6	2.6	2.6
0.9	0.9	0.9	0.9	0.9	0.9
1.2	1.2	1.4	1.4	1.6	1.6
2.0	2.0	2.2	2.2	2.5	2.5
13.0	18.0	21.0	25.5	29.0	31.0
16.5	22.0	25.5	31.0	34.5	37.5
300	590	920	1210	1570	1860
99	132	153	186	207	225
165	220	255	310	345	375
165	220	255	310	345	375
198	264	306	372	414	450
165	220	255	310	345	375
248	330	383	465	518	563
200	350	600	900	1200	1500
10.5	10.5	10.5	10.5	10.5	10.5
8.21	8.21	8.21	8.21	8.21	8.21
3	3	3	3	3	3
30	30	30	30	30	30
63	63	63	63	63	63
0.305	0.305	0.305	0.305	0.305	0.305

NSHTÖU ... ×2.5+5×1(C) 0.6/1 kV

					19	25
	1	Number of cores			19	25
Mechanical properties	20	Minimum bending radius for fixed installation	mm		210	225
	22	Minimum bending radius for free flexing	mm		350	375
	23	Minimum bending radius for cable entry	mm		350	375
	24	Minimum bending radius for reeling operation	mm		420	450
	26	Minimum bending radius for power track system	mm		350	375
	27	Minimum bending radius for guide rollers	mm		525	563
	29	Perm. pulling force for install. and operation	N		950	1250
Electrical properties	30	d.c. resistance/unit length at 90 °C	Ω/km		10.5	10.5
	31	d.c. resistance/unit length at 20 °C	Ω/km		8.21	8.21
Current-carrying capacity[1]	37	Number of loaded cores			3	3
	41	Installation on walls	A		30	30
	42	Minimum time value	s		63	63
Short-circuit	43	Rated short-time current of conductor (1s)	kA		0.305	0.305

[1] Current-carrying capacity not according to DIN VDE 0298 Part 4, see Clause 1.2

Table 2.8.2 c
CORDAFLEX(K) rubber-sheathed flexible cables (4 cores) for travel speeds up to 120 m/min

Design				4	6	10	16
Design	2	Nom. cross-sectional area of conductor	mm²	4	6	10	16
	6	Type of conductor		FS	FS	FS	FS
	7	Coating of strands		tinned	tinned	tinned	tinned
	8	Number of strands		80	120	210	230
	9	Diameter of strand	mm	0.26	0.26	0.26	0.31
	10	Diameter of conductor	mm	3.2	3.9	5.1	6.3
	11	Thickness of insulation	mm	1.0	1.0	1.2	1.2
	13	Thickness of inner sheath	mm	1.2	1.2	1.4	1.6
	14	Thickness of outer sheath	mm	2.0	2.0	2.2	2.5
	15	Overall diameter of cable (min. value)	mm	17.0	18.5	20.0	26.0
	16	Overall diameter of cable (max. value)	mm	20.5	22.5	26.5	31.5
	19	Weight of cable	kg/km	490	610	920	1300
Mechanical properties	20	Minimum bending radius for fixed installation	mm	123	135	159	189
	22	Minimum bending radius for free flexing	mm	205	225	265	315
	23	Minimum bending radius for cable entry	mm	205	225	265	315
	24	Minimum bending radius for reeling operation	mm	246	270	318	378
	26	Minimum bending radius for power track system	mm	205	225	265	315
	27	Minimum bending radius for guide rollers	mm	308	338	398	473
	29	Perm. pulling force for install. and operation	N	320	480	800	1280
Electrical properties	30	d.c. resistance/unit length at 90 °C	Ω/km	6.49	4.32	2.49	1.58
	31	d.c. resistance/unit length at 20 °C	Ω/km	5.09	3.39	1.95	1.24
Current-carrying capacity[1]	37	Number of loaded cores		3	3	3	3
	41	Installation on walls	A	41	53	74	99
	42	Minimum time value	s	87	117	166	237
Short-circuit	43	Rated short-time current of conductor (1s)	kA	0.488	0.732	1.22	1.95

[1] Current-carrying capacity not according to DIN VDE 0298 Part 4, see Clause 1.2

25	35	50	70	95	120
FS tinned	FS tinned	FS tinned	FS tinned	FS tinned	FS tinned
360	510	530	730	780	1000
0.31	0.31	0.36	0.36	0.41	0.41
7.8	9.2	11.0	13.1	15.1	17.0
1.4	1.4	1.6	1.6	1.8	1.8
1.8	1.8	2.0	2.0	2.4	2.8
3.0	3.0	3.5	3.5	4.0	4.5
31.5	34.0	40.0	44.5	51.5	55.5
38.0	44.5	47.5	52.5	60.5	65.0
2000	2520	3480	4560	6030	7340
228	267	285	315	363	390
380	445	475	525	605	650
380	445	475	525	605	650
456	534	570	630	726	780
380	445	475	525	605	650
570	668	713	788	908	975
2000	2800	4000	5600	7600	9600
1.01	0.720	0.501	0.353	0.268	0.209
0.795	0.565	0.393	0.277	0.210	0.164
3	3	3	3	3	3
131	162	202	250	301	352
331	424	557	713	906	1057
3.05	4.27	6.10	8.54	11.6	14.6

Table 2.8.3
CORDAFLEX(SM) rubber-sheathed flexible cables for very high dynamic stresses

Design							
	1	Number of cores		46	12	24	7
	2	Nom. cross-sectional area of conductor	mm²	1	1.5	1.5	2.5
	6	Type of conductor		FS	FS	FS	FS
	7	Coating of strands		tinned	tinned	tinned	tinned
	8	Number of strands		60	50	50	80
	9	Diameter of strand	mm	0.16	0.21	0.21	0.21
	10	Diameter of conductor	mm	1.5	1.8	1.8	2.6
	11	Thickness of insulation	mm	0.8	0.8	0.8	0.9
	13	Thickness of inner sheath	mm	1.6	1.4	1.4	1.2
	14	Thickness of outer sheath	mm	2.5	2.2	2.2	2.0
	15	Overall diameter of cable (min. value)	mm	31.5	22.5	25.5	18.5
	16	Overall diameter of cable (max. value)	mm	38.0	27.5	31.0	22.5
	19	Weight of cable	kg/km	1700	830	1130	620
Mechanical properties	20	Minimum bending radius for fixed installation	mm	228	165	186	135
	22	Minimum bending radius for free flexing	mm	380	275	310	225
	23	Minimum bending radius for cable entry	mm	380	275	310	225
	24	Minimum bending radius for reeling operation	mm	456	330	372	270
	26	Minimum bending radius for power track system	mm	380	275	310	225
	27	Minimum bending radius for guide rollers	mm	570	413	465	338
	29	Perm. pulling force for install. and operation	N	920	360	720	350
Electrical properties	30	d.c. resistance/unit length at 90 °C	Ω/km	25.5	17.5	17.5	10.5
	31	d.c. resistance/unit length at 20 °C	Ω/km	20.0	13.7	13.7	8.21
Current-carrying capacity[1]	37	Number of loaded cores		3	3	3	3
	41	Installation on walls	A	23	23	23	30
	42	Minimum time value	s	17	39	39	63
Short-circuit	43	Rated short-time current of conductor (1s)	kA	0.122	0.183	0.183	0.305

[1] Current-carrying capacity not according to DIN VDE 0298 Part 4, see Clause 1.2

Table 2.8.4
SPREADERFLEX basket cables

Design							
	1	Number of cores		48	30	36	42
	2	Nom. cross-sectional area of conductor	mm²	1	2.5	2.5	2.5
	6	Type of conductor		FS	FS	FS	FS
	7	Coating of strands		plain	plain	plain	plain
	8	Number of strands		55	76	76	76
	9	Diameter of strand	mm	0.16	0.21	0.21	0.21
	10	Diameter of conductor	mm	1.5	2.6	2.6	2.6
	11	Thickness of insulation	mm	0.6	0.7	0.7	0.7
	12	Thickness of sheath	mm	2.0	2.0	2.0	2.0
	15	Overall diameter of cable (min. value)	mm	29.5	29.5	33.0	35.5
	16	Overall diameter of cable (max. value)	mm	35.5	35.5	39.0	42.5
	19	Weight of cable	kg/km	2190	2300	2740	3460

NSHTÖU ... × ... 0.6/1 kV

12	24	30	4	4	4	4	4
2.5	2.5	2.5	10	16	25	35	50
FS	FS	FS	FS	FS	FS	FS	FS
tinned	tinned	tinned	tinned	tinned	tinned	tinned	tinned
80	80	80	210	230	360	510	530
0.21	0.21	0.21	0.26	0.31	0.31	0.31	0.36
2.6	2.6	2.6	5.1	6.3	7.8	9.2	11.0
0.9	0.9	0.9	1.2	1.2	1.4	1.4	1.6
1.4	1.6	1.8	1.4	1.6	1.8	1.8	2.0
2.2	2.5	3.0	2.2	2.5	3.0	3.0	3.5
25.5	29.5	33.5	23.0	26.5	32.5	35.5	41.5
31.0	35.5	41.0	28.0	32.0	38.5	42.0	49.0
1130	1640	2200	1060	1500	2230	2850	4000
186	213	246	168	192	231	252	294
310	355	410	280	320	385	420	490
310	355	410	280	320	385	420	490
372	426	492	336	384	462	504	588
310	355	410	280	320	385	420	490
465	533	615	420	480	578	630	735
600	1200	1500	800	1280	2000	2800	4000
10.5	10.5	10.5	2.49	1.58	1.01	0.720	0.501
8.21	8.21	8.21	1.95	1.24	0.795	0.565	0.393
3	3	3	3	3	3	3	3
30	30	30	74	99	131	162	202
63	63	63	166	237	331	424	557
0.305	0.305	0.305	1.22	1.95 ,	3.05	4.27	6.10

YSLTÖ ... × ... 300/500 V

					48	30	36	42
	1	Number of cores						
	2	Nom. cross-sectional area of conductor	mm^2		1	2.5	2.5	2.5
echanical operties	20	Minimum bending radius for fixed installation	mm		142	142	156	170
	22	Minimum bending radius for free flexing	mm		178	178	195	213
	23	Minimum bending radius for cable entry	mm		178	178	195	213
	29	Perm. pulling force for install. and operation	N		720	1125	1350	1575
ectrical operties	30	d.c. resistance/unit length at 70 °C	Ω/km		23.3	9.55	9.55	9.55
	31	d.c. resistance/unit length at 20 °C	Ω/km		19.5	7.98	7.98	7.98
urrent-carrying pacity	37	Number of loaded cores			3	3	3	3
	41	Installation on walls	A		15.0	26	26	26
	42	Minimum time value	s		40	84	84	84
ort-circuit	43	Rated short-time current of conductor (1s)	kA		0.109	0.272	0.272	0.272

Table 2.8.5 a
PLANOFLEX rubber-sheathed flat flexible cables nominal cross-sectional area of conductor = 1.5 and 2.5 mm^2

Design						
	1	Number of cores		4	7	8
	2	Nom. cross-sectional area of conductor	mm^2	1.5	1.5	1.5
	6	Type of conductor		FF	FF	FF
	7	Coating of strands		plain	plain	plain
	8	Number of strands		77	77	77
	9	Diameter of strand	mm	0.16	0.16	0.16
	10	Diameter of conductor	mm	1.8	1.8	1.8
	11	Thickness of insulation	mm	0.8	0.8	0.8
	12	Thickness of sheath	mm	1.2	1.2	1.2
	17	Overall dimensions of cable (min. value)	mm × mm	5.4×15.0	5.4×25.0	5.4×27.5
	18	Overall dimensions of cable (max. value)	mm × mm	6.4×18.5	6.4×30.0	6.4×32.0
	19	Weight of cable	kg/km	160	280	310
Mechanical properties	20	Minimum bending radius for fixed installation	mm	26	26	26
	25	Minimum bending radius for festoon system	mm	19	19	19
	29	Perm. pulling force for install. and operation	N	90	158	180
Electrical properties	30	d.c. resistance/unit length at 90 °C	Ω/km	17.0	17.0	17.0
	31	d.c. resistance/unit length at 20 °C	Ω/km	13.3	13.3	13.3
Current-carrying capacity[1]	37	Number of loaded cores		3	3	3
	41	Installation on walls	A	23	23	23
	42	Minimum time value	s	39	39	39
Short-circuit	43	Rated short-time current of conductor (1s)	kA	0.214	0.214	0.214

[1] Current-carrying capacity not according to DIN VDE 0298 Part 4, see Clause 1.2

Table 2.8.5 b
PLANOFLEX rubber-sheathed flat flexible cables 4 cores

Design						
	2	Nom. cross-sectional area of conductor	mm^2	4	6	10
	6	Type of conductor		FF	FF	FF
	7	Coating of strands		plain	plain	plain
	8	Number of strands		210	175	300
	9	Diameter of strand	mm	0.16	0.21	0.21
	10	Diameter of conductor	mm	3.2	3.9	5.1
	11	Thickness of insulation	mm	1.0	1.0	1.2
	12	Thickness of sheath	mm	1.8	1.8	1.8
	17	Overall dimensions of cable (min. value)	mm × mm	7.9×22.5	8.5×24.5	9.9×30.0
	18	Overall dimensions of cable (max. value)	mm × mm	9.6×28.0	10.4×31.0	12.1×38.0
	19	Weight of cable	kg/km	400	500	760
Mechanical properties	20	Minimum bending radius for fixed installation	mm	38	42	48
	25	Minimum bending radius for festoon system	mm	38	42	60
	29	Perm. pulling force for install. and operation	N	240	360	600
Electrical properties	30	d.c. resistance/unit length at 90 °C	Ω/km	6.31	4.21	2.44
	31	d.c. resistance/unit length at 20 °C	Ω/km	4.95	3.30	1.91
Current-carrying capacity[1]	37	Number of loaded cores		3	3	3
	41	Installation on walls	A	41	53	74
	42	Minimum time value	s	87	117	166
Short-circuit	43	Rated short-time current of conductor (1s)	kA	0.572	0.858	1.43

[1] Current-carrying capacity not according to DIN VDE 0298 Part 4, see Clause 1.2

NGFLGÖU ...×... 300/500 V

10	12	24	4	7	8	12	24
1.5	1.5	1.5	2.5	2.5	2.5	2.5	2.5
FF	FF	FF	FF	FF	FF	FF	FF
plain	plain	plain	plain	plain	plain	plain	plain
77	77	77	130	130	130	130	130
0.16	0.16	0.16	0.16	0.16	0.16	0.16	0.16
1.8	1.8	1.8	2.6	2.6	2.6	2.6	2.6
0.8	0.8	0.8	0.9	0.9	0.9	0.9	0.9
1.5	1.5	1.8	1.5	1.5	1.5	1.8	2.4
6.0×35.5	6.0×42.0	11.5×51.0	6.6×18.5	6.6×31.0	6.6×33.5	7.2×50.5	15.0×66.0
7.0×41.5	7.0×48.5	13.5×56.0	8.2×24.0	8.2×39.5	8.2×42.5	8.8×63.5	17.0×72.5
430	510	1050	270	460	520	830	1730
28	28	54	33	33	33	35	68
21	21	68	33	33	33	35	85
225	270	540	150	263	300	450	900
17.0	17.0	17.0	10.2	10.2	10.2	10.2	10.2
13.3	13.3	13.3	7.98	7.98	7.98	7.98	7.98
3	3	3	3	3	3	3	3
23	23	23	30	30	30	30	30
39	39	39	63	63	63	63	63
0.214	0.214	0.214	0.357	0.357	0.357	0.357	0.357

NGFLGÖU 4×... 300/500 V

16	25	35	50	70	95
FF	FF	F	F	F	F
plain	plain	plain	plain	plain	plain
480	750	276	396	560	740
0.21	0.21	0.41	0.41	0.41	0.41
6.3	7.8	9.2	11.0	13.1	15.1
1.2	1.4	1.4	1.6	1.6	1.8
2.1	2.1	2.4	2.7	3.0	3.3
11.5×35.0	13.1×41.5	14.8×47.0	17.2×55.0	19.5×62.5	22.0×71.0
14.0×44.0	15.7×51.0	18.0×58.5	21.0×69.0	23.7×78.0	26.9×89.0
1100	1580	2150	2960	4000	5470
56	63	72	84	95	108
70	78	90	105	118	134
960	1500	2100	3000	4200	5700
1.54	0.995	0.706	0.492	0.347	0.263
1.21	0.780	0.554	0.386	0.272	0.206
3	3	3	3	3	3
99	131	162	202	250	301
237	331	424	557	713	906
2.29	3.57	5.00	7.15	10.0	13.6

Table 2.8.5c
PLANOFLEX rubber-sheathed flat flexible cables 7 cores

Design				4	6	10
	2	Nom. cross-sectional area of conductor	mm²	4	6	10
	6	Type of conductor		FF	FF	FF
	7	Coating of strands		plain	plain	plain
	8	Number of strands		210	175	300
	9	Diameter of strand	mm	0.16	0.21	0.21
	10	Diameter of conductor	mm	3.2	3.9	5.1
	11	Thickness of insulation	mm	1.0	1.0	1.2
	12	Thickness of sheath	mm	1.8	1.8	1.8
	17	Overall dimensions of cable (min. value)	mm × mm	7.9×37.0	8.5×41.0	9.9×50.5
	18	Overall dimensions of cable (max. value)	mm × mm	9.6×46.5	10.4×51.5	12.1×63.5
	19	Weight of cable	kg/km	680	860	1320
Mechanical properties	20	Minimum bending radius for fixed installation	mm	38	42	48
	25	Minimum bending radius for festoon system	mm	38	42	60
	29	Perm. pulling force for install. and operation	N	420	630	1050
Electrical properties	30	d.c. resistance/unit length at 90 °C	Ω/km	6.31	4.21	2.44
	31	d.c. resistance/unit length at 20 °C	Ω/km	4.95	3.30	1.91
Current-carrying capacity[1]	37	Number of loaded cores		3	3	3
	41	Installation on walls	A	41	53	74
	42	Minimum time value	s	87	117	166
Short-circuit	43	Rated short-time current of conductor (1s)	kA	0.572	0.858	1.43

[1] Current-carrying capacity not according to DIN VDE 0298 Part 4, see Clause 1.2

Table 2.8.6
PVC lift control cables for suspension lengths up to 50 m   **YSLTK ... ×1 300/500 V**

Design				7	12	18	24	30
	1	Number of cores		7	12	18	24	30
	6	Type of conductor		F	F	F	F	F
	7	Coating of strands		plain	plain	plain	plain	plain
	8	Number of strands		32	32	32	32	32
	9	Diameter of strand	mm	0.21	0.21	0.21	0.21	0.21
	10	Diameter of conductor	mm	1.5	1.5	1.5	1.5	1.5
	11	Thickness of insulation	mm	0.6	0.6	0.6	0.6	0.6
	12	Thickness of sheath	mm	1.0	1.2	1.2	1.4	1.4
	15	Overall diameter of cable (min. value)	mm	11.3	15.3	15.5	18.5	20.8
	16	Overall diameter of cable (max. value)	mm	13.7	18.5	18.8	21.5	23.8
	19	Weight of cable	kg/km	190	340	370	540	680
Mechanical properties	20	Minimum bending radius for fixed installation	mm	55	74	75	86	95
Electrical properties	30	d.c. resistance/unit length at 70 °C	Ω/km	23.3	23.3	23.3	23.3	23.3
	31	d.c. resistance/unit length at 20 °C	Ω/km	19.5	19.5	19.5	19.5	19.5
Current-carrying capacity	37	Number of loaded cores		3	3	3	3	3
	41	Installation on walls	A	15.0	15.0	15.0	15.0	15.0
	42	Minimum time value	s	29	29	29	29	29
Short-circuit	43	Rated short-time current of conductor (1s)	kA	0.109	0.109	0.109	0.109	0.109

16	25	35
FF	FF	F
plain	plain	plain
480	750	280
0.21	0.21	0.41
6.3	7.8	9.2
1.2	1.4	1.4
2.4	2.7	2.7
12.1×59.5	14.3×71.5	15.4×79.5
14.6×74.0	17.0×88.5	18.6×99.5
1970	2810	3830
58	68	74
73	85	93
1680	2625	3675
1.54	0.995	0.706
1.21	0.780	0.554
3	3	3
99	131	162
237	331	424
2.29	3.57	5.00

Table 2.8.7
PVC lift control cables for suspension lengths up to 150 m, 2 screened cores   **YSLYTK 28×1+2×0.5FM(C) 300/500 V**

esign	6	Type of conductor		F
	7	Coating of strands		plain
	8	Number of strands		32
	9	Diameter of strand	mm	0.21
	10	Diameter of conductor	mm	1.5
	11	Thickness of insulation	mm	0.6
	12	Thickness of sheath	mm	1.4
	15	Overall diameter of cable (min. value)	mm	23.7
	16	Overall diameter of cable (max. value)	mm	28.2
	19	Weight of cable	kg/km	780
Mechanical Properties	20	Minimum bending radius for fixed installation	mm	113
Electrical Properties	30	d.c. resistance/unit length at 70 °C	Ω/km	23.3
	31	d.c. resistance/unit length at 20 °C	Ω/km	19.5
Current-carrying capacity	37	Number of loaded cores		3
	41	Installation on walls	A	15.0
	42	Minimum time value	s	29
Short-circuit	43	Rated short-time current of conductor (1s)	kA	0.109

Table 2.8.8   **YSLYCYTK 28×1+2×0.5FM(C) 300/500 V**
PVC lift control cable for suspension lengths up to 150 m, 2 screened cores and common screen

Design	6	Type of conductor		F
	7	Coating of strands		plain
	8	Number of strands		32
	9	Diameter of strand	mm	0.21
	10	Diameter of conductor	mm	1.5
	11	Thickness of insulation	mm	0.6
	13	Thickness of inner sheath	mm	1.0
	14	Thickness of outer sheath	mm	1.4
	15	Overall diameter of cable (min. value)	mm	24.9
	16	Overall diameter of cable (max. value)	mm	29.4
	19	Weight of cable	kg/km	910
Mechanical properties	20	Minimum bending radius for fixed installation	mm	118
Electrical properties	30	d.c. resistance/unit length at 70 °C	Ω/km	23.3
	31	d.c. resistance/unit length at 20 °C	Ω/km	19.5
Current-carrying capacity	37	Number of loaded cores		3
	41	Installation on walls	A	15.0
	42	Minimum time value	s	29
Short-circuit	43	Rated short-time current of conductor (1s)	kA	0.109

Table 2.8.9
PROTOLON reeling and trailing cables for travel speeds up to 60 m/min   NTSWÖU 3×...+3×.../3 0.6/1 kV

esign	3	Nom. cross-sectional area of phase conductor	mm²	120	150	185
	4	Nom. cross-sectional area of PE conductor	mm²	70	70	95
	6	Type of conductor		F	F	F
	7	Coating of strands		tinned	tinned	tinned
	8	Number of strands		962	1191	1457
	9	Diameter of strand	mm	0.41	0.41	0.41
	10	Diameter of conductor	mm	17.0	1941.0	21.0
	11	Thickness of insulation	mm	2.4	2.4	2.4
	13	Thickness of inner sheath	mm	2.4	2.4	2.8
	14	Thickness of outer sheath	mm	4.0	4.0	4.5
	15	Overall diameter of cable (min. value)	mm	52.0	55.5	61.0
	16	Overall diameter of cable (max. value)	mm	61.0	65.0	71.0
	19	Weight of cable	kg/km	6770	7880	9600
echanical operties	20	Minimum bending radius for fixed installation	mm	366	390	426
	22	Minimum bending radius for free flexing	mm	610	650	710
	23	Minimum bending radius for cable entry	mm	610	650	710
	24	Minimum bending radius for reeling operation	mm	732	780	852
	27	Minimum bending radius for guide rollers	mm	915	975	1065
	29	Perm. pulling force for install. and operation	N	5400	6750	8325
ectrical operties	30	d.c. resistance/unit length at 90 °C	Ω/km	0.209	0.168	0.138
	31	d.c. resistance/unit length at 20 °C	Ω/km	0.164	0.132	0.108
	32	Inductance/unit length per conductor	mH/km	0.240	0.190	0.230
urrent-carrying pacity	37	Number of loaded cores		3	3	3
	41	Installation on walls	A	352	404	461
	42	Minimum time value	s	1057	1253	1464
ort-circuit	43	Rated short-time current of conductor (1s)	kA	14.6	18.3	22.6

Table 2.8.10
PROTOLON reeling and trailing cables for travel speeds up to 60 m/min

Design	3	Nom. cross-sectional area of phase conductor	mm^2	25	35	50	70
	4	Nom. cross-sectional area of PE conductor	mm^2	25	25	25	35
	6	Type of conductor		F	F	F	F
	7	Coating of strands		tinned	tinned	tinned	tinned
	8	Number of strands		196	276	396	565
	9	Diameter of strand	mm	0.41	0.41	0.41	0.41
	10	Diameter of conductor	mm	7.8	9.2	11.0	13.1
	11	Thickness of insulation	mm	3.0	3.0	3.0	3.0
	13	Thickness of inner sheath	mm	2.0	2.0	2.4	2.4
	14	Thickness of outer sheath	mm	3.5	3.5	4.0	4.0
	15	Overall diameter of cable (min. value)	mm	41.5	44.5	49.0	53.0
	16	Overall diameter of cable (max. value)	mm	48.5	52.5	58.6	62.0
	19	Weight of cable	kg/km	3000	3580	4480	5490
Mechanical properties	20	Minimum bending radius for fixed installation	mm	291	315	352	372
	22	Minimum bending radius for free flexing	mm	485	525	586	620
	23	Minimum bending radius for cable entry	mm	485	525	586	620
	24	Minimum bending radius for reeling operation	mm	582	630	703	744
	27	Minimum bending radius for guide rollers	mm	728	788	879	930
	29	Perm. pulling force for install. and operation	N	1125	1575	2250	3150
Electrical properties	30	d.c. resistance/unit length at 90 °C	Ω/km	1.01	0.720	0.501	0.353
	31	d.c. resistance/unit length at 20 °C	Ω/km	0.795	0.565	0.393	0.277
	32	Inductance/unit length per conductor	mH/km	0.290	0.280	0.270	0.260
	33	Operating capacitance/unit length	μF/km	0.370	0.420	0.470	0.540
	34	Charging current	A/km	0.400	0.460	0.510	0.590
	35	Earth fault current	A/km	1.21	1.37	1.53	1.76
	36	Electric field strength at the conductor	kV/mm	1.49	1.45	1.41	1.38
Current-carrying capacity	37	Number of loaded cores		3	3	3	3
	41	Installation on walls	A	131	162	202	250
	42	Minimum time value	s	331	424	557	713
Short-circuit	43	Rated short-time current of conductor (1s)	kA	3.05	4.27	6.10	8.54

95	120
50	70
F	F
tinned	tinned
746	962
0.41	0.41
15.1	17.0
3.0	3.0
2.8	2.8
4.5	4.5
56.5	61.5
66.5	72.0
6610	8140
399	432
665	720
665	720
798	864
998	1080
4275	5400
0.268	0.209
0.210	0.164
0.250	0.240
0.600	0.660
0.650	0.720
1.96	2.15
1.35	1.33
3	3
301	352
906	1057
11.6	14.6

Table 2.8.11
PROTOLON reeling and trailing cables for travel speeds up to 60 m/min

Design							
	3	Nom. cross-sectional area of phase conductor	mm^2	25	35	50	70
	4	Nom. cross-sectional area of PE conductor	mm^2	25	25	25	35
	6	Type of conductor		F	F	F	F
	7	Coating of strands		tinned	tinned	tinned	tinned
	8	Number of strands		196	276	396	565
	9	Diameter of strand	mm	0.41	0.41	0.41	0.41
	10	Diameter of conductor	mm	7.8	9.2	11.0	13.1
	11	Thickness of insulation	mm	3.4	3.4	3.4	3.4
	13	Thickness of inner sheath	mm	2.0	2.0	2.4	2.4
	14	Thickness of outer sheath	mm	3.5	3.5	4.0	4.0
	15	Overall diameter of cable (min. value)	mm	43.0	46.0	51.0	54.5
	16	Overall diameter of cable (max. value)	mm	50.5	55.0	60.0	64.0
	19	Weight of cable	kg/km	3170	3780	4710	5830
Mechanical properties	20	Minimum bending radius for fixed installation	mm	303	330	360	384
	22	Minimum bending radius for free flexing	mm	505	550	600	640
	23	Minimum bending radius for cable entry	mm	505	550	600	640
	24	Minimum bending radius for reeling operation	mm	606	660	720	768
	27	Minimum bending radius for guide rollers	mm	758	825	900	960
	29	Perm. pulling force for install. and operation	N	1125	1575	2250	3150
Electrical properties	30	d.c. resistance/unit length at 90 °C	Ω/km	1.01	0.720	0.501	0.353
	31	d.c. resistance/unit length at 20 °C	Ω/km	0.795	0.565	0.393	0.277
	32	Inductance/unit length per conductor	mH/km	0.300	0.290	0.270	0.260
	33	Operating capacitance/unit length	μF/km	0.340	0.380	0.430	0.480
	34	Charging current	A/km	0.620	0.690	0.780	0.870
	35	Earth fault current	A/km	1.85	2.07	2.34	2.61
	36	Electric field strength at the conductor	kV/mm	2.26	2.19	2.12	2.07
Current-carrying capacity	37	Number of loaded cores		3	3	3	3
	41	Installation on walls	A	131	162	202	250
	42	Minimum time value	s	331	424	557	713
Short-circuit	43	Rated short-time current of conductor (1s)	kA	3.05	4.27	6.10	8.54

95	120
50	70
F	F
tinned	tinned
746	962
0.41	0.41
15.1	17.0
3.4	3.4
2.8	2.8
4.5	4.5
60.0	63.5
70.0	74.0
7120	8410
420	444
700	740
700	740
840	888
1050	1110
4275	5400
0.268	0.209
0.210	0.164
0.260	0.250
0.540	0.590
0.980	1.07
2.94	3.21
2.02	1.99
3	3
301	352
906	1057
11.6	14.6

Table 2.8.12 PROTOLON(SM) reeling and trailing cables travel speeds up to 120 m/min			NTSCGEWÖU 3×...+3×.../3 3.6/6 kV				
Design	3	Nom. cross-sectional area of phase conductor	mm^2	25	35	50	70
	4	Nom. cross-sectional area of PE conductor	mm^2	25	25	25	35
	6	Type of conductor		FS	FS	FS	FS
	7	Coating of strands		tinned	tinned	tinned	tinned
	8	Number of strands		349	494	516	736
	9	Diameter of strand	mm	0.31	0.31	0.36	0.36
	10	Diameter of conductor	mm	7.8	9.2	11.0	13.1
	11	Thickness of insulation	mm	3.0	3.0	3.0	3.0
	13	Thickness of inner sheath	mm	3.4	3.6	4.0	4.2
	14	Thickness of outer sheath	mm	4.9	5.3	5.8	6.3
	15	Overall diameter of cable (min. value)	mm	46.5	50.0	55.0	61.0
	16	Overall diameter of cable (max. value)	mm	55.0	59.0	64.5	71.0
	19	Weight of cable	kg/km	3780	4510	5780	7050
Mechanical properties	20	Minimum bending radius for fixed installation	mm	330	354	387	426
	22	Minimum bending radius for free flexing	mm	550	590	645	710
	23	Minimum bending radius for cable entry	mm	550	590	645	710
	24	Minimum bending radius for reeling operation	mm	660	708	774	852
	27	Minimum bending radius for guide rollers	mm	825	885	968	1065
	29	Perm. pulling force for install. and operation	N	1125	1575	2250	3150
Electrical properties	30	d.c. resistance/unit length at 90 °C	Ω/km	1.01	0.720	0.501	0.353
	31	d.c. resistance/unit length at 20 °C	Ω/km	0.795	0.565	0.393	0.277
	32	Inductance/unit length per conductor	mH/km	0.290	0.280	0.270	0.260
	33	Operating capacitance/unit length	μF/km	0.370	0.420	0.470	0.540
	34	Charging current	A/km	0.400	0.460	0.510	0.590
	35	Earth fault current	A/km	1.21	1.37	1.53	1.76
	36	Electric field strength at the conductor	kV/mm	1.49	1.45	1.41	1.38
Current-carrying capacity	37	Number of loaded cores		3	3	3	3
	41	Installation on walls	A	131	162	202	250
	42	Minimum time value	s	331	424	557	713
Short-circuit	43	Rated short-time current of conductor (1s)	kA	3.05	4.27	6.10	8.54

Table 2.8.13
PROTOLON(SM) reeling and trailing cables travel speeds up to 120 m/min   **NTSCGEWÖU 3×…+3×…/3 6/10 kV**

Design	3	Nom. cross-sectional area of phase conductor	mm^2	25	35	50	70
	4	Nom. cross-sectional area of PE conductor	mm^2	25	25	25	35
	6	Type of conductor		FS	FS	FS	FS
	7	Coating of strands		tinned	tinned	tinned	tinned
	8	Number of strands		349	494	516	736
	9	Diameter of strand	mm	0.31	0.31	0.36	0.36
	10	Diameter of conductor	mm	7.8	9.2	11.0	13.1
	11	Thickness of insulation	mm	3.4	3.4	3.4	3.4
	13	Thickness of inner sheath	mm	3.5	3.8	4.1	4.4
	14	Thickness of outer sheath	mm	5.2	5.5	6.0	6.5
	15	Overall diameter of cable (min. value)	mm	48.5	52.5	57.0	63.0
	16	Overall diameter of cable (max. value)	mm	57.2	61.5	67.0	73.0
	19	Weight of cable	kg/km	4060	4850	5870	7410
Mechanical properties	20	Minimum bending radius for fixed installation	mm	343	369	402	438
	22	Minimum bending radius for free flexing	mm	572	615	670	730
	23	Minimum bending radius for cable entry	mm	572	615	670	730
	24	Minimum bending radius for reeling operation	mm	686	738	804	876
	27	Minimum bending radius for guide rollers	mm	858	923	1005	1095
	29	Perm. pulling force for install. and operation	N	1125	1575	2250	3150
Electrical properties	30	d.c. resistance/unit length at 90 °C	Ω/km	1.01	0.720	0.501	0.353
	31	d.c. resistance/unit length at 20 °C	Ω/km	0.795	0.565	0.393	0.277
	32	Inductance/unit length per conductor	mH/km	0.300	0.290	0.270	0.260
	33	Operating capacitance/unit length	μF/km	0.340	0.380	0.430	0.480
	34	Charging current	A/km	0.620	0.690	0.780	0.870
	35	Earth fault current	A/km	1.85	2.07	2.34	2.61
	36	Electric field strength at the conductor	kV/mm	2.26	2.19	2.12	2.07
Current-carrying capacity	37	Number of loaded cores		3	3	3	3
	41	Installation on walls	A	131	162	202	250
	42	Minimum time value	s	331	424	557	713
Short-circuit	43	Rated short-time current of conductor (1s)	kA	3.05	4.27	6.10	8.54

Table 2.8.14 PROTOLON(SB) trailing cables					NTSCGEWÖU 3×...+3×50/3 6/10 kV			
Design	3	Nom. cross-sectional area of phase conductor	mm²	35	50	70	95
	6	Type of conductor		F	F	F	F
	7	Coating of strands		tinned	tinned	tinned	tinned
	8	Number of strands		276	396	565	746
	9	Diameter of strand	mm	0.41	0.41	0.41	0.41
	10	Diameter of conductor	mm	9.2	11.0	13.1	15.1
	11	Thickness of insulation	mm	3.4	3.4	3.4	3.4
	12	Thickness of sheath	mm	5.5	6.4	6.4	7.3
	15	Overall diameter of cable (min. value)	mm	46.0	51.0	54.0	60.0
	16	Overall diameter of cable (max. value)	mm	55.5	60.0	64.0	70.0
	19	Weight of cable	kg/km	3920	4840	5810	7090
Mechanical properties	20	Minimum bending radius for fixed installation	mm	333	360	384	420
	22	Minimum bending radius for free flexing	mm	555	600	640	700
	23	Minimum bending radius for cable entry	mm	555	600	640	700
	29	Perm. pulling force for install. and operation	N	1575	2250	3150	4275
Electrical properties	30	d.c. resistance/unit length at 90 °C	Ω/km	0.720	0.501	0.353	0.209
	31	d.c. resistance/unit length at 20 °C	Ω/km	0.565	0.393	0.277	0.164
	32	Inductance/unit length per conductor	mH/km	0.290	0.270	0.260	0.260
	33	Operating capacitance/unit length	μF/km	0.380	0.430	0.480	0.540
	34	Charging current	A/km	0.690	0.780	0.870	0.980
	35	Earth fault current	A/km	2.07	2.34	2.61	2.94
	36	Electric field strength at the conductor	kV/mm	2.19	2.12	2.07	2.02
Current-carrying capacity	37	Number of loaded cores		3	3	3	3
	41	Installation on walls	A	162	202	250	301
	42	Minimum time value	s	424	557	713	906
Short-circuit	43	Rated short-time current of conductor (1s)	kA	4.27	6.10	8.54	11.6

Table 2.8.15 PROTOLON(SB) trailing cables				NTSCGEWÖU 3×...+3×50/3 12/20 kV			
esign	3	Nom. cross-sectional area of phase conductor	mm^2	35	50	70	95
	6	Type of conductor		F	F	F	F
	7	Coating of strands		tinned	tinned	tinned	tinned
	8	Number of strands		276	396	565	746
	9	Diameter of strand	mm	0.41	0.41	0.41	0.41
	10	Diameter of conductor	mm	9.2	11.0	13.1	15.1
	11	Thickness of insulation	mm	5.5	5.5	5.5	5.5
	12	Thickness of sheath	mm	6.4	6.8	6.8	6.8
	15	Overall diameter of cable (min. value)	mm	56.0	60.5	64.0	70.0
	16	Overall diameter of cable (max. value)	mm	65.5	70.5	75.0	81.0
	19	Weight of cable	kg/km	5300	6330	7380	8830
echanical operties	20	Minimum bending radius for fixed installation	mm	393	423	450	486
	22	Minimum bending radius for free flexing	mm	655	705	750	810
	23	Minimum bending radius for cable entry	mm	655	705	750	810
	29	Perm. pulling force for install. and operation	N	1575	2250	3150	4275
ectrical operties	30	d.c. resistance/unit length at 90 °C	Ω/km	0.720	0.501	0.353	0.268
	31	d.c. resistance/unit length at 20 °C	Ω/km	0.565	0.393	0.277	0.210
	32	Inductance/unit length per conductor	mH/km	0.330	0.310	0.300	0.290
	33	Operating capacitance/unit length	μF/km	0.260	0.290	0.330	0.360
	34	Charging current	A/km	0.940	1.05	1.20	1.31
	35	Earth fault current	A/km	2.83	3.16	3.59	3.92
	36	Electric field strength at the conductor	kV/mm	3.04	2.92	2.81	2.73
urrent-carrying pacity	37	Number of loaded cores		3	3	3	3
	41	Installation on walls	A	172	215	265	319
	42	Minimum time value	s	376	492	634	806
ort-circuit	43	Rated short-time current of conductor (1s)	kA	4.27	6.10	8.54	11.6

Table 2.8.16 PROTOLON(SF) trailing cables for fixed installation				NTSCGEWÖU 3×...+3×.../3 6/10 kV				
Design	3	Nom. cross-sectional area of phase conductor	mm²	35	50	70	95	120
	4	Nom. cross-sectional area of PE conductor	mm²	25	25	35	50	70
	6	Type of conductor		F	F	F	F	F
	7	Coating of strands		tinned	tinned	tinned	tinned	tinned
	8	Number of strands		276	396	565	746	962
	9	Diameter of strand	mm	0.41	0.41	0.41	0.41	0.41
	10	Diameter of conductor	mm	9.2	11.0	13.1	15.1	17.0
	11	Thickness of insulation	mm	3.4	3.4	3.4	3.4	3.4
	13	Thickness of inner sheath	mm	2.0	2.4	2.4	2.8	2.8
	14	Thickness of outer sheath	mm	3.5	4.0	4.0	4.5	4.5
	15	Overall diameter of cable (min. value)	mm	45.5	50.0	54.0	59.0	62.5
	16	Overall diameter of cable (max. value)	mm	54.0	59.0	63.5	69.0	73.0
	19	Weight of cable	kg/km	3690	4590	5660	6970	8270
Mechanical properties	20	Minimum bending radius for fixed installation	mm	324	354	381	414	438
	23	Minimum bending radius for cable entry	mm	540	590	635	690	730
	29	Perm. pulling force for install. and operation	N	1575	2250	3150	4275	5400
Electrical properties	30	d.c. resistance/unit length at 90 °C	Ω/km	0.720	0.501	0.353	0.268	0.209
	31	d.c. resistance/unit length at 20 °C	Ω/km	0.565	0.393	0.277	0.210	0.164
	32	Inductance/unit length per conductor	mH/km	0.290	0.270	0.260	0.260	0.250
	33	Operating capacitance/unit length	µF/km	0.380	0.430	0.480	0.540	0.590
	34	Charging current	A/km	0.690	0.780	0.870	0.980	1.07
	35	Earth fault current	A/km	2.07	2.34	2.61	2.94	3.21
	36	Electric field strength at the conductor	kV/mm	2.19	2.12	2.07	2.02	1.99
Current-carrying capacity	37	Number of loaded cores		3	3	3	3	3
	41	Installation on walls	A	162	202	250	301	352
	42	Minimum time value	s	424	557	713	906	1057
Short-circuit	43	Rated short-time current of conductor (1s)	kA	4.27	6.10	8.54	11.6	14.6

Table 2.8.17

PROTOLON(SF) trailing cables for fixed installation

NTSCGEWÖU 3×...+3×.../3 12/20 kV

Design	3	Nom. cross-sectional area of phase conductor	mm²	35	50	70	95	120
	4	Nom. cross-sectional area of PE conductor	mm²	25	25	35	50	70
	6	Type of conductor		F	F	F	F	F
	7	Coating of strands		tinned	tinned	tinned	tinned	tinned
	8	Number of strands		276	396	565	746	962
	9	Diameter of strand	mm	0.41	0.41	0.41	0.41	0.41
	10	Diameter of conductor	mm	9.2	11.0	13.1	15.1	17.0
	11	Thickness of insulation	mm	5.5	5.5	5.5	5.5	5.5
	13	Thickness of inner sheath	mm	2.4	2.8	2.8	3.2	3.2
	14	Thickness of outer sheath	mm	4.0	4.5	4.5	5.0	5.0
	15	Overall diameter of cable (min. value)	mm	55.5	60.0	63.5	69.0	72.5
	16	Overall diameter of cable (max. value)	mm	65.0	70.0	74.0	80.0	84.0
	19	Weight of cable	kg/km	5030	6060	7210	8700	10120
Mechanical properties	20	Minimum bending radius for fixed installation	mm	390	420	444	480	504
	23	Minimum bending radius for cable entry	mm	650	700	740	800	840
	29	Perm. pulling force for install. and operation	N	1575	2250	3150	4275	5400
Electrical properties	30	d.c. resistance/unit length at 90 °C	Ω/km	0.720	0.501	0.353	0.268	0.209
	31	d.c. resistance/unit length at 20 °C	Ω/km	0.565	0.393	0.277	0.210	0.164
	32	Inductance/unit length per conductor	mH/km	0.330	0.310	0.300	0.290	0.280
	33	Operating capacitance/unit length	μF/km	0.260	0.290	0.330	0.360	0.390
	34	Charging current	A/km	0.940	1.05	1.20	1.31	1.41
	35	Earth fault current	A/km	2.83	3.16	3.59	3.92	4.24
	36	Electric field strength at the conductor	kV/mm	3.04	2.92	2.81	2.73	2.67
Current-carrying capacity	37	Number of loaded cores		3	3	3	3	3
	41	Installation on walls	A	172	215	265	319	371
	42	Minimum time value	s	376	492	634	806	951
Short-circuit	43	Rated short-time current of conductor (1s)	kA	4.27	6.10	8.54	11.6	14.6

Table 2.8.18 PROTOLON(ST) trailing cables for use in water			NTSCGEWÖU 3×...+3×.../3 6/10 kV			
Design	3	Nom. cross-sectional area of phase conductor	mm²	35	50	70	95
	4	Nom. cross-sectional area of PE conductor	mm²	25	25	35	50
	6	Type of conductor		F	F	F	F
	7	Coating of strands		tinned	tinned	tinned	tinned
	8	Number of strands		276	396	565	746
	9	Diameter of strand	mm	0.41	0.41	0.41	0.41
	10	Diameter of conductor	mm	9.2	11.0	13.1	15.1
	11	Thickness of insulation	mm	3.4	3.4	3.4	3.4
	13	Thickness of inner sheath	mm	2.0	2.4	2.4	2.8
	14	Thickness of outer sheath	mm	3.5	4.0	4.0	4.5
	15	Overall diameter of cable (min. value)	mm	45.5	50.0	54.0	59.0
	16	Overall diameter of cable (max. value)	mm	54.0	59.0	63.5	69.5
	19	Weight of cable	kg/km	3780	4670	5730	7090
Mechanical properties	20	Minimum bending radius for fixed installation	mm	324	354	381	417
	22	Minimum bending radius for free flexing	mm	540	590	635	695
	23	Minimum bending radius for cable entry	mm	540	590	635	695
	29	Perm. pulling force for install. and operation	N	1575	2250	3150	4275
Electrical properties	30	d.c. resistance/unit length at 90 °C	Ω/km	0.720	0.501	0.353	0.268
	31	d.c. resistance/unit length at 20 °C	Ω/km	0.565	0.393	0.277	0.210
	32	Inductance/unit length per conductor	mH/km	0.290	0.270	0.260	0.260
	33	Operating capacitance/unit length	μF/km	0.380	0.430	0.480	0.540
	34	Charging current	A/km	0.690	0.780	0.870	0.980
	35	Earth fault current	A/km	2.07	2.34	2.61	2.94
	36	Electric field strength at the conductor	kV/mm	2.19	2.12	2.07	2.02
Current-carrying capacity	37	Number of loaded cores		3	3	3	3
	41	Installation on walls	A	162	202	250	301
	42	Minimum time value	s	424	557	713	906
Short-circuit	43	Rated short-time current of conductor (1s)	kA	4.27	6.10	8.54	11.6

Table 2.8.19
PROTOLON(ST) trailing cables for use in water

NTSCGEWÖU 3×...+3×.../3 12/20 kV

sign							
	3	Nom. cross-sectional area of phase conductor	mm²	25	35	50	70
	4	Nom. cross-sectional area of PE conductor	mm²	25	25	25	35
	6	Type of conductor	.	F	F	F	F
	7	Coating of strands		tinned	tinned	tinned	tinned
	8	Number of strands		196	276	396	565
	9	Diameter of strand	mm	0.41	0.41	0.41	0.41
	10	Diameter of conductor	mm	7.8	9.2	11.0	13.1
	11	Thickness of insulation	mm	5.5	5.5	5.5	5.5
	13	Thickness of inner sheath	mm	2.4	2.4	2.8	2.8
	14	Thickness of outer sheath	mm	4.0	4.0	4.5	4.5
	15	Overall diameter of cable (min. value)	mm	52.0	55.5	60.0	63.5
	16	Overall diameter of cable (max. value)	mm	61.5	65.0	70.0	74.0
	19	Weight of cable	kg/km	4430	5125	6170	7360
echanical operties	20	Minimum bending radius for fixed installation	mm	369	390	420	444
	22	Minimum bending radius for free flexing	mm	615	650	700	740
	23	Minimum bending radius for cable entry	mm	615	650	700	740
	29	Perm. pulling force for install. and operation	N	1125	1575	2250	3150
ectrical operties	30	d.c. resistance/unit length at 90 °C	Ω/km	1.01	0.720	0.501	0.353
	31	d.c. resistance/unit length at 20 °C	Ω/km	0.795	0.565	0.393	0.277
	32	Inductance/unit length per conductor	mH/km	0.340	0.330	0.310	0.300
	33	Operating capacitance/unit length	μF/km	0.240	0.260	0.290	0.330
	34	Charging current	A/km	0.870	0.940	1.05	1.20
	35	Earth fault current	A/km	2.61	2.83	3.16	3.59
	36	Electric field strength at the conductor	kV/mm	3.17	3.04	2.92	2.81
rrent-carrying pacity	37	Number of loaded cores		3	3	3	3
	41	Installation on walls	A	139	172	215	265
	42	Minimum time value	s	294	376	492	634
ort-circuit	43	Rated short-time current of conductor (1s)	kA	3.05	4.27	6.10	8.54

Table 2.8.20
PROTOLON reeling and trailing cables 1 core

NTMCGCWÖU 1×... 12/20 kV

				25	50	95
Design	2	Nom. cross-sectional area of conductor	mm²	25	50	95
	6	Type of conductor		F	F	F
	7	Coating of strands		tinned	tinned	tinned
	8	Number of strands		196	396	746
	9	Diameter of strand	mm	0.41	0.41	0.41
	10	Diameter of conductor	mm	7.8	11.0	15.1
	11	Thickness of insulation	mm	5.5	5.5	5.5
	12	Thickness of sheath	mm	2.5	3.0	3.0
	15	Overall diameter of cable (min. value)	mm	25.5	29.0	33.5
	16	Overall diameter of cable (max. value)	mm	31.0	35.0	40.0
	19	Weight of cable	kg/km	1080	1530	2190
Mechanical properties	20	Minimum bending radius for fixed installation	mm	186	210	240
	22	Minimum bending radius for free flexing	mm	310	350	400
	23	Minimum bending radius for cable entry	mm	310	350	400
	29	Perm. pulling force for install. and operation	N	375	750	1425
Electrical properties	30	d.c. resistance/unit length at 90 °C	Ω/km	1.01	0.501	0.268
	31	d.c. resistance/unit length at 20 °C	Ω/km	0.795	0.393	0.210
	33	Operating capacitance/unit length	μF/km	0.240	0.290	0.360
	34	Charging current	A/km	0.870	1.05	1.31
	35	Earth fault current	A/km	2.61	3.16	3.92
	36	Electric field strength at the conductor	kV/mm	3.17	2.92	2.73
Current-carrying capacity	37	Number of loaded cores		3	3	3
	41	Installation on walls	A	139	215	319
	42	Minimum time value	s	294	492	806
Short-circuit	43	Rated short-time current of conductor (1s)	kA	3.05	6.10	11.6

Table 2.8.21 **PROTOLON reeling and trailing cables** steel braid reinforcement				**NTSCGERLWÖU** 3×...+3×.../3E 6/10 kV				
esign	3 Nom. cross-sectional area of phase conductor	mm²	25	35	50	70	95
	4 Nom. cross-sectional area of PE conductor	mm²	16	16	25	35	50
	6 Type of conductor		F	F	F	F	F
	7 Coating of strands		tinned	tinned	tinned	tinned	tinned
	8 Number of strands		196	276	396	565	746
	9 Diameter of strand	mm	0.41	0.41	0.41	0.41	0.41
	10 Diameter of conductor	mm	7.8	9.2	11.0	13.1	15.1
	11 Thickness of insulation	mm	3.4	3.4	3.4	3.4	3.4
	13 Thickness of inner sheath	mm	2.0	2.0	2.4	2.4	2.8
	14 Thickness of outer sheath	mm	3.5	3.5	4.0	4.0	4.5
	15 Overall diameter of cable (min. value)	mm	54.5	63.5	60.0	63.5	76.5
	16 Overall diameter of cable (max. value)	mm	64.0	68.0	70.0	74.0	81.0
	19 Weight of cable	kg/km	5300	6200	6600	8200	9500
echanical operties	20 Minimum bending radius for fixed installation	mm	384	408	420	444	486
	22 Minimum bending radius for free flexing	mm	640	680	700	740	810
	23 Minimum bending radius for cable entry	mm	640	680	700	740	810
	29 Perm. pulling force for install. and operation	N	1125	1575	2250	3150	4275
ectrical operties	30 d.c. resistance/unit length at 90 °C	Ω/km	1.01	0.720	0.501	0.353	0.268
	31 d.c. resistance/unit length at 20 °C	Ω/km	0.795	0.565	0.393	0.277	0.210
	32 Inductance/unit length per conductor	mH/km	0.300	0.290	0.270	0.260	0.260
	33 Operating capacitance/unit length	μF/km	0.340	0.380	0.430	0.480	0.540
	34 Charging current	A/km	0.620	0.690	0.780	0.870	0.980
	35 Earth fault current	A/km	1.85	2.07	2.34	2.61	2.94
	36 Electric field strength at the conductor	kV/mm	2.26	2.19	2.12	2.07	2.02
urrent-carrying pacity	37 Number of loaded cores		3	3	3	3	3
	41 Installation on walls	A	131	162	202	250	301
	42 Minimum time value	s	331	424	557	713	906
ort-circuit	43 Rated short-time current of conductor (1s)	kA	3.05	4.27	6.10	8.54	11.6

Table 2.8.22
PROTOLON reeling and trailing cables steel braid reinforcement

Design						
Design	3	Nom. cross-sectional area of phase conductor	mm^2	25	35	50
	4	Nom. cross-sectional area of PE conductor	mm^2	16	16	25
	6	Type of conductor		F	F	F
	7	Coating of strands		tinned	tinned	tinned
	8	Number of strands		196	276	396
	9	Diameter of strand	mm	0.41	0.41	0.41
	10	Diameter of conductor	mm	8.0	9.2	11.0
	11	Thickness of insulation	mm	5.5	5.5	5.5
	13	Thickness of inner sheath	mm	2.4	2.4	2.8
	14	Thickness of outer sheath	mm	4.0	4.0	4.5
	15	Overall diameter of cable (min. value)	mm	70.5	74.5	77.5
	16	Overall diameter of cable (max. value)	mm	75.0	78.5	82.0
	19	Weight of cable	kg/km	6900	7300	8600
Mechanical properties	20	Minimum bending radius for fixed installation	mm	450	471	492
	22	Minimum bending radius for free flexing	mm	750	785	820
	23	Minimum bending radius for cable entry	mm	750	785	820
	29	Perm. pulling force for install. and operation	N	1125	1575	2250
Electrical properties	30	d.c. resistance/unit length at 90 °C	Ω/km	1.01	0.720	0.501
	31	d.c. resistance/unit length at 20 °C	Ω/km	0.795	0.565	0.393
	32	Inductance/unit length per conductor	mH/km	0.340	0.330	0.310
	33	Operating capacitance/unit length	μF/km	0.240	0.260	0.290
	34	Charging current	A/km	0.870	0.940	1.05
	35	Earth fault current	A/km	2.61	2.83	3.16
	36	Electric field strength at the conductor	kV/mm	3.15	3.04	2.92
Current-carrying capacity	37	Number of loaded cores		3	3	3
	41	Installation on walls	A	139	172	215
	42	Minimum time value	s	294	376	492
Short-circuit	43	Rated short-time current of conductor (1s)	kA	3.05	4.27	6.10

70	95
35	50
F	F
tinned	tinned
565	746
0.41	0.41
13.1	15.1
5.5	5.5
2.8	3.2
4.5	5.0
83.5	87.5
89.0	93.0
10200	11500
534	558
890	930
890	930
3150	4275
0.353	0.268
0.277	0.210
0.300	0.290
0.330	0.360
1.20	1.31
3.59	3.92
2.81	2.73
3	3
265	319
634	806
8.54	11.6

Table 2.9.1a
Heavy duty PROTOMONT rubber-sheathed flexible cables for power distribution, concentric PE conductor

Design	3	Nom. cross-sectional area of phase conductor	mm^2	2.5	6	10	16
	4	Nom. cross-sectional area of PE conductor	mm^2	2.5	6	10	16
	6	Type of conductor		F	F	F	F
	7	Coating of strands		tinned	tinned	tinned	tinned
	8	Number of strands		50	84	80	126
	9	Diameter of strand	mm	0.26	0.31	0.41	0.41
	10	Diameter of conductor	mm	2.6	3.9	5.1	6.3
	11	Thickness of insulation	mm	0.9	1.0	1.2	1.2
	13	Thickness of inner sheath	mm	1.0	1.2	1.4	1.4
	14	Thickness of outer sheath	mm	1.6	2.0	2.2	2.2
	15	Overall diameter of cable (min. value)	mm	15.0	18.0	22.0	28.0
	16	Overall diameter of cable (max. value)	mm	18.0	21.0	26.0	32.0
	19	Weight of cable	kg/km	370	640	920	1350
Mechanical properties	20	Minimum bending radius for fixed installation	mm	108	126	156	192
	22	Minimum bending radius for free flexing	mm	180	210	260	320
	23	Minimum bending radius for cable entry	mm	180	210	260	320
	29	Perm. pulling force for install. and operation	N	113	270	450	720
Electrical properties	30	d.c. resistance/unit length at 90 °C	Ω/km	10.5	4.32	2.49	1.58
	31	d.c. resistance/unit length at 20 °C	Ω/km	8.21	3.39	1.95	1.24
	32	Inductance/unit length per conductor	mH/km	0.290	0.270	0.270	0.250
	33	Operating capacitance/unit length	μF/km	0.350	0.440	0.480	0.570
	34	Charging current	A/km	0.0600	0.0800	0.0900	0.1000
	35	Earth fault current	A/km	0.190	0.240	0.260	0.310
Current-carrying capacity	37	Number of loaded cores		3	3	3	3
	41	Installation on walls	A	30	53	74	99
	42	Minimum time value	s	63	117	166	237
Short-circuit	43	Rated short-time current of conductor (1s)	kA	0.305	0.732	1.22	1.95

25	35	50	70	95	120
16	16	25	35	50	70
F	F	F	F	F	F
tinned	tinned	tinned	tinned	tinned	tinned
196	276	396	560	740	950
0.41	0.41	0.41	0.41	0.41	0.41
7.8	9.2	11.0	13.1	15.1	17.0
1.4	1.4	1.6	1.6	1.8	1.8
1.6	1.8	2.0	2.0	2.4	2.4
2.5	3.0	3.5	3.5	4.0	4.0
30.0	34.0	41.0	44.0	52.0	62.0
34.0	38.0	46.0	49.0	57.0	68.0
1800	2400	3320	4230	5590	6880
204	228	276	294	342	408
340	380	460	490	570	680
340	380	460	490	570	680
1125	1575	2250	3150	4275	5400
1.01	0.720	0.501	0.353	0.268	0.209
0.795	0.565	0.393	0.277	0.210	0.164
0.250	0.240	0.240	0.230	0.230	0.230
0.600	0.690	0.720	0.840	0.860	0.960
0.110	0.130	0.130	0.150	0.160	0.170
0.330	0.380	0.390	0.460	0.470	0.520
3	3	3	3	3	3
131	162	202	250	301	352
331	424	557	713	906	1057
3.05	4.27	6.10	8.54	11.6	14.6

Table 2.9.1b
Heavy duty PROTOMONT rubber-sheathed flexible cables for power distribution, PE conductor over each core, control core

Design	3	Nom. cross-sectional area of phase conductor	mm^2	2.5	6	10	16
	4	Nom. cross-sectional area of PE conductor	mm^2	2.5	6	10	16
	5	Nom. cross-sectional area of control core	mm^2	1.5	1.5	2.5	2.5
	6	Type of conductor		F	F	F	F
	7	Coating of strands		tinned	tinned	tinned	tinned
	8	Number of strands		50	84	80	126
	9	Diameter of strand	mm	0.26	0.31	0.41	0.41
	10	Diameter of conductor	mm	2.6	3.9	5.1	6.3
	11	Thickness of insulation	mm	0.9	1.0	1.2	1.2
	13	Thickness of inner sheath	mm	1.0	1.2	1.4	1.4
	14	Thickness of outer sheath	mm	1.6	2.0	2.2	2.2
	15	Overall diameter of cable (min. value)	mm	16.0	20.0	22.0	28.0
	16	Overall diameter of cable (max. value)	mm	20.0	24.0	26.0	32.0
	19	Weight of cable	kg/km	490	680	1050	1340
Mechanical properties	20	Minimum bending radius for fixed installation	mm	120	144	156	192
	22	Minimum bending radius for free flexing	mm	200	240	260	320
	23	Minimum bending radius for cable entry	mm	200	240	260	320
	29	Perm. pulling force for install. and operation	N	113	270	450	720
Electrical properties	30	d.c. resistance/unit length at 90 °C	Ω/km	10.5	4.32	2.49	1.58
	31	d.c. resistance/unit length at 20 °C	Ω/km	8.21	3.39	1.95	1.24
	32	Inductance/unit length per conductor	mH/km	0.290	0.270	0.270	0.250
	33	Operating capacitance/unit length	μF/km	0.350	0.440	0.480	0.570
	34	Charging current	A/km	0.0600	0.0800	0.0900	0.1000
	35	Earth fault current	A/km	0.190	0.240	0.260	0.310
Current-carrying capacity	37	Number of loaded cores		3	3	3	3
	41	Installation on walls	A	30	53	74	99
	42	Minimum time value	s	63	117	166	237
Short-circuit	43	Rated short-time current of conductor (1s)	kA	0.305	0.732	1.22	1.95

25	35	50	70	95
16	16	25	35	50
2.5	2.5	2.5	2.5	2.5
F	F	F	F	F
tinned	tinned	tinned	tinned	tinned
196	276	396	560	740
0.41	0.41	0.41	0.41	0.41
7.8	9.2	11.0	13.1	15.2
1.4	1.4	1.6	1.6	1.8
1.6	1.8	2.0	2.0	2.4
2.5	3.0	3.5	3.5	4.0
30.0	34.0	41.0	44.0	52.0
34.0	38.0	46.0	49.0	57.0
1850	2430	3360	4260	5610
204	228	276	294	342
340	380	460	490	570
340	380	460	490	570
1125	1575	2250	3150	4275
1.01	0.720	0.501	0.353	0.268
0.795	0.565	0.393	0.277	0.210
0.250	0.240	0.240	0.230	0.230
0.600	0.690	0.720	0.840	0.860
0.110	0.130	0.130	0.150	0.160
0.330	0.380	0.390	0.460	0.470
3	3	3	3	3
131	162	202	250	301
331	424	557	713	906
3.05	4.27	6.10	8.54	11.6

Table 2.9.1c						NSSHÖU ... × ... / ... KON 0.6/1 kV
Heavy duty PROTOMONT rubber-sheathed flexible cables for power distribution, and overall concentric PE conductor						

Design	1	Number of cores		3	5	5	5
	3	Nom. cross-sectional area of phase conductor	mm^2	2.5	2.5	4	6
	4	Nom. cross-sectional area of PE conductor	mm^2	2.5	2.5	4	6
	6	Type of conductor		F	F	F	F
	7	Coating of strands		tinned	tinned	tinned	tinned
	8	Number of strands		50	50	56	84
	9	Diameter of strand	mm	0.26	0.26	0.31	0.31
	10	Diameter of conductor	mm	2.6	2.6	3.2	3.9
	11	Thickness of insulation	mm	0.9	0.9	1.0	1.0
	13	Thickness of inner sheath	mm	1.0	1.2	1.2	1.4
	14	Thickness of outer sheath	mm	1.6	2.0	2.0	2.2
	15	Overall diameter of cable (min. value)	mm	14.0	18.0	20.0	21.0
	16	Overall diameter of cable (max. value)	mm	17.0	21.0	24.0	24.0
	19	Weight of cable	kg/km	350	510	650	830
Mechanical properties	20	Minimum bending radius for fixed installation	mm	102	126	144	144
	22	Minimum bending radius for free flexing	mm	170	210	240	240
	23	Minimum bending radius for cable entry	mm	170	210	240	240
	29	Perm. pulling force for install. and operation	N	113	188	300	450
Electrical properties	30	d.c. resistance/unit length at 90 °C	Ω/km	10.5	10.5	6.49	4.32
	31	d.c. resistance/unit length at 20 °C	Ω/km	8.21	8.21	5.09	3.39
Current-carrying capacity	37	Number of loaded cores		3	3	3	3
	41	Installation on walls	A	30	30	41	53
	42	Minimum time value	s	63	63	87	117
Short-circuit	43	Rated short-time current of conductor (1s)	kA	0.305	0.305	0.488	0.732

Table 2.9.2

NSSHCGEÖU 3×...+3×(1.5STKON+.../3KON)SM 0.6/1 kV

Heavy duty PROTOMONT rubber-sheathed flexible cables for coal cutters

Design	3	Nom. cross-sectional area of phase conductor	mm²	35	50	70	95
	4	Nom. cross-sectional area of PE conductor	mm²	16	25	35	50
	6	Type of conductor		FS	FS	FS	FS
	7	Coating of strands		plain	plain	plain	plain
	8	Number of strands		494	516	736	776
	9	Diameter of strand	mm	0.31	0.36	0.36	0.41
	10	Diameter of conductor	mm	9.2	11.0	13.1	15.1
	11	Thickness of insulation	mm	1.4	1.6	1.6	1.8
	13	Thickness of inner sheath	mm	2.0	2.0	2.4	2.4
	14	Thickness of outer sheath	mm	3.5	3.5	4.0	4.0
	15	Overall diameter of cable (min. value)	mm	46.0	48.0	56.0	57.0
	16	Overall diameter of cable (max. value)	mm	51.0	53.0	61.5	62.5
	19	Weight of cable	kg/km	3330	3860	5460	6030
Mechanical properties	20	Minimum bending radius for fixed installation	mm	306	318	369	375
	22	Minimum bending radius for free flexing	mm	510	530	615	625
	23	Minimum bending radius for cable entry	mm	510	530	615	625
	24	Minimum bending radius for reeling operation	mm	612	636	738	750
	27	Minimum bending radius for guide rollers	mm	765	795	923	938
	29	Perm. pulling force for install. and operation	N	1575	2250	3150	4275
Electrical properties	30	d.c. resistance/unit length at 90 °C	Ω/km	0.706	0.492	0.347	0.263
	31	d.c. resistance/unit length at 20 °C	Ω/km	0.554	0.386	0.272	0.206
	32	Inductance/unit length per conductor	mH/km	0.240	0.240	0.230	0.230
	33	Operating capacitance/unit length	μF/km	0.690	0.720	0.840	0.860
	34	Charging current	A/km	0.130	0.130	0.150	0.160
	35	Earth fault current	A/km	0.380	0.390	0.460	0.470
	36	Operational diel. strength at the conductor	kV/mm	0.472	0.411	0.403	0.358
Current-carrying capacity	37	Number of loaded cores		3	3	3	3
	41	Installation on walls	A	162	202	250	301
	42	Minimum time value	s	424	557	713	906
Short-circuit	43	Rated short-time current of conductor (1s)	kA	5.00	7.15	10.0	13.6

Table 2.9.3				NSSHCGEÖU 3×.../... KON+3×(1.5STKON/1.5ÜLKON)V 0.6/1 kV					
Heavy duty PROTOMONT rubber-sheathed flexible cables for coal cutters									
Design	3	Nom. cross-sectional area of phase conductor	mm²	25	35	50	70	95
	4	Nom. cross-sectional area of PE conductor	mm²	16	16	25	35	50
	6	Type of conductor		FS	FS	FS	FS	FS
	7	Coating of strands		tinned	tinned	tinned	tinned	tinned
	8	Number of strands		349	494	516	736	776
	9	Diameter of strand	mm	0.31	0.31	0.36	0.36	0.41
	10	Diameter of conductor	mm	7.8	9.2	11.0	13.1	15.1
	11	Thickness of insulation	mm	1.4	1.4	1.6	1.6	1.8
	13	Thickness of inner sheath	mm	2.8	2.8	3.0	3.0	3.4
	14	Thickness of outer sheath	mm	3.0	3.0	3.5	3.5	4.0
	15	Overall diameter of cable (min. value)	mm	45.2	45.2	47.6	51.6	58.6
	16	Overall diameter of cable (max. value)	mm	48.2	48.2	50.6	55.6	62.6
	19	Weight of cable	kg/km	2510	3150	4120	5150	6700
Mechanical properties	20	Minimum bending radius for fixed installation	mm	289	289	304	334	376
	22	Minimum bending radius for free flexing	mm	482	482	506	556	626
	23	Minimum bending radius for cable entry	mm	482	482	506	556	626
	29	Perm. pulling force for install. and operation	N	1125	1575	2250	3150	4275
Electrical properties	30	d.c. resistance/unit length at 90 °C	Ω/km	1.01	0.720	0.501	0.353	0.268
	31	d.c. resistance/unit length at 20 °C	Ω/km	0.795	0.565	0.393	0.277	0.210
	32	Inductance/unit length per conductor	mH/km	0.250	0.240	0.240	0.230	0.230
	33	Operating capacitance/unit length	μF/km	0.600	0.690	0.720	0.840	0.860
	34	Charging current	A/km	0.110	0.130	0.130	0.150	0.160
	35	Earth fault current	A/km	0.330	0.380	0.390	0.460	0.470
Current-carrying capacity	37	Number of loaded cores		3	3	3	3	3
	41	Installation on walls	A	131	162	202	250	301
	42	Minimum time value	s	331	424	557	713	906
Short-circuit	43	Rated short-time current of conductor (1s)	kA	3.05	4.27	6.10	8.54	11.6

Table 2.9.4

NSSHCGEÖU 3×.../... KON+3×(1.5STKON/1.5ÜLKON)Z 0.6/1 kV

Heavy duty PROTOMONT rubber-sheathed flexible cables for coal cutters

sign	3	Nom. cross-sectional area of phase conductor	mm²	25	35	50	70	95
	4	Nom. cross-sectional area of PE conductor	mm²	16	16	25	35	50
	6	Type of conductor		FS	FS	FS	FS	FS
	7	Coating of strands		plain	plain	plain	plain	plain
	8	Number of strands		349	494	516	736	776
	9	Diameter of strand	mm	0.31	0.31	0.36	0.36	0.41
	10	Diameter of conductor	mm	7.8	9.2	11.0	13.1	15.1
	11	Thickness of insulation	mm	1.4	1.4	1.6	1.6	1.8
	13	Thickness of inner sheath	mm	2.8	2.8	3.0	3.0	3.4
	14	Thickness of outer sheath	mm	3.5	3.5	3.5	3.5	4.5
	15	Overall diameter of cable (min. value)	mm	47.5	47.5	48.7	53.9	60.4
	16	Overall diameter of cable (max. value)	mm	50.5	50.5	51.7	57.9	64.4
	19	Weight of cable	kg/km	3010	3250	4180	5300	7100
echanical perties	20	Minimum bending radius for fixed installation	mm	303	303	310	347	386
	22	Minimum bending radius for free flexing	mm	505	505	517	579	644
	23	Minimum bending radius for cable entry	mm	505	505	517	579	644
	24	Minimum bending radius for reeling operation	mm	606	606	620	695	773
	27	Minimum bending radius for guide rollers	mm	758	758	775	868	966
	29	Perm. pulling force for install. and operation	N	1125	1575	2250	3150	4275
ctrical perties	30	d.c. resistance/unit length at 90 °C	Ω/km	0.995	0.706	0.492	0.347	0.263
	31	d.c. resistance/unit length at 20 °C	Ω/km	0.780	0.554	0.386	0.272	0.206
	32	Inductance/unit length per conductor	mH/km	0.250	0.240	0.240	0.230	0.230
	33	Operating capacitance/unit length	μF/km	0.600	0.690	0.720	0.840	0.860
	34	Charging current	A/km	0.110	0.130	0.130	0.150	0.160
	35	Earth fault current	A/km	0.330	0.380	0.390	0.460	0.470
rrent-carrying acity	37	Number of loaded cores		3	3	3	3	3
	41	Installation on walls	A	131	162	202	250	301
	42	Minimum time value	s	331	424	557	713	906
rt-circuit	43	Rated short-time current of conductor (1s)	kA	3.57	5.00	7.15	10.0	13.6

Table 2.9.5 Heavy duty PROTOMONT rubber-sheathed flexible cables for hoists				NTMTWÖU ... ×2.5ST+... ×1FM(C) 0.6/1 kV	
Design	1	Number of cores		8+2	14+6
	6	Type of conductor		F	F
	7	Coating of strands		tinned	tinned
	8	Number of strands		80	80
	9	Diameter of strand	mm	0.21	0.21
	10	Diameter of conductor	mm	2.6	2.6
	11	Thickness of insulation	mm	1.5	1.5
	12	Thickness of sheath	mm	2.0	2.5
	15	Overall diameter of cable (min. value)	mm	21.8	27.5
	16	Overall diameter of cable (max. value)	mm	25.0	31.0
	19	Weight of cable	kg/km	760	1190
Mechanical properties	20	Minimum bending radius for fixed installation	mm	150	186
	22	Minimum bending radius for free flexing	mm	250	310
	23	Minimum bending radius for cable entry	mm	250	310
Electrical properties	30	d.c. resistance/unit length at 90 °C	Ω/km	10.5	10.5
	31	d.c. resistance/unit length at 20 °C	Ω/km	8.21	8.21
Current-carrying capacity	37	Number of loaded cores		3	3
	41	Installation on walls	A	30	30
	42	Minimum time value	s	63	63
Short-circuit	43	Rated short-time current of conductor (1s)	kA	0.305	0.305

Table 2.9.6
Screened SUPROMONT PVC cables NYHSSYCY 3×...+3×.../3E+3×2.5ST+ÜL 3.6/6 kV

Design								
Design	3	Nom. cross-sectional area of phase conductor	mm²	25	35	50	70	95
	4	Nom. cross-sectional area of PE conductor	mm²	16	16	25	35	50
	6	Type of conductor		F	F	F	F	F
	7	Coating of strands		plain	plain	plain	plain	plain
	8	Number of strands		196	276	396	580	740
	9	Diameter of strand	mm	0.41	0.41	0.41	0.41	0.41
	10	Diameter of conductor	mm	7.8	9.2	11.0	13.1	15.1
	11	Thickness of insulation	mm	3.4	3.4	3.4	3.4	3.4
	13	Thickness of inner sheath	mm	3.0	3.0	3.0	3.0	3.0
	14	Thickness of outer sheath	mm	3.0	3.0	3.0	3.0	3.0
	15	Overall diameter of cable (min. value)	mm	48.0	52.0	55.0	59.0	62.0
	16	Overall diameter of cable (max. value)	mm	53.0	57.0	61.0	65.0	68.0
	19	Weight of cable	kg/km	3900	4600	5500	6700	7900
Mechanical properties	20	Minimum bending radius for fixed installation	mm	318	342	366	390	408
	21	Minimum bending radius, formed bend	mm	212	228	244	260	272
	29	Perm. pulling force for install. and operation	N	1125	1575	2250	3150	4275
Electrical properties	30	d.c. resistance/unit length at 70 °C	Ω/km	0.933	0.663	0.462	0.325	0.246
	31	d.c. resistance/unit length at 20 °C	Ω/km	0.780	0.554	0.386	0.272	0.206
	32	Inductance/unit length per conductor	mH/km	0.310	0.300	0.280	0.270	0.260
	33	Operating capacitance/unit length	μF/km	0.620	0.700	0.810	0.930	1.05
	34	Charging current	A/km	0.670	0.760	0.880	1.01	1.14
	35	Earth fault current	A/km	2.02	2.29	2.64	3.04	3.43
Current-carrying capacity	37	Number of loaded cores		3	3	3	3	3
	41	Installation on walls	A	101	126	153	196	238
	42	Minimum time value	s	396	498	690	824	1029
Short-circuit	43	Rated short-time current of conductor (1s)	kA	2.88	4.02	5.75	8.05	10.9

Table 2.10.1a
Light-sheathed SIENOPYR cables 1, 2 and 3 cores

Design				1	1	1	1
	1	Number of cores		1	1	1	1
	2	Nom. cross-sectional area of conductor	mm^2	1.5	2.5	4	6
	6	Type of conductor		E	E	E	E
	7	Coating of strands		plain	plain	plain	plain
	8	Number of strands		1	1	1	1
	10	Diameter of conductor	mm	1.5	1.9	2.4	2.9
	11	Thickness of insulation	mm	0.5	0.5	0.6	0.6
	12	Thickness of sheath	mm	1.4	1.4	1.4	1.4
	15	Overall diameter of cable (min. value)	mm	5.0	5.4	6.0	6.4
	16	Overall diameter of cable (max. value)	mm	8.8	9.4	10.0	10.5
	19	Weight of cable	kg/km	92	110	135	160
Mechanical properties	20	Minimum bending radius for fixed installation	mm	35	38	40	42
	21	Minimum bending radius, formed bend	mm	9	9	10	21
	28	Perm. pulling force for installation	N	75	125	200	300
Electrical properties	30	d.c. resistance/unit length at 70 °C	Ω/km	14.5	8.87	5.52	3.69
	31	d.c. resistance/unit length at 20 °C	Ω/km	12.1	7.41	4.61	3.08
Current-carrying capacity	37	Number of loaded cores		3	3	3	3
	39	Install. method C, on/in walls or under plaster	A	17.5	24	32	41
	42	Minimum time value	s	47	70	101	138
Short-circuit	43	Rated short-time current of conductor (1s)	kA	0.172	0.287	0.460	0.690

Table 2.10.1b
Light-sheathed SIENOPYR cables 4, 5 and 7 cores

Design				4	4	4	4	4
	1	Number of cores		4	4	4	4	4
	2	Nom. cross-sectional area of conductor	mm^2	1.5	2.5	4	6	10
	6	Type of conductor		E	E	E	E	E
	7	Coating of strands		plain	plain	plain	plain	plain
	8	Number of strands		1	1	1	1	1
	10	Diameter of conductor	mm	1.5	1.9	2.4	2.9	3.7
	11	Thickness of insulation	mm	0.5	0.5	0.6	0.6	0.7
	12	Thickness of sheath	mm	1.4	1.4	1.6	1.6	1.6
	15	Overall diameter of cable (min. value)	mm	9.0	10.0	12.0	13.0	15.5
	16	Overall diameter of cable (max. value)	mm	10.5	11.5	14.0	15.5	18.0
	19	Weight of cable	kg/km	150	200	300	395	595
Mechanical properties	20	Minimum bending radius for fixed installation	mm	42	46	56	62	72
	21	Minimum bending radius, formed bend	mm	21	23	28	31	36
	28	Perm. pulling force for installation	N	300	500	800	1200	2000
Electrical properties	30	d.c. resistance/unit length at 70 °C	Ω/km	14.5	8.87	5.52	3.69	2.19
	31	d.c. resistance/unit length at 20 °C	Ω/km	12.1	7.41	4.61	3.08	1.83
Current-carrying capacity	37	Number of loaded cores		3	3	3	3	3
	39	Install. method C, on/in walls or under plaster	A	17.5	24	32	41	57
	42	Minimum time value	s	47	70	101	138	199
Short-circuit	43	Rated short-time current of conductor (1s)	kA	0.172	0.287	0.460	0.690	1.15

NHXMH ... × ... 300/500 V

1	1	2	3	3	3	3	3
10	16	1.5	1.5	2.5	4	6	10
E	M	E	E	E	E	E	E
plain	plain	plain	plain	plain	plain	plain	plain
1	7	1	1	1	1	1	1
3.7	5.3	1.5	1.5	1.9	2.4	2.9	3.7
0.7	0.7	0.5	0.5	0.5	0.6	0.6	0.7
1.4	1.4	1.4	1.4	1.4	1.4	1.6	1.6
7.4	8.6	8.0	8.4	9.3	10.5	12.0	14.5
12.0	13.5	9.4	9.8	11.0	12.5	14.0	16.5
215	295	110	130	165	235	320	480
48	54	38	39	44	50	56	66
24	27	9	10	22	25	28	33
500	800	150	225	375	600	900	1500
2.19	1.38	14.5	14.5	8.87	5.52	3.69	2.19
1.83	1.15	12.1	12.1	7.41	4.61	3.08	1.83
3	3	2	2	2	2	2	2
57	76	19.5	19.5	26	35	46	63
199	286	38	38	60	84	110	163
1.15	1.84	0.172	0.172	0.287	0.460	0.690	1.15

NHXMH ... × ... 300/500 V

4	4	4	5	5	5	5	5	5	7	7
16	25	35	1.5	2.5	4	6	10	16	1.5	2.5
M	M	M	E	E	E	E	E	M	E	E
plain	plain	plain	plain	plain	plain	plain	plain	plain	plain	plain
7	7	7	1	1	1	1	1	7	1	1
5.3	6.6	7.9	1.5	1.9	2.4	2.9	3.7	5.3	1.5	1.5
0.7	0.9	0.9	0.5	0.5	0.6	0.6	0.7	0.7	0.5	0.5
1.6	1.8	1.8	1.4	1.4	1.6	1.6	1.6	1.8	1.4	1.6
19.0	23.5	26.5	9.6	10.5	13.0	14.5	17.0	21.0	10.0	12.0
22.5	28.0	31.0	11.5	12.5	15.5	16.5	19.5	25.0	12.0	14.0
935	1420	1910	175	235	350	480	710	1140	210	300
90	112	124	46	50	62	66	78	100	48	56
45	84	93	23	25	31	33	39	50	24	28
3200	5000	7000	375	625	1000	1500	2500	4000	525	875
1.38	0.870	0.627	14.5	8.87	5.52	3.69	2.19	1.38	14.5	8.87
1.15	0.727	0.524	12.1	7.41	4.61	3.08	1.83	1.15	12.1	7.41
3	3	3	3	3	3	3	3	3	3	3
76	96	119	17.5	24	32	41	57	76	17.5	24
286	438	559	47	70	101	138	199	286	47	70
1.84	2.88	4.02	0.172	0.287	0.460	0.690	1.15	1.84	0.172	0.287

Table 2.10.2
SIENOPYR-Rubber-insulated single-core cables for special purposes finely stranded conductor

Design					4	6	10	16
	2	Nom. cross-sectional area of conductor		mm^2	4	6	10	16
	6	Type of conductor			F	F	F	F
	7	Coating of strands			tinned	tinned	tinned	tinned
	8	Number of strands			56	84	80	126
	9	Diameter of strand		mm	0.31	0.31	0.41	0.41
	10	Diameter of conductor		mm	3.2	3.9	5.1	6.3
	11	Thickness of insulation		mm	1.3	1.3	1.5	1.5
	12	Thickness of sheath		mm	0.8	0.8	0.8	0.8
	15	Overall diameter of cable (min. value)		mm	6.2	6.8	8.0	9.8
	16	Overall diameter of cable (max. value)		mm	9.0	9.5	11.0	13.0
	19	Weight of cable		kg/km	90	120	180	250
Mechanical properties	20	Minimum bending radius for fixed installation		mm	54	57	66	78
	21	Minimum bending radius, formed bend		mm	36	38	44	52
	29	Perm. pulling force for install. and operation		N	60	90	150	240
Electrical properties	30	d.c. resistance/unit length at 90 °C		Ω/km	6.49	4.32	2.49	1.58
	31	d.c. resistance/unit length at 20 °C		Ω/km	5.09	3.39	1.95	1.24
Current-carrying capacity	37	Number of loaded cores			1	1	1	1
	38	Installation in free air		A	62	80	111	149
	42	Minimum time value		s	38	51	74	105
Short-circuit	43	Rated short-time current of conductor (1s)		kA	0.488	0.732	1.22	1.95

Table 2.10.3
SIENOPYR heat-resistant single-core non-sheathed cables solid conductor   **(N)HX4GA 1×... 450/750 V**

Design					0.75	1	1.5	2.5	4
	2	Nom. cross-sectional area of conductor		mm^2	0.75	1	1.5	2.5	4
	6	Type of conductor			E	E	E	E	E
	7	Coating of strands			tinned	tinned	tinned	tinned	tinned
	8	Number of strands			1	1	1	1	1
	10	Diameter of conductor		mm	1.0	1.2	1.5	1.9	2.4
	11	Thickness of insulation		mm	0.8	0.8	0.8	0.9	1.0
	15	Overall diameter of cable (min. value)		mm	2.5	2.7	2.9	3.5	3.6
	16	Overall diameter of cable (max. value)		mm	3.0	3.2	3.4	4.1	5.2
	19	Weight of cable		kg/km	13	15	20	35	55
Mechanical properties	20	Minimum bending radius for fixed installation		mm	12	13	14	16	21
	21	Minimum bending radius, formed bend		mm	3	3	3	4	5
	28	Perm. pulling force for installation		N	38	50	75	125	200
Electrical properties	30	d.c. resistance/unit length at 120 °C		Ω/km	34.5	25.4	17.0	10.5	6.55
	31	d.c. resistance/unit length at 20 °C		Ω/km	24.8	18.2	12.2	7.56	4.70
Current-carrying capacity	37	Number of loaded cores			1	1	1	1	1
	38	Installation in free air		A	15.0	19.0	24	32	42
	42	Minimum time value		s	31	35	49	76	113
Short-circuit	43	Rated short-time current of conductor (1s)		kA	0.0765	0.102	0.153	0.255	0.408

25	35	50	70	95	120	150	185	240
F	F	F	F	F	F	F	F	F
tinned	tinned	tinned	tinned	tinned	tinned	tinned	tinned	tinned
196	276	396	360	475	608	756	925	1221
0.41	0.41	0.41	0.51	0.51	0.51	0.51	0.51	0.51
7.8	9.2	11.0	13.1	15.1	17.0	19.0	21.0	24.0
1.8	1.8	1.8	1.8	2.2	2.2	2.2	2.4	2.6
1.0	1.0	1.0	1.0	1.0	1.0	1.2	1.2	1.2
11.9	13.3	14.8	16.6	18.9	21.2	22.8	24.9	28.4
15.0	16.5	18.0	20.5	24.0	26.0	28.0	31.0	34.5
380	480	620	870	1180	1420	1750	2150	2800
90	99	108	123	144	156	168	186	207
60	66	72	82	96	104	112	124	138
375	525	750	1050	1425	1800	2250	2775	3600
1.01	0.720	0.501	0.353	0.268	0.209	0.168	0.138	0.104
0.795	0.565	0.393	0.277	0.210	0.164	0.132	0.108	0.0817
1	1	1	1	1	1	1	1	1
197	244	304	376	453	529	608	693	823
146	187	246	315	400	468	553	648	773
3.05	4.27	6.10	8.54	11.6	14.6	18.3	22.6	29.3

Table 2.10.4
SIENOPYR heat-resistant single-core non-sheathed cables finely stranded conductor

Design	2	Nom. cross-sectional area of conductor	mm^2	0.75	1	1.5	2.5
	6	Type of conductor		F	F	F	F
	7	Coating of strands		tinned	tinned	tinned	tinned
	8	Number of strands		24	32	30	50
	9	Diameter of strand	mm	0.21	0.21	0.26	0.26
	10	Diameter of conductor	mm	1.3	1.5	1.8	2.6
	11	Thickness of insulation	mm	0.8	0.8	0.8	0.9
	15	Overall diameter of cable (min. value)	mm	2.6	2.7	3.0	3.6
	16	Overall diameter of cable (max. value)	mm	3.2	3.4	3.7	4.4
	19	Weight of cable	kg/km	15	20	25	40
Mechanical properties	20	Minimum bending radius for fixed installation	mm	13	14	15	18
	21	Minimum bending radius, formed bend	mm	3	3	4	4
	29	Perm. pulling force for install. and operation	N	11	15	23	38
Electrical properties	30	d.c. resistance/unit length at 120 °C	Ω/km	37.2	27.9	19.1	11.4
	31	d.c. resistance/unit length at 20 °C	Ω/km	26.7	20.0	13.7	8.21
Current-carrying capacity	37	Number of loaded cores		1	1	1	1
	38	Installation in free air	A	15.0	19.0	24	32
	42	Minimum time value	s	31	35	49	76
Short-circuit	43	Rated short-time current of conductor (1s)	kA	0.0765	0.102	0.153	0.255

Table 2.10.5a
SIENOPYR(X) rubber-sheathed flexible cables 3 cores

Design	2	Nom. cross-sectional area of conductor	mm^2	1.5	2.5	4	6
	6	Type of conductor		F	F	F	F
	7	Coating of strands		tinned	tinned	tinned	tinned
	8	Number of strands		30	50	56	84
	9	Diameter of strand	mm	0.26	0.26	0.31	0.31
	10	Diameter of conductor	mm	1.8	2.6	3.2	3.9
	11	Thickness of insulation	mm	0.8	0.9	1.0	1.0
	13	Thickness of inner sheath	mm	1.0	1.0	1.2	1.2
	14	Thickness of outer sheath	mm	1.6	1.6	2.0	2.0
	15	Overall diameter of cable (min. value)	mm	10.5	12.0	14.5	15.5
	16	Overall diameter of cable (max. value)	mm	15.0	16.5	20.0	22.0
	19	Weight of cable	kg/km	220	280	420	520
Mechanical properties	20	Minimum bending radius for fixed installation	mm	90	99	120	132
	22	Minimum bending radius for free flexing	mm	150	165	200	220
	23	Minimum bending radius for cable entry	mm	150	165	200	220
	29	Perm. pulling force for install. and operation	N	68	113	180	270
Electrical properties	30	d.c. resistance/unit length at 90 °C	Ω/km	17.5	10.5	64.9	4.32
	31	d.c. resistance/unit length at 20 °C	Ω/km	13.7	8.21	50.9	3.39
	32	Inductance/unit length per conductor	mH/km	0.320	0.290	0.290	0.270
Current-carrying capacity	37	Number of loaded cores		2	2	2	2
	41	Installation on walls	A	23	30	41	53
	42	Minimum time value	s	39	63	87	117
Short-circuit	43	Rated short-time current of conductor (1s)	kA	0.183	0.305	0.488	0.732

(N)HX4GAF 1×... 450/750 V

4	6	10	16	25	35	50	70	95
F	F	F	F	F	F	F	F	F
tinned	tinned	tinned	tinned	tinned	tinned	tinned	tinned	tinned
56	84	80	126	196	276	396	360	475
0.31	0.31	0.41	0.41	0.41	0.41	0.41	0.51	0.51
3.2	3.9	5.1	6.3	7.8	9.2	11.0	13.1	15.1
1.0	1.0	1.2	1.2	1.4	1.4	1.6	1.6	1.8
4.3	4.8	6.0	8.0	9.7	10.5	12.7	14.5	16.5
5.5	6.3	7.9	9.0	11.0	12.4	14.6	16.7	19.1
60	80	130	210	320	420	540	750	985
22	25	32	36	44	50	58	67	76
6	6	8	9	22	25	29	33	38
60	90	150	240	375	525	750	1050	1425
709	4.72	2.72	1.73	1.11	0.787	0.547	0.386	0.293
509	3.39	1.95	1.24	0.795	0.565	0.393	0.277	0.210
1	1	1	1	1	1	1	1	1
42	54	73	98	129	158	198	245	292
113	154	234	333	469	613	796	1019	1321
0.408	0.612	1.02	1.63	2.55	3.57	5.10	7.14	9.69

(N)HXSHXÖ 3×... 0.6/1 kV

10	16	25	35	50
F	F	F	F	F
tinned	tinned	tinned	tinned	tinned
80	126	196	276	369
0.41	0.41	0.41	0.41	0.41
5.1	6.3	7.8	9.2	11.0
1.2	1.2	1.4	1.4	1.6
1.4	1.4	1.6	1.8	2.0
2.2	2.2	2.5	3.0	3.5
19.5	22.5	28.1	32.0	35.0
25.5	28.5	34.5	39.0	45.0
820	1150	1680	2290	3150
153	171	207	234	270
255	285	345	390	450
255	285	345	390	450
450	720	1125	1575	2250
2.49	1.58	1.01	0.720	0.501
1.95	1.24	0.795	0.565	0.393
0.270	0.250	0.250	0.240	0.240
2	2	2	2	2
74	99	131	162	202
166	237	331	424	557
1.22	1.95	3.05	4.27	6.10

Table 2.10.5 b
SIENOPYR(X) rubber-sheathed flexible cables 4 and 5 cores

Design								
	1	Number of cores		4	4	4	4	4
	2	Nom. cross-sectional area of conductor	mm²	1.5	2.5	4	6	10
	6	Type of conductor		F	F	F	F	F
	7	Coating of strands		tinned	tinned	tinned	tinned	tinned
	8	Number of strands		30	50	56	84	80
	9	Diameter of strand	mm	0.26	0.26	0.31	0.31	0.41
	10	Diameter of conductor	mm	1.8	2.6	3.2	3.9	5.1
	11	Thickness of insulation	mm	0.8	0.9	1.0	1.0	1.2
	13	Thickness of inner sheath	mm	1.0	1.2	1.2	1.2	1.4
	14	Thickness of outer sheath	mm	1.6	2.0	2.0	2.0	2.2
	15	Overall diameter of cable (min. value)	mm	11.0	13.5	15.5	17.0	20.5
	16	Overall diameter of cable (max. value)	mm	16.0	19.0	21.5	23.0	27.5
	19	Weight of cable	kg/km	250	370	490	620	940
Mechanical properties	20	Minimum bending radius for fixed installation	mm	96	114	129	138	165
	22	Minimum bending radius for free flexing	mm	160	190	215	230	275
	23	Minimum bending radius for cable entry	mm	160	190	215	230	275
	29	Perm. pulling force for install. and operation	N	90	150	240	360	600
Electrical properties	30	d.c. resistance/unit length at 90 °C	Ω/km	17.5	10.5	6.49	4.32	2.49
	31	d.c. resistance/unit length at 20 °C	Ω/km	13.7	8.21	5.09	3.39	1.95
	32	Inductance/unit length per conductor	mH/km	0.340	0.320	0.310	0.290	0.290
Current-carrying capacity	37	Number of loaded cores		3	3	3	3	3
	41	Installation on walls	A	23	30	41	53	74
	42	Minimum time value	s	39	63	87	117	166
Short-circuit	43	Rated short-time current of conductor (1s)	kA	0.183	0.305	0.488	0.732	1.22

Table 2.10.5 c
SIENOPYR(X) rubber-sheathed flexible cables 5 cores and more

Design							
	1	Number of cores		7	7	7	7
	2	Nom. cross-sectional area of conductor	mm²	1.5	2.5	4	6
	6	Type of conductor		F	F	F	F
	7	Coating of strands		tinned	tinned	tinned	tinned
	8	Number of strands		30	50	56	84
	9	Diameter of strand	mm	0.26	0.26	0.31	0.31
	10	Diameter of conductor	mm	1.8	2.6	3.2	3.9
	11	Thickness of insulation	mm	0.8	0.9	1.0	1.0
	13	Thickness of inner sheath	mm	1.2	1.2	1.4	1.4
	14	Thickness of outer sheath	mm	2.0	2.0	2.2	2.2
	15	Overall diameter of cable (min. value)	mm	15.0	16.5	18.5	20.5
	16	Overall diameter of cable (max. value)	mm	19.5	21.5	25.5	28.5
	19	Weight of cable	kg/km	450	540	790	1060
Mechanical properties	20	Minimum bending radius for fixed installation	mm	117	129	153	171
	22	Minimum bending radius for free flexing	mm	195	215	255	285
	23	Minimum bending radius for cable entry	mm	195	215	255	285
	29	Perm. pulling force for install. and operation	N	158	263	420	630
Electrical properties	30	d.c. resistance/unit length at 90 °C	Ω/km	17.5	10.5	6.49	4.32
	31	d.c. resistance/unit length at 20 °C	Ω/km	13.7	8.21	5.09	3.39
Current-carrying capacity	37	Number of loaded cores		3	3	3	3
	41	Installation on walls	A	23	30	41	53
	42	Minimum time value	s	39	63	87	117
Short-circuit	43	Rated short-time current of conductor (1s)	kA	0.183	0.305	0.488	0.732

(N)HXSHXÖ ... ×... 0.6/1 kV

4	4	4	4	5	5	5	5	5	5	5
16	25	35	50	1.5	2.5	4	6	10	16	25
F	F	F	F	F	F	F	F	F	F	F
tinned	tinned	tinned	tinned	tinned	tinned	tinned	tinned	tinned	tinned	tinned
126	196	276	369	30	50	56	84	80	126	196
0.41	0.41	0.41	0.41	0.26	0.26	0.31	0.31	0.41	0.41	0.41
6.3	7.8	9.2	11.0	1.8	2.6	3.2	3.9	5.1	6.3	7.8
1.2	1.4	1.4	1.6	0.8	0.9	1.0	1.0	1.2	1.2	1.4
1.6	1.8	1.8	2.0	1.0	1.2	1.2	1.4	1.4	1.6	1.8
2.5	3.0	3.0	3.5	1.6	2.0	2.0	2.2	2.2	2.5	3.0
25.0	30.0	33.5	39.0	12.0	14.5	16.0	18.5	22.0	27.0	32.5
32.0	39.0	42.5	49.0	17.0	20.0	23.0	26.5	30.0	34.0	42.0
1420	2120	2750	3780	280	410	560	740	1070	1670	2460
192	234	255	294	102	120	138	159	180	204	252
320	390	425	490	170	200	230	265	300	340	420
320	390	425	490	170	200	230	265	300	340	420
960	1500	2100	3000	113	188	300	450	750	1200	1875
1.58	1.01	0.720	0.501	17.5	10.5	6.49	4.32	2.49	1.58	1.01
1.24	0.795	0.565	0.393	13.7	8.21	5.09	3.39	1.95	1.24	0.795
0.280	0.270	0.260	0.260	0.350	0.330	0.320	0.300	0.300	0.290	0.280
3	3	3	3	3	3	3	3	3	3	3
99	131	162	202	23	30	41	53	74	99	131
237	331	424	557	39	63	87	117	166	237	331
1.95	3.05	4.27	6.10	0.183	0.305	0.488	0.732	1.22	1.95	3.05

(N)HXSHXÖ ... ×... 0.6/1 kV

8	16	20	36	33
1.5	1.5	1.5	1.5	2.5
F	F	F	F	F
tinned	tinned	tinned	tinned	tinned
30	30	30	30	50
0.26	0.26	0.26	0.26	0.26
1.8	1.8	1.8	1.8	2.6
0.8	0.8	0.8	0.8	0.9
1.2	1.2	1.4	1.6	1.8
2.0	2.0	2.2	2.5	3.0
16.0	18.5	22.0	31.5	31.5
20.5	24.0	27.5	37.0	37.0
510	730	930	1410	1910
123	144	165	222	222
205	240	275	370	370
205	240	275	370	370
180	360	450	810	1238
17.5	17.5	17.5	17.5	10.5
13.7	13.7	13.7	13.7	8.21
3	3	3	3	3
23	23	23	23	30
39	39	39	39	63
0.183	0.183	0.183	0.183	0.305

Table 2.10.5d SIENOPYR(X) rubber-sheathed flexible cables additional control cores			(N)HXSHXÖ ... ×... +2×1.5 0.6/1 kV				
Design	1	Number of cores		4	7	4	7	4
	2	Nom. cross-sectional area of conductor	mm²	2.5	2.5	4	4	6
	6	Type of conductor		F	F	F	F	F
	7	Coating of strands		tinned	tinned	tinned	tinned	tinned
	8	Number of strands		50	50	56	56	84
	9	Diameter of strand	mm	0.26	0.26	0.31	0.31	0.31
	10	Diameter of conductor	mm	2.6	2.6	3.2	3.2	3.9
	11	Thickness of insulation	mm	0.9	0.9	1.0	1.0	1.0
	13	Thickness of inner sheath	mm	1.2	1.4	1.2	1.4	1.4
	14	Thickness of outer sheath	mm	2.0	2.2	2.0	2.2	2.2
	15	Overall diameter of cable (min. value)	mm	17.0	22.0	19.5	25.0	22.0
	16	Overall diameter of cable (max. value)	mm	19.5	25.0	22.0	28.0	25.0
	19	Weight of cable	kg/km	505	825	635	1080	865
Mechanical properties	20	Minimum bending radius for fixed installation	mm	117	150	132	168	150
	22	Minimum bending radius for free flexing	mm	195	250	220	280	250
	23	Minimum bending radius for cable entry	mm	195	250	220	280	250
	29	Perm. pulling force for install. and operation	N	150	263	240	420	360
Electrical properties	30	d.c. resistance/unit length at 90 °C	Ω/km	10.5	10.5	6.49	64.9	4.32
	31	d.c. resistance/unit length at 20 °C	Ω/km	8.21	8.21	5.09	50.9	3.39
Current-carrying capacity	37	Number of loaded cores		3	3	3	3	3
	41	Installation on walls	A	30	30	41	41	53
	42	Minimum time value	s	63	63	87	87	117
Short-circuit	43	Rated short-time current of conductor (1s)	kA	0.305	0.305	0.488	0.488	0.732

3 Explanations to the Project Planning Data of Cables

General

PE conductor (green/yellow marked core)

Cables with PE conductor have the same design data as cables without PE conductor provided that they have the same number of cores. Therefore, in the tables of Clause 2, for non-harmonized cables, no distinction is made between cables with PE conductor (J) and cables without PE conductor (O).

For harmonized cables, however, both designs are given because the information whether or not there is a PE conductor (G or X) is relevant for the designation of the number of cores and the nominal cross-sectional area.

Installation methods

The current-carrying capacities listed in the tables of Clause 2 are always given for a certain installation method. The current-carrying capacities for further installation methods are shown in Clause 1.2.

Current-carrying capacity of multi-core cables

In the tables of Clause 2, the current-carrying capacity for multi-core cables is always given for three loaded cores. If there are more than three loaded cores, the current-carrying capacity shall be converted with the factors mentioned in Clause 1.2.

Frequency

The current-carrying capacities apply to a system frequency of 50 Hz, by approximation also for operation with 60 Hz.

Symbols

The symbols given basically comply with those in Part 1, with the exception of the symbol F for the permissible pulling force (Items 28 and 29).

Explanations to the Items of the Project Planning Data

(Table 2.1.1 to 2.10.5)

Item	Designation	Symbol	Unit	Explanation	Further explanations see Part 1	Details in DIN VDE
Design						
1	Number of cores			The number of cores with the same nominal cross-sectional area of conductor is given, including the green/yellow marked core, if available, and provided that it also has the same nominal cross-sectional area. The number of cores is mentioned in the heading of those tables only containing cables with the same number of cores.		
2	Nominal cross-sectional area of conductor	q_n	mm^2	A d.c. resistance per unit length at 20 °C according to DIN VDE 0295 (see Item 31) is allocated to the nominal cross-sectional area of conductor. It is proved by measurement. The cross-sectional area is mentioned in the heading of those tables containing cables with the same nominal cross-sectional area.	Clause 20.1, Page 320 Clause 37, Page 454 to 455	0295 0472 Part 501
3	Nominal cross-sectional area of phase conductor	q_n	mm^2	See explanations to Item 2		
4	Nominal cross-sectional area of PE conductor	q_n	mm^2	See explanations to Item 2		
5	Nominal cross-sectional area of control core	q_n	mm^2	See explanations to Item 2		
6	Type of conductor			Only circular copper conductors are used. Depending on the design, the following distinction is made according to DIN VDE 0295: E Solid conductor (Class 1) M Stranded conductor (Class 2) F Finely stranded conductor (Class 5) FF Extra finely stranded conductor (Class 6 or as well extra finely stranded conductor for welding cables according to Table 5 in DIN VDE 0295) FS Finely stranded special conductor (not included in DIN VDE 0295)		

Item	Designation	Symbol	Unit	Explanation	Further explanations see Part 1	Details in DIN VDE
7	Coating of strand			Depending on the cable design, plain or tinned strands are specified.	Clause 37 Page 454 to 455	0250 0281 0282 0295
8	Number of strands			For stranded conductors, the minimum number of strands is given; for finely stranded conductors, finely stranded special conductors and extra finely stranded conductors, a recommended value is given.		
9	Diameter of strand	d_D	mm	A maximum value is given. For finely stranded and extra finely stranded conductors, it is specified in DIN VDE 0295. For finely stranded special conductors, the maximum value is given by the company.	Clause 37 Page 454 to 455	0295
10	Diameter of conductor	d_L	mm	A maximum value is given.	Clause 37 Page 454 to 455	0295
11	Thickness of insulation	$\delta^{1)}$	mm	A nominal value is given. Conductive layers, if available, are not included in the thickness of insulation.		0250 0281 0282
12	Thickness of sheath	$\delta^{1)}$	mm	A nominal value is given. This item applies to cables with one sheath only. For cables having two sheaths Items 13 and 14 apply.		0250 0281 0282
13	Thickness of inner sheath	$\delta^{1)}$	mm	A nominal value is given.		0250 0281 0282
14	Thickness of outer sheath	$\delta^{1)}$	mm	A nominal value is given.		0250 0281 0282
15	Overall diameter of cable (min. value)	d	mm	The values for circular cables laid down in the VDE specifications are given in the tables. For cables for which no VDE specifications exist, the values were determined by calculation. This also applies to the minimum value if only a maximum value is given in the VDE specifications.		0250 0281 0282

[1] As in Part 1, no difference is made in the symbol δ

$\triangleright$

Explanations to the Items of Project Planning Data (Continued)

(Table 2.1.1 to 2.10.5)

Item	Designation	Symbol	Unit	Explanation	Further explanations see Part 1	Details in DIN VDE
16	Overall diameter of cable (max. value)	d	mm	See explanations to Item 15.		
17	Overall dimensions of cable (min. value)	$d \times b$	mm $\times$ mm	The dimensions for flat cables specified in the VDE specifications are given in the tables. The values were determined by calculation for those cables for which no VDE specifications exist.		0250 0281
18	Overall dimensions of cable (max. value)	$d \times b$	mm $\times$ mm	See explanations to Item 17.		
19	Weight of cable	m'	kg/km	The net weight determined by calculation is given in the tables. The value given is a recommended value.		

Mechanical properties

Item	Designation	Symbol	Unit	Explanation	Further explanations see Part 1	Details in DIN VDE
20 to 27	Minimum bending radius...	r	mm	The minimum bending radii were calculated by means of the factors given in DIN VDE 0298 Part 3, Table 2. When falling below these radii, the service life of the cable can be reduced.	Clause 11, Page 87	0298 Part 3
28	Permissible pulling force for installation	F	N	For permanently installed cables, the highest permissible pulling force is given. The permissible pulling force is calculated by means of the sum of the nominal cross-sectional areas of conductors. Screens, concentric conductors and split PE conductors as well as additional control and pilot cores must not be considered. In DIN VDE 0298 Part 3, a permissible pulling stress of 50 N/mm^2 is given. When exceeding the permissible pulling force, the service life of the cable can be reduced.		0298 Part 3

Item	Designation	Symbol	Unit	Explanation	Further explanations see Part 1	Details in DIN VDE
29	Permissible pulling force for installation and operation	F	N	For flexible cables the highest permissible pulling force is given. The permissible pulling force is calculated by means of the sum of the nominal cross-sectional areas of conductors. Screens, concentric conductors and split PE conductors as well as additional control and pilot cores must not be considered. DIN VDE 0298 Part 3 gives a permissible pulling stress of 15 N/mm². For CORDAFLEX(K) and CORDAFLEX(SM) cables a value of 20 N/mm² is permitted because of the special cable design. When exceeding the permissible tensile force, the service life of the cable can be reduced.		0298 Part 3

Electrical properties

Item	Designation	Symbol	Unit	Explanation	Further explanations see Part 1	Details in DIN VDE
30	d.c. resistance per unit length at ... °C	R'_ϑ	Ω/km	The d.c. resistance per unit length at the permissible operating temperature is given. This value was determined by conversion of the d.c. resistance per unit length at 20 °C (Item 31).	Clause 1, Page 11	
31	d.c. resistance per unit length at 20 °C	R'_{20}	Ω/km	The maximum value for the d.c. resistance per unit length at 20 °C is given in the tables. This value is specified in DIN VDE 0295. For deviating conductor temperatures, the d.c. resistance per unit length can be converted with the temperature coefficients given in Part 1.	Clause 20.1, Page 320 Clause 37, Page 454 and 455	0295
32	Inductance per unit length per conductor	L'_m	mH/km	The mean inductance per unit length is given.	Clause 21, Page 322 to 330	
33	Operating capacitance per unit length	C'_b	μF/km	The operating capacitance per unit length is given for cables with screened cores only. It applies to the permissible operating temperature and is a recommended value.	Clause 22.15, Page 331 to 334	

$\triangleright$

Explanations to the Items of Project Planning Data (Continued)

(Table 2.1.1 to 2.10.5)

Item	Designation	Symbol	Unit	Explanation	Further explanations see Part 1	Details in DIN VDE
34	Charging current	I_c'	A/km	The charging current is a reactive current which depends on the voltage, the frequency and the capacitance of the cable. The charging current given was calculated with the operating capacitance per unit length (see Item 33), the voltage $U/\sqrt{3}$ and a system frequency of 50 Hz.	Clause 22.3 Page 334 to 335	
35	Earth fault current	I_e'	A/km	A fault between a phase conductor and earth arising in an isolated neutral system is designated as earth fault. The current which results from this fault is called earth-fault current. In this table the earth-fault current is given for three-phase operation either for a multi-core cable with screened cores or a cable system consisting of three screened single-core cables.	Clause 22.3 Page 334 to 335	
36	Electric field strength at the conductor	E_i	kV/mm	The electric field strength on the inner conductive layer is given for the voltage $U/\sqrt{3}$ for all cables with a radial field distribution.		

Current-carrying capacity

Item	Designation	Symbol	Unit	Explanation	Further explanations see Part 1	Details in DIN VDE
37	Number of loaded cores			The number of cores loaded at the same time is given in the tables. The current-carrying capacities given in Items 38 to 41 refer to the number of loaded cores.		
38 to 41	Current-carrying capacity[1]	I_r	A	The current-carrying capacities recommended in DIN VDE 0298 Part 4 are given (see also Clause 1.2). They apply to the number of loaded cores (Item 37) at an ambient temperature of 30 °C, continuous operation and the installation methods mentioned. These installation methods and further ones are described in DIN VDE 0298 Part 4 and in Clause 1.2. The values were allocated for those cables which are not included in DIN VDE 0298 Part 4.	DIN VDE 0298 Part 4 supersedes the details in Clause 1.2	0298 Part 4

[1] Further details for heat-resistant cables can be taken from Clause 1.2.2

176

Item	Designation	Symbol	Unit	Explanation	Further explanations see Part 1	Details in DIN VDE
42	Minimum time value	τ	s	The minimum time value τ is one fifth of the time taken from the temperature curve to almost reach the permissible final temperature at constant current. It is relevant for the calculation of the current-carrying capacity at short-time operation and intermittent operation.	Clause 18.6 Page 239 to 244	

Short-circuit

Item	Designation	Symbol	Unit	Explanation	Further explanations see Part 1	Details in DIN VDE
43	Rated short-time current of conductor (1 s)	I_{thr}	kA	The rated short-time current I_{thr} of a conductor is defined for the rated short-circuit duration $t_{kr} = 1$ s (rated value of the short-circuit current-carrying capacity). Permissible short-circuit capacities I_{thz} for other times t_k up to 5 s are obtained from the following equation: $$I_{thz} = I_{thr}\sqrt{\frac{t_{kr}}{t_k}}.$$		0298 Part 4

Power Cables

4 Selection of Cables

4.1 Summary and Application

Designation of Cables

According to DIN VDE the following details are given for the correct designation of cables:

▷ type designation code,
▷ number of cores × nominal cross-sectional area in mm²,
▷ designation code for conductor shape and type,
▷ nominal cross-sectional area of screen or of concentric conductor in mm², if necessary,
▷ rated voltages in kV.

The type designation code is a combination of defined letters after the initial letter "N". The code is formed according to the order of construction beginning with the conductor. The letter "N" for "standard type" identifies cable types complying with the relevant VDE specifications. For cable types for which no VDE specifications exist but which fulfil all requirements, the letter N is put in parantheses (N). The most important type designation codes for cables are shown in the table (page 179).

Designation codes for conductors:

RE Solid circular conductor
RM Stranded circular conductor
SE Solid sector-shaped conductor
SM Stranded sector-shaped conductor

The following items are not specified:

▷ copper conductor,
▷ insulation of mass-impregnated paper (core, belt),
▷ inner and outer semi-conductive layers for cables with an extruded insulation,
▷ inner covering,
▷ inner protective coverings of fibrous material.

The nominal cross-sectional area of copper screens is given after an oblique ('/') located after the designation code for the phase conductors, e.g.: NYSEY 3 × 95 RM/16 6/10 kV. This also applies to the nominal cross-sectional area of the concentric conductor, e.g.: NYCWY 3 × 95 SM/50 0.6/1 kV.

Examples for the construction of cable designations (rated voltages are not given)

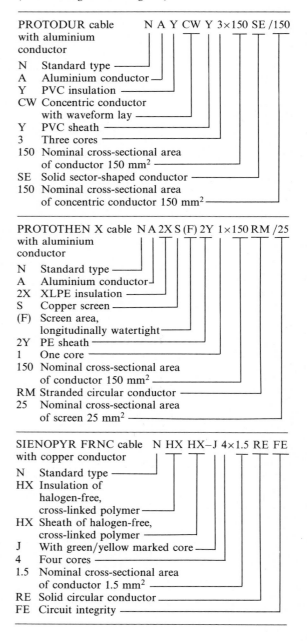

PROTODUR cable   N A Y CW Y 3×150 SE /150
with aluminium conductor

N   Standard type
A   Aluminium conductor
Y   PVC insulation
CW Concentric conductor with waveform lay
Y   PVC sheath
3   Three cores
150 Nominal cross-sectional area of conductor 150 mm²
SE Solid sector-shaped conductor
150 Nominal cross-sectional area of concentric conductor 150 mm²

PROTOTHEN X cable N A 2X S (F) 2Y 1×150 RM /25
with aluminium conductor

N   Standard type
A   Aluminium conductor
2X XLPE insulation
S   Copper screen
(F) Screen area, longitudinally watertight
2Y PE sheath
1   One core
150 Nominal cross-sectional area of conductor 150 mm²
RM Stranded circular conductor
25 Nominal cross-sectional area of screen 25 mm²

SIENOPYR FRNC cable N HX HX–J 4×1.5 RE FE
with copper conductor

N   Standard type
HX Insulation of halogen-free, cross-linked polymer
HX Sheath of halogen-free, cross-linked polymer
J   With green/yellow marked core
4   Four cores
1.5 Nominal cross-sectional area of conductor 1.5 mm²
RE Solid circular conductor
FE Circuit integrity

Letters of the designation codes for cables used in Clauses 4 and 5

Constructional elements	Cables with mass-impregnated paper insulation DIN VDE 0255	Cables with extruded insulation DIN VDE 0265, 0266, 0271, 0272, 0273
Standard type	N	N
Conductor - of copper - of aluminium	no letter A	no letter A
Insulation - paper with mass impregnation - PVC, polyvinyl chloride - XLPE, cross-linked polyethylene - cross-linked, halogen-free polymer	no letter – – –	– Y 2X HX
Concentric copper conductor - with helical layer - with waveform layer	– –	C CW
Copper screen - for single-core cables or multi-core cables with common screen - for multi-core cables with individual screens	– –	S SE

Metal sheath of lead
- for single-core cables and multi-core cables
  with common sheath
- for three-core separately lead-sheathed (S.L.) cables with corrosion
  protection on each sheath
	K
	
	EK
		K
		
		–

Inner coverings - longitudinally watertight screen area - lapped protective covering with embedded layer of plastic foils	– E	(F) –
Armour - steel tape - galvanized flat steel wire - counter helix or reinforcement helix of galvanized steel tape	B – –	– F G
Outer protective covering - fibrous material (jute) in compound - PVC sheath - PVC protective covering - cross-linked, halogen-free polymer - non-cross-linked, halogen-free polymer - PE sheath	A Y – – –	– Y – HX H 2Y
Cable with $U_0/U = 0.6/1$ kV without concentric conductor - with green/yellow core - without green/yellow core	– –	-J -O
Circuit integrity (after the designation code for the conductor)	–	-FE

4.1 Summary and Application

Design	Type designation code	Number of cores	Nominal cross-sectional area of conductors mm^2	Rated voltage U_0/U (U_m) kV	Project planning data	
					Table	Page
4.1.1 PROTODUR cables (PVC insulation) with copper conductor	NYY	single-core	4 – 500	0.6/1 (1.2)	5.1.1	216, 218
	NYY-J	three-core	1.5 – 400		5.1.2	220
	NYY-O	three-core	1.5 – 400		5.1.3	222
	NYY	three-core with one reduced conductor	25 – 400		5.1.4	224
		four-core	1.5 – 400		5.1.5	226
		five-core	1.5 – 25		5.1.6	228
		multi-core (control cable)	... × 1.5 and ... × 2.5		5.1.7	230
	NYCWY	three-core with reduced concentric conductor	1.5 – 400	0.6/1 (1.2)	5.1.8	232
		four-core with reduced concentric conductor	1.5 – 400		5.1.9	234
	NYCY	multi-core (control cable)	... × 1.5 and ... × 2.5	0.6/1 (1.2)	5.1.10	236
	NYKY-J	three-core	1.5 – 400	0.6/1 (1.2)	5.1.11	238
	NYKY	three-core with one reduced conductor	25 – 400		5.1.12	240
		four-core	1.5 – 400		5.1.13	242
		five-core	1.5 – 25		5.1.14	244
		multi-core (control cable)	... × 1.5 and ... × 2.5		5.1.15	246

eferred plication	Information on installation and laying	Information on the current-carrying capacity	Relevant accessories		
			Designation	Table	Page

or installation indoors, cable channels, tdoors and in ground, mechanical damages e not to be expected; ainly in power plants, dustry and switchgear d for the needs of in-allers.	In case of high mechanical stress during installation and operation, an armour is recom-mended.	The current-carrying capacities of single-core cables for d.c. operation and of two-core cables for single-phase a.c. operation shall be taken from Part 1, Page 160.	Termination
			– Indoor SKSA
			           PEA
			– Outdoor SKSE
			
			Straight joint SKSM
			            SKSM M
			            SKSM ST
			            SKEM
			            PV
			            PV/U
			
			Branch joint GNKA 1
			           PA
			           PAK
			           PAD
			           GMSA
			
			Transition joint SKSM PK
				8.1.1
				8.1.2
				8.2.1
				
				8.5.1
				8.5.2
				8.5.4
				8.5.5
				8.5.6
				8.5.7
				
				8.6.1
				8.6.2
				8.6.3
				8.6.4
				8.6.5
				
				8.7.1
					358
					359
					373
					
					389
					390
					392
					393
					394
					395
					
					409
					410
					411
					412
					413
					
					415

or installation indoors, cable channels, out-oors and in ground, if ubsequent mechanical amage is likely to ccur; for urban etworks, house services nd street lighting. f the cable is spiked, e concentric conductor nsures that the over-urrent protective de-ice, connected in series, perates.	The concentric conductor with wave-form layer (CW) is not cut at branches. The concentric conductor shall not be considered as armour.	The electrical and thermal data given in the tables (Item 23 to 49) also apply to cables with helical concentric conductor (C) and to cables with a concentric conduc-tor having the same nominal cross sectional area as the main conductor.	Termination
			– Indoor SKSA
			           PEA
			– Outdoor SKSE
			
			Straight joint SKSM C
			            SKEM
			            PV
			            PV/U
			
			Branch joint GNKA-1
			           PA
			           PAD
			           GMSA
				8.1.1
				8.1.2
				8.2.1
				
				8.5.3
				8.5.5
				8.5.6
				8.5.7
				
				8.6.1
				8.6.2
				8.6.4
				8.6.5
					358
					359
					373
					
					391
					393
					394
					395
					
					409
					410
					412
					413

or installation indoors, n cable channels, utdoors and in ground, f there is a risk that olvents and fuel may enetrate and mechani-al damage is not likely o occur.	The lead sheath shall not be used as neutral conductor (N). The sheath wire, if available, is used for earthing the lead sheath (DIN VDE 0165).		Termination – Indoor SKSA – Outdoor SKSE Straight joint SKEM PV	8.1.1 8.2.1 8.5.5 8.5.8	358 373 393 396

4.1 Summary and Application

Design	Type designation code	Number of cores	Nominal cross-sectional area of conductors mm^2	Rated voltage $U_0/U\,(U_m)$ kV	Project planning data	
					Table	Page
4.1.1 PROTODUR cables (PVC insulation) with copper conductor (continued)	NYFGY	three-core	25 – 300	3.6/6 (7.2)	5.1.16	248
	NYSY	single-core	25 – 400	6/10 (12)	5.1.17	250, 252
	NYSEY	three-core	25 – 300	6/10 (12)	5.1.18	254

Preferred application	Information on installation and laying	Information on the current-carrying capacity	Relevant accessories		
			Designation	Table	Page
For power plants, industry and switchgear; for installation indoors, in cable channels, outdoors and in ground, if increased mechanical protection is required or where high tensile stress may occur during installation and operation.			Termination – Indoor PEA PEB – Outdoor FFK 10 Straight joint PV	8.1.3 8.1.4 8.2.2 8.5.9	360 361 374 397
For power plants and switchgear as well as for installation of sub-stations; for installation indoors in confined spaces and cable channels because of small bending radii. As buried cable, because of its light weight preferred in situations where installation is difficult (e.g. steep slopes in pipes)	When installing in air adequate fixing shall be provided with regard to the dynamic effect of short-circuit currents	When selecting the cross-sectional areas of the screens the earth fault or double line-to-earth fault conditions of the network shall be considered. For calculation of annual cost in addition to the ohmic losses the dielectric losses are relevant.	Termination – Indoor IAES 10 – Outdoor FEP 10 Separable elbow connector WS, BWS Straight joint WPS 10 WP Transition joint ÜMP 10	8.1.5 8.2.4 8.4.1 8.5.11 8.5.13 8.7.2	362 377 386 400 403 416
For power plants, industry and switchgear; for installation indoors, in cable channels, outdoors and in ground; as generator cable. The peak short-circuit withstand can be increased, e.g. up to 110 kA, by means of an additional bandage.		When selecting the cross-sectional areas of the screens the earth fault or double line-to-earth fault conditions of the network shall be considered. For calculation of annual cost in addition to the ohmic losses the dielectric losses are relevant.	Termination – Indoor IAES 10 – Outdoor FEP 10 Straight joint WP Transition joint ÜMP 10	8.1.6 8.2.4 8.5.12 8.7.2	364 377 402 416

4.1 Summary and Application

Design	Type designation code	Number of cores	Nominal cross-sectional area of conductors mm^2	Rated voltage $U_0/U\,(U_\mathrm{m})$ kV	Project planning data	
					Table	Page
4.1.2 PROTOTHEN X cables (XLPE insulation) with copper conductor	N2XS2Y	single-core	25 – 500	6/10 (12)	5.2.1	256, 258
	N2XS2Y	single-core	25 – 500	12/20 (24)	5.2.2	260, 262
	N2XS2Y	single-core	25 – 500	18/30 (36)	5.2.3	264, 266
4.1.3 Halogen-free cables with improved performance in the case of fire (SIENOPYR FRNC cables) with copper conductor	(N)2XH	single-core	4 – 300	0.6/1 (1.2)	5.3.1	268
	(N)2XH-J	three-core	1.5 – 240		5.3.2	270
	(N)2XH	three-core with one reduced conductor	25 – 240		5.3.3	270
		four-core	1.5 – 240		5.3.4	272
		five-core	1.5 – 16		5.3.5	272
		multi-core (control cable)	. . . × 1.5 and . . . × 2.5		5.3.6	274
	(N)2XCH	three-core with reduced concentric conductor	1.5 – 240	0.6/1 (1.2)	5.3.7	274
		four-core with reduced concentric conductor	1.5 – 95		5.3.8	276
		multi-core (control cable)	. . . × 1.5, . . . × 2.5 and . . . × 4		5.3.9	276

Preferred application	Information on installation and laying	Information on the current-carrying capacity	Relevant accessories		
			Designation	Table	Page
...a ground for distribution ...tworks because of the ...xtremely low dielectric ...sses. ...Germany cables with ...VC sheath which have ...een used up to now for ...stribution networks are ...creasingly superseded ...y mechanically more ...sistant designs with PE ...heath. ...the penetration of ...ater is possible after ...echanical damage, ...bles with a longitudi-...ally watertight screen ...rea are recommended.	When installing these cables indoors and in cable chan-nels it shall be taken into account that the PE sheath is not flame retardant. When installing single-core cables in air adequate fixing shall be provided with regard to the dynamic effect of short-circuit currents	When selecting the cross-sectional areas of the screens the earth fault or double line-to-earth fault conditions of the network shall be considered. The electrical and thermal data given in the tables (Item 23 to 49) also apply to cables with longitudinally watertight screen area (F) as well as to cables with PVC outer sheath (Y).	Termination		
			– Indoor IAES 10	8.1.5	362
			IAES 20	8.1.5	362
			IAES 30	8.1.5	362
			IAE 20	8.1.8	369
			– Outdoor FAE 10	8.2.3	375
			FAE 20	8.2.3	375
			FAE 30	8.2.3	375
			FEP 10	8.2.4	377
			FAL 20	8.2.6	379
			FEL-2Y	8.2.8	381
			Separable el-bow connector WS, BWS	8.4.1	386
			Straight joint AMS 10	8.5.10	398
			AMS 20	8.5.10	398
			WPS 10	8.5.11	400
			WPS 20	8.5.11	400
			WPS 30	8.5.11	400
			WP	8.5.13	403
			Transition joint ÜMP 10	8.7.2	416
			SM-WP	8.7.3	417
...buildings or installa-...ons with increased ...afety requirements for ...e protection of persons ...r high-quality equipment ...here the performance in ...e case of fire regarding ...re propagation, corrosiv-...y of combustion gases ...nd smoke density is of ...e utmost importance.	Direct laying in ground is not recommended for these cables.	These cables are designed for a permissible operat-ing temperature of 90 °C and a permissi-ble short-circuit temperature of 250 °C.			

4.1 Summary and Application

Design	Type designation code	Number of cores	Nominal cross-sectional area of conductors mm²	Rated voltage U_0/U (U_m) kV	Project planning data	
					Table	Page
4.1.3 Halogen-free cables with improved performance in the case of fire (SIENOPYR FRNC cables) with copper conductor (continued)	NHXHX..FE	single-core	4 – 300	0.6/1 (1.2)	5.3.10	278
	NHXHX-J..FE	three-core	1.5 – 240		5.3.11	280
	NHXHX..FE	three-core with one reduced conductor	25 – 240		5.3.12	280
		four-core	1.5 – 240		5.3.13	282
		five-core	1.5 – 16		5.3.14	282
		multi-core (control cable)	... × 1.5 and ... × 2.5		5.3.15	284
	NHXCHX..FE	three-core with reduced concentric conductor	1.5 – 240	0.6/1 (1.2)	5.3.16	284
		four-core with reduced concentric conductor	1.5 – 95		5.3.17	286
		multi-core (control cable)	... × 1.5, ... × 2.5 and ... × 4		5.3.18	286
4.1.4 Cables with paper insulation and copper conductor	NKBA	three-core	35 – 300	6/10 (12)	5.4.1	288
	NEKEBA	three-core	25 – 300	12/20 (24)	5.4.2	290
	NEKEBA	three-core	35 – 300	18/30 (36)	5.4.3	292

Preferred application	Information on installation and laying	Information on the current-carrying capacity	Relevant accessories		
			Designation	Table	Page
In buildings or installations with increased safety requirements for the protection of persons or high-quality equipment where the circuit integrity as well as the performance in the case of fire regarding reduced fire propagation, corrosivity of combustion gases and smoke density are of the utmost importance. Cables with circuit integrity are preferably used in installations where electrical equipment shall function for a certain time in the case of fire, e.g. emergency lighting. According to DIN VDE 0266, the cables listed in this table are designed for a circuit integrity of at least 20 minutes in the case of fire. In contrast to these cables, cables with a circuit integrity of 180 minutes (FE 180) have an additional bandage over the conductor.	According to DIN VDE 0266 direct laying in ground is not permissible for these cables.	These cables are designed for a permissible operating temperature of 70 °C and a permissible short-circuit temperature of 160 °C. For cables with a circuit integrity of 180 minutes (FE 180) the same electrical and thermal data apply.	–		
For distribution networks in ground	For differences in level exceeding those in Part 1, Clause 29.3, Page 401, a polymeric-insulated cable should be used.	The electrical and thermal data given in the tables (Item 23 to 49) also apply to cables with PVC outer sheath (Y).	Termination – Indoor IKM 10 EoD – Outdoor FF 10 FEL 20 FEL 30 Spreader box AS SS Straight VS joint SMI Transition ÜMP 10 joint SM-WP	8.1.7 8.1.9 8.2.5 8.2.7 8.2.7 8.3.1 8.3.2 8.5.14 8.5.15 8.7.2 8.7.3	368 371 378 380 380 383 384 405 406 416 417

4.1 Summary and Application

Design	Type designation code	Number of cores	Nominal cross-sectional area of conductors mm^2	Rated voltage $U_0/U\,(U_m)$ kV	Project planning data	
					Table	Page
4.1.5 PROTODUR cables (PVC insulation) with aluminium conductor	NAYY	single-core	50 – 240	0.6/1 (1.2)	5.5.1	294, 296
		four-core	25 – 240		5.5.2	298
	NAYCWY	three-core with concentric conductor	50 – 185	0.6/1 (1.2)	5.5.3	300
4.1.6 PROTOTHEN X cables (XLPE insulation) with aluminium conductor	NA2XY	single-core	50 – 240	0.6/1 (1.2)	5.6.1	302, 304
		four-core	25 – 240		5.6.2	306

Preferred application	Information on installation and laying	Information on the current-carrying capacity	Relevant accessories			
			Designation		Table	Page
For installation indoors, in cable channels, outdoors and in ground; for urban networks if mechanical damage is not likely to occur. Also for power plants, industry and switchgear		The current-carrying capacities of single-core cables for d.c. operation and of two-core cables for single-phase a.c. operation shall be taken from Part 1, Page 160.	Termination – Indoor	SKSA	8.1.1	358
				PEA	8.1.2	359
			– Outdoor	SKSE	8.2.1	373
			Straight joint	SKSM	8.5.1	389
				SKSM M	8.5.2	390
				SKEM	8.5.5	393
				PV	8.5.6	394
				PV/U	8.5.7	395
			Branch joint	GNKA-1	8.6.1	409
				PA	8.6.2	410
				PAK	8.6.3	411
				PAD	8.6.4	412
				GMSA	8.6.5	413
			Transition joint	SKSM PK	8.7.1	415
For installation indoors, in cable channels, outdoors and in ground if subsequent mechanical damages are likely to occur; mainly for urban networks, house services and street lighting. If the cable is spiked, the concentric conductor ensures that the overcurrent protective device, connected in series, operates.	The concentric conductor with wave-form layer (CW) is not cut at branches. The concentric conductor shall not be considered as armour.	The electrical and thermal data given in the tables (Items 23 to 49) also apply to cables with helical concentric conductor (C).	Termination – Indoor	SKSA	8.1.1	358
				PEA	8.1.2	359
			– Outdoor	SKSE	8.2.1	373
			Straight joint	SKSM C	8.5.3	391
				SKEM	8.5.5	393
				PV	8.5.6	394
				PV/U	8.5.7	395
			Branch joint	GNKA-1	8.6.1	409
				PA	8.6.2	410
				PAD	8.6.4	412
				GMSA	8.6.5	413
For installation indoors, in cable channels, outdoors and in ground; for urban networks if mechanical damage is not to be expected. Especially for urban networks with high peak loads, for excessive cable grouping and other extreme ambient conditions		The current-carrying capacities of single-core cables for d.c. operation shall be taken from Part 1, Page 160.	Termination – Indoor	SKSA	8.1.1	358
				PEA	8.1.2	359
			– Outdoor	SKSE	8.2.1	373
			Straight joint	SKSM	8.5.1	389
				SKSM M	8.5.2	390
				SKEM	8.5.5	393
				PV	8.5.6	394
			Branch joint	GNKA-1	8.6.1	409
				PA	8.6.2	410
				PAK	8.6.3	411
				GMSA	8.6.5	413
			Transition joint	SKSM PK	8.7.1	415

4.1 Summary and Application

Design	Type designation code	Number of cores	Nominal cross-sectional area of conductors mm^2	Rated voltage $U_0/U\,(U_m)$ kV	Project planning data	
					Table	Page
4.1.6 PROTOTHEN X cables (XLPE insulation) with aluminium conductor (continued)	NA2XS2Y	single-core	25 – 500	6/10 (12)	5.6.3	308, 310
	NA2XS(F)2Y	single-core	25 – 500		5.6.4	312, 314
	NA2XS2Y	three-core	50 – 185		5.6.5	316
	NA2XS2Y	single-core	25 – 500	12/20 (24)	5.6.6	318, 320
	NA2XS(F)2Y	single-core	25 – 500		5.6.7	322, 324
	NA2XS2Y	single-core	25 – 500	18/30 (36)	5.6.8	326, 328
	NA2XS(F)2Y	single-core	25 – 500		5.6.9	330, 332
4.1.7 Cables with paper-insulation and aluminium conductor	NAKBA	three-core	50 – 300	6/10 (12)	5.7.1	334
	NAEKEBA	three-core	35 – 300	12/20 (24)	5.7.2	336
	NAEKEBA	three-core	35 – 300	18/30 (36)	5.7.3	338

eferred plication	Information on installation and laying	Information on the current-carrying capacity	Relevant accessories		
			Designation	Table	Page

r distribution networks ground because of the tremely low dielectric sses. Germany cables with /C sheath previously ed for distribution tworks are increasingly perseded by mecanically ore resistant designs th PE sheath. If the netration of water is ssible after mechanical mage, cables with a ngitudinally watertight reen area are commended.	When installing these cables indoors and in cable channels it shall be taken into account that the PE sheath is not flame retardant. Multi-core polymeric-insulated cables are not available in a longitudinally water-tight design. When installing single-core cables in air adequate fixing shall be provided with regard to the dynamic effect to short-circuit currents.	When selecting the cross-sectional areas of the screens the earth fault or double line-to-earth fault conditions of the networks shall be considered. The electrical and thermal data given in the tables (Item 23 to 49) also apply to cables with PVC outer sheath (Y).	Termination – Indoor IAES 10
			       IAES 20
			       IAES 30
			       IAES 10[1)]
			       IAE 20
			– Outdoor FAE 10
			       FAE 20
			       FAE 30
			       FEP 10
			       FAL 20
			       FEL-2Y
			
			Separable WS, BWS
			elbow
			connector
			
			Straight AMS 10
			joint AMS 20
			       WPS 10
			       WPS 20
			       WPS 30
			       WP[1)]
			       WP
			
			Transition ÜMP 10
			joint SM-WP
				8.1.5
				8.1.5
				8.1.5
				8.1.6
				8.1.8
				8.2.3
				8.2.3
				8.2.3
				8.2.4
				8.2.6
				8.2.8
				
				8.4.1
				
				8.5.10
				8.5.10
				8.5.11
				8.5.11
				8.5.11
				8.5.12
				8.5.13
				
				8.7.2
				8.7.3
					362
					362
					362
					364
					369
					375
					375
					375
					377
					379
					381
					
					386
					
					398
					398
					400
					400
					400
					402
					403
					
					416
					417

r distribution networks ground	For differences in level exceeding those in Part 1, Clause 29.3, Page 401, a polymeric-insulated cable should be used.	The electrical and thermal data given in the tables (Item 23 to 49) also apply to cables with PVC outer sheath (Y).	Termination – Indoor IKM 10
			       EoD
			– Outdoor FFK 10
			       FEL 20
			       FEL 30
			
			Spreader box AS
			       SS
			
			Straight VS
			joint SMI
			
			Transition ÜMP 10
			joint SM-WP
				8.1.7
				8.1.9
				8.2.2
				8.2.7
				8.2.7
				
				8.3.1
				8.3.2
				
				8.5.14
				8.5.15
				
				8.7.2
				8.7.3
					368
					371
					374
					380
					380
					
					383
					384
					
					405
					406
					
					416
					417

or three-core cables

4.2 Determination of the Cross-Sectional Area of Cables

4.2.1 General

In order to achieve a safe project planning of cable installations the cross-sectional area of conductor shall be determined such that the requirement

current-carrying capacity $I_z \geq$ loading I_b

is fulfilled for all operating conditions which can occur. A distinction is made between the current-carrying capacity

▷ for normal operation (Clause 4.2.2)
▷ and for short-circuit (operation under fault conditions) (Clause 4.2.3).

Especially in low-voltage systems, the cross-sectional area of conductor shall be additionally determined in respect of the permitted voltage drop ΔU (Clause 4.2.4) and in order to avoid thermal overloading of the cable a suitable protective device is to be selected (Clause 4.2.5). Besides that the relevant installation rules shall be observed.

With regard to these criteria brief instructions are given for project planning. They are sufficient for most cases when using the values listed in Clause 5. The procedure is shown by examples.

More comprehensive calculation methods with detailed project planning data shall be taken from Part 1 of the book "Power Cables and their Application".

Instructions for project planning of cables on ships are given in Clause 4.2.6.

4.2.2 Determination of the Cross-Sectional Area for Normal Operation

Laying in Ground

The calculation procedure for the determination of the cross-sectional area for installation in ground is given in Table 4.2.1. The "reference operating conditions", shown in this table, in most cases represent the relevant project planning data; the current-carrying capacities I_r (rated currents) for these "reference operating conditions" are given in Clause 5, Item 32.

For "site operating conditions" the rated current shall be multiplied by the relevant rating factors given in Tables 4.2.2 to 4.2.8. Instructions for this procedure are shown in the right column of Table 4.2.1.

The rating factors for *multi-core cables* given in Table 4.2.2 shall be applied to the current-carrying capacities of multi-core cables with three loaded cores. These rating factors have already been considered in the tables of Clause 5, on the assumption that all cores are loaded. However, if not all cores of a multi-core cable are loaded, the current-carrying capacity determined in this way may then be converted to the number of loaded cores by means of the values listed in Table 4.2.2.

Examples for Laying in Ground

1st Example: Current-Carrying Capacity

Three PROTODUR cables NYFGY 3 × 240 SM 3.6/6 kV are laid in ground. Which power can be transmitted in this case?

Permissible operating temperature:	70 °C
Type of operation	
– supply utility load, load factor:	0.7
Laying conditions	
– depth of lay:	0.7 m
– clearance between cables:	7 cm
– cover of bricks without trapped air	
Ambient conditions	
– ground temperature:	30 °C
– soil thermal resistivity of the moist area:	1.5 Km/W
Rating factors	
– f_1 (Table 4.2.3):	0.80
– f_2 (Table 4.2.7):	0.77
Total rating factor Πf:	0.62

The cover has no influence on the current-carrying capacity as there is no trapped air. According to Table 5.1.16 the rated value is $I_r = 460$ A. Thus the current-carrying capacity becomes

$$I_z = I_r \cdot \Pi f = 460 \text{ A} \cdot 0.62 = 285 \text{ A}.$$

The connection of three cables can therefore transmit

$$I_{zges} = 3 \cdot 285 \text{ A} = 855 \text{ A}$$

or a three-phase power of

$$S_{ges} = \sqrt{3} \cdot 855 \text{ A} \cdot 6 \text{ kV} = 8.89 \text{ MVA}.$$

Table 4.2.1   Operating conditions for laying in ground

Reference operating conditions for the determination of rated currents I_r	Site operating conditions[1] and calculation of the current-carrying capacity $I_z = I_r \, \Pi \, f$
Type of operation (see Part 1, Clause 18.2.1)	
Load factor 0.7 and maximum load according to the current-carrying capacities given in the tables for laying in ground	Rating factors f_1 according to Table 4.2.3 or 4.2.4 f_2 according to Table 4.2.5 to 4.2.8
Installation conditions (see Part 1, Clause 18.2.1)	
Depth of lay 0.7 m	For depths of lay up to 1.2 m no conversion is necessary
Arrangement: 1 multi-core cable 1 single-core cable in d.c. system 3 single-core cables in three-phase system, lying in a single layer with a clearance of 7 cm 3 single-core cables in three-phase system, bunched[2]	Rating factors for multi-core cables according to Table 4.2.2, for grouping: f_1 according to Table 4.2.3 or 4.2.4 f_2 according to Table 4.2.5 to 4.2.8
Embedded in sand or soil backfill and, if necessary, with a cover of bricks, concrete plates or flat to slightly curved thin plastic plates	Rating factors for covers with trapped air $f = 0.9$, for laying in pipes $f = 0.85$
Ambient conditions (see Part 1, Clause 18.2.1)	
Ground temperature 20 °C at depth of lay	Rating factors
Soil thermal resistivity of the moist area 1 Km/W	f_1 according to Table 4.2.3 or 4.2.4 f_2 according to Table 4.2.5 to 4.2.8
Soil thermal resistivity of the dry area 2.5 Km/W	
Protection from external heating, e.g heating ducts	see Part 1, Clause 16, Table 16.2
Jointing and earthing of the metal sheaths or screens at both ends (see Part 1, Clause 21)	
System frequency 50 to 60 Hz	

[1] Site operating conditions for laying in ground shall always be calculated by means of the two rating factors f_1 and f_2, as both rating factors depend on the soil thermal resistivity and on the load factor: $\Pi f = f_1 \cdot f_2$

[2] Three cables laid touching in trefoil formation are designated as "bunched"

193

2nd Example: *Determination of the Cross-Sectional Area*

In a 10 kV system a continuous load of 5 MVA shall be transmitted with single-core PROTOTHEN X cables having a cross-sectional area of not more than 70 mm². The cables are laid in bunched arrangement. How many cable systems have to be installed in parallel for this purpose?

Permissible operating temperature: 90 °C

Type of operation
– continuous operation, load factor: 1.0

Installation conditions
– depth of lay   0.7 m
– clearance between the systems: 25 cm
– cover of bricks
    without trapped air

Ambient conditions
– ground temperature: 25 °C
– soil thermal resistivity
    of the moist area: 1.0 Km/W

Load current:

$$I_b = \frac{5 \text{ MVA}}{\sqrt{3} \cdot 10 \text{ kV}} = 289 \text{ A}$$

Estimation: Two cable systems are necessary.

Rating factors

– f_1 (Table 4.2.3):   0.90
– f_2 (Table 4.2.6):   0.75

Total rating factor: 0.68

Thus the "fictitious load current" I_{bf} per cable system converted to the "reference operating conditions" is

$$I_{bf} = \frac{289 \text{ A}}{2 \cdot 0.68} = 213 \text{ A}.$$

Table 5.2.1a shows that the conductor cross-sectional area of 50 mm² is sufficient in this case because the rated current for a PROTOTHEN cable N2XS2Y 1 × 50 RM/16 6/10 kV is given with

$$I_r = 221 \text{ A}.$$

3rd Example: *Current-Carrying Capacity of a Multi-Core Cable*

Only 10 cores of a multi-core PROTODUR cable NYY 19 × 1.5 RE 0.6/1 kV are loaded. The cable is laid in ground under "reference operating conditions". The current-carrying capacity of the cable is calculated as follows:

Rated value I_r
for 19 loaded cores
(Table 5.1.7):   10.4 A

Rating factors for
– 19 loaded cores
    (Table 4.2.2):   0.40
– 10 loaded cores
    (Table 4.2.2):   0.50

Total rating factor
($f = 0.50/0.40$):   1.25

Current-carrying
capacity $I_z = I_r \cdot \Pi f$:   13 A

The same result can be obtained for a four-core cable NYY 4 × 1.5 RE 0.6/1 kV.

Rated value I_r
(Table 5.1.5):   26 A

Rating factor for
– 10 loaded cores
    (Table 4.2.2):   0.50

Current-carrying
capacity $I_z = I_r \cdot \Pi f$:   13 A

Table 4.2.2
Rating factors[1], multi-core cables with conductor cross-sectional areas of 1.5 to 10 mm². Laying in ground or installation in air

Number of loaded cores	Laying in ground	Installation in air
5	0.70	0.75
7	0.60	0.65
10	0.50	0.55
14	0.45	0.50
19	0.40	0.45
24	0.35	0.40
40	0.30	0.35
61	0.25	0.30

[1] The rating factors shall be applied to the current-carrying capacities of four-core cables $U_0/U = 0.6/1$ kV in three-phase operation

Table 4.2.3 Rating factors f_1 for laying in ground
(these rating factors do *not* apply to PVC cables with $U_0/U = 6/10$ kV).
f_1 shall only be applied together with the rating factor f_2 according to Tables 4.2.5 to 4.2.8

Permissible operating temperature °C	Ground tempe-rature °C	Soil thermal resistivity of the moist area															
		0.7 Km/W Load factor					1.0 Km/W Load factor					1.5 Km/W Load factor					2.5 Km/W Load factor
		0.50	0.60	0.70	0.83	1.00	0.50	0.60	0.70	0.85	1.00	0.50	0.60	0.70	0.85	1.00	0.5 to 1.00
90	5	1.24	1.21	1.18	1.13	1.07	1.11	1.09	1.07	1.03	1.00	0.99	0.98	0.97	0.96	0.94	0.89
	10	1.23	1.19	1.16	1.11	1.05	1.09	1.07	1.05	1.01	0.98	0.97	0.96	0.95	0.93	0.91	0.86
	15	1.21	1.17	1.14	1.08	1.03	1.07	1.05	1.02	0.99	0.95	0.95	0.93	0.92	0.91	0.89	0.84
	20	1.19	1.15	1.12	1.06	1.00	1.05	1.02	1.00	0.96	0.93	0.92	0.91	0.90	0.88	0.86	0.81
	25						1.02	1.00	0.98	0.94	0.90	0.90	0.88	0.87	0.85	0.84	0.78
	30								0.95	0.91	0.88	0.87	0.86	0.84	0.83	0.81	0.75
	35													0.82	0.80	0.78	0.72
	40																0.68
70	5	1.29	1.26	1.22	1.15	1.09	1.13	1.11	1.08	1.04	1.00	0.99	0.98	0.97	0.95	0.93	0.86
	10	1.27	1.23	1.19	1.13	1.06	1.11	1.08	1.06	1.01	0.97	0.96	0.95	0.94	0.92	0.89	0.83
	15	1.25	1.21	1.17	1.10	1.03	1.08	1.06	1.03	0.99	0.94	0.93	0.92	0.91	0.88	0.86	0.79
	20	1.23	1.18	1.14	1.08	1.01	1.06	1.03	1.00	0.96	0.91	0.90	0.89	0.87	0.85	0.83	0.76
	25						1.03	1.00	0.97	0.93	0.88	0.87	0.85	0.84	0.82	0.79	0.72
	30								0.94	0.89	0.85	0.84	0.82	0.80	0.78	0.76	0.68
	35													0.77	0.74	0.72	0.63
	40																0.59
65	5	1.31	1.27	1.23	1.16	1.09	1.14	1.11	1.09	1.04	1.00	0.99	0.98	0.96	0.94	0.92	0.85
	10	1.29	1.24	1.20	1.14	1.06	1.11	1.09	1.06	1.02	0.97	0.96	0.95	0.93	0.91	0.89	0.82
	15	1.26	1.22	1.18	1.11	1.04	1.09	1.06	1.03	0.98	0.94	0.93	0.91	0.90	0.88	0.85	0.78
	20	1.24	1.20	1.15	1.08	1.01	1.06	1.03	1.00	0.95	0.90	0.90	0.88	0.86	0.84	0.82	0.74
	25						1.03	1.00	0.97	0.92	0.87	0.86	0.84	0.83	0.80	0.78	0.70
	30								0.94	0.89	0.83	0.82	0.81	0.79	0.77	0.74	0.65
	35													0.75	0.72	0.70	0.60
	40																0.55
60	5	1.33	1.28	1.24	1.17	1.10	1.15	1.12	1.09	1.05	1.00	0.99	0.98	0.96	0.94	0.92	0.84
	10	1.30	1.26	1.21	1.14	1.07	1.12	1.09	1.06	1.02	0.97	0.96	0.94	0.93	0.90	0.88	0.80
	15	1.28	1.23	1.19	1.12	1.04	1.09	1.06	1.03	0.98	0.93	0.92	0.91	0.89	0.87	0.84	0.76
	20	1.25	1.21	1.16	1.09	1.01	1.06	1.03	1.00	0.95	0.90	0.89	0.87	0.86	0.83	0.80	0.72
	25						1.03	1.00	0.97	0.92	0.86	0.85	0.83	0.82	0.79	0.76	0.67
	30								0.93	0.88	0.82	0.81	0.79	0.78	0.75	0.72	0.62
	35													0.73	0.70	0.67	0.57
	40																0.51

Note:

For paper-insulated cables the temperature rise is limited at
ground temperatures below 20 °C (see Part 1, Table 18.1)

Table 4.2.4 Rating factors f_1 for laying in ground
(*only* for PVC cables with $U_0/U = 6/10$ kV).
f_1 shall only be applied together with the rating factor f_2 according to Tables 4.2.5 to 4.2.8

Arrangement a	b	c	Ground temperature °C	0.7 Km/W					1.0 Km/W					1.5 Km/W					2.5 Km/W
Number of systems	cables		(Load factor →)	0.50	0.60	0.70	0.85	1.00	0.50	0.60	0.70	0.85	1.00	0.50	0.60	0.70	0.85	1.00	0.5 to 1.00
1	1	1	5	1.31	1.27	1.23	1.16	1.09	1.14	1.12	1.09	1.05	1.00	0.99	0.98	0.96	0.94	0.92	0.85
			10	1.29	1.26	1.21	1.14	1.07	1.12	1.09	1.06	1.02	0.97	0.96	0.95	0.93	0.91	0.89	0.81
			15	1.27	1.22	1.18	1.11	1.04	1.09	1.06	1.03	0.98	0.94	0.93	0.91	0.90	0.87	0.85	0.77
			20	1.24	1.20	1.15	1.08	1.01	1.06	1.03	1.00	0.95	0.90	0.89	0.88	0.86	0.84	0.81	0.73
			25						1.03	1.00	0.97	0.92	0.87	0.85	0.84	0.83	0.80	0.77	0.69
			30						0.94	0.89	0.83			0.82	0.80	0.79	0.76	0.73	0.64
			35											0.75	0.72	0.70			0.59
			40																0.54
4	3	3	5	1.29	1.24	1.20	1.13	1.06	1.11	1.08	1.05	1.01	0.96	0.95	0.94	0.93	0.90	0.88	0.81
			10	1.26	1.22	1.17	1.11	1.03	1.08	1.05	1.03	0.98	0.93	0.92	0.91	0.89	0.87	0.84	0.77
			15	1.24	1.19	1.15	1.08	1.00	1.05	1.03	0.99	0.95	0.90	0.89	0.87	0.86	0.83	0.81	0.73
			20	1.21	1.17	1.12	1.05	0.97	1.03	0.99	0.96	0.91	0.86	0.85	0.84	0.82	0.79	0.77	0.68
			25						0.99	0.96	0.93	0.88	0.83	0.82	0.80	0.78	0.76	0.73	0.64
			30						0.90	0.84	0.79			0.78	0.76	0.74	0.71	0.68	0.59
			35											0.70	0.67	0.64			0.53
			40																0.47
10	5	6	5	1.26	1.21	1.17	1.10	1.03	1.08	1.05	1.02	0.97	0.93	0.92	0.90	0.89	0.86	0.84	0.76
			10	1.23	1.19	1.14	1.07	1.00	1.05	1.02	0.99	0.94	0.89	0.88	0.87	0.85	0.83	0.80	0.72
			15	1.21	1.16	1.12	1.04	0.96	1.02	0.99	0.96	0.91	0.86	0.85	0.83	0.81	0.79	0.76	0.68
			20	1.18	1.14	1.09	1.01	0.93	0.99	0.96	0.93	0.87	0.82	0.81	0.79	0.77	0.75	0.72	0.63
			25						0.96	0.93	0.89	0.84	0.78	0.77	0.75	0.73	0.70	0.68	0.58
			30						0.86	0.80	0.74			0.73	0.71	0.69	0.66	0.63	0.52
			35											0.64	0.61	0.58			0.46
			40																0.38
–	8	10	5	1.23	1.19	1.14	1.07	0.99	1.05	1.02	0.99	0.94	0.89	0.88	0.86	0.85	0.82	0.80	0.72
			10	1.21	1.16	1.11	1.04	0.96	1.02	0.99	0.96	0.91	0.85	0.84	0.83	0.81	0.78	0.76	0.67
			15	1.18	1.13	1.09	1.01	0.93	0.99	0.96	0.92	0.87	0.82	0.81	0.79	0.77	0.74	0.72	0.63
			20	1.15	1.11	1.06	0.98	0.90	0.96	0.92	0.89	0.84	0.78	0.77	0.75	0.73	0.70	0.67	0.57
			25						0.92	0.89	0.85	0.80	0.74	0.73	0.71	0.69	0.66	0.63	0.52
			30						0.82	0.76	0.70			0.68	0.66	0.64	0.61	0.57	0.45
			35											0.60	0.56	0.52			0.38
			40																0.29
–	10	–	5	1.22	1.17	1.13	1.05	0.98	1.03	1.00	0.97	0.92	0.87	0.86	0.84	0.83	0.80	0.78	0.69
			10	1.19	1.15	1.10	1.02	0.94	1.00	0.97	0.94	0.89	0.83	0.82	0.81	0.79	0.76	0.73	0.65
			15	1.17	1.12	1.07	0.99	0.91	0.97	0.94	0.90	0.85	0.79	0.78	0.77	0.75	0.72	0.69	0.60
			20	1.14	1.09	1.04	0.96	0.88	0.94	0.90	0.87	0.81	0.76	0.74	0.73	0.71	0.68	0.65	0.54
			25						0.90	0.87	0.83	0.78	0.71	0.70	0.68	0.66	0.63	0.60	0.48
			30						0.79	0.73	0.67			0.66	0.63	0.61	0.58	0.54	0.41
			35											0.56	0.52	0.48			0.33
			40																0.22

Arrangement a   Arrangement b   Arrangement c

Table 4.2.5 Rating factors f_2 for laying in ground.
Single-core cables in three-phase systems, bunched arrangement,
f_1 shall only be applied together with the rating factor f_2 according to Tables 4.2.3 and 4.2.4

7cm

Design	Number of systems	Soil thermal resistivity of the moist area											
		0.7 Km/W			1.0 Km/W			1.5 Km/W			2.5 Km/W		
		Load factor			Load factor			Load factor			Load factor		
		0.5	0.6	0.7	0.5	0.6	0.7	0.5	0.6	0.7	0.5	0.6	0.7
XLPE cables 0.6/1 to 18/30 kV	1	1.09	1.04	0.99	1.11	1.05	1.00	1.13	1.07	1.01	1.17	1.09	1.03
	2	0.97	0.90	0.84	0.98	0.91	0.85	1.00	0.92	0.86	1.02	0.94	0.87
	3	0.88	0.80	0.74	0.89	0.82	0.75	0.90	0.82	0.76	0.92	0.83	0.76
	4	0.83	0.75	0.69	0.84	0.76	0.70	0.85	0.77	0.70	0.82	0.78	0.71
	5	0.79	0.71	0.65	0.80	0.72	0.66	0.80	0.73	0.66	0.81	0.73	0.67
	6	0.76	0.68	0.62	0.77	0.69	0.63	0.77	0.70	0.63	0.78	0.70	0.64
	8	0.72	0.64	0.58	0.72	0.65	0.59	0.73	0.65	0.59	0.74	0.66	0.59
	10	0.69	0.61	0.56	0.69	0.62	0.56	0.70	0.62	0.56	0.70	0.63	0.57
PVC cables 0.6/1 kV to 6/10 kV	1	1.01	1.02	0.99	1.04	1.05	1.00	1.07	1.06	1.01	1.11	1.08	1.01
	2	0.94	0.89	0.84	0.97	0.91	0.85	0.99	0.92	0.86	1.01	0.93	0.87
	3	0.86	0.79	0.74	0.89	0.81	0.75	0.90	0.83	0.76	0.91	0.83	0.77
	4	0.82	0.75	0.69	0.84	0.76	0.70	0.85	0.77	0.71	0.86	0.78	0.71
	5	0.78	0.71	0.65	0.80	0.72	0.66	0.80	0.73	0.66	0.81	0.73	0.67
	6	0.75	0.68	0.62	0.77	0.69	0.63	0.77	0.70	0.64	0.78	0.70	0.64
	8	0.71	0.64	0.58	0.72	0.65	0.59	0.73	0.65	0.59	0.73	0.66	0.60
	10	0.68	0.61	0.55	0.69	0.62	0.56	0.69	0.62	0.56	0.70	0.63	0.57

Design	Number of systems	Load factor		Load factor		Load factor		Load factor	
		0.85	1.0	0.85	1.0	0.85	1.0	0.85	1.0
All types	1	0.93	0.87	0.93	0.87	0.94	0.87	0.94	0.87
	2	0.77	0.71	0.77	0.71	0.77	0.71	0.78	0.71
	3	0.67	0.61	0.67	0.61	0.68	0.61	0.68	0.61
	4	0.62	0.56	0.62	0.56	0.62	0.56	0.63	0.56
	5	0.58	0.52	0.58	0.52	0.58	0.52	0.59	0.62
	6	0.55	0.50	0.55	0.50	0.56	0.50	0.56	0.50
	8	0.51	0.46	0.52	0.46	0.52	0.46	0.52	0.46
	10	0.49	0.44	0.49	0.44	0.49	0.44	0.49	0.44

Table 4.2.6 Rating factors f_2 for laying in ground.
Single-core cables in three-phase systems, bunched arrangement.
f_2 shall only be applied together with the rating factor f_1 according to Tables 4.2.3 and 4.2.4

Design	Number of systems	Soil thermal resistivity of the moist area											
		0.7 Km/W			1.0 Km/W			1.5 Km/W			2.5 Km/W		
		Load factor			Load factor			Load factor			Load factor		
		0.5	0.6	0.7	0.5	0.6	0.7	0.5	0.6	0.7	0.5	0.6	0.7
XLPE cables 0.6/1 to 18/30 kV	1	1.09	1.04	0.99	1.11	1.05	1.00	1.13	1.07	1.01	1.17	1.09	1.03
	2	1.01	0.94	0.89	1.02	0.95	0.89	1.04	0.97	0.90	1.06	0.98	0.91
	3	0.94	0.87	0.81	0.95	0.88	0.82	0.97	0.89	0.82	0.99	0.90	0.83
	4	0.91	0.84	0.78	0.92	0.84	0.78	0.93	0.85	0.79	0.95	0.86	0.79
	5	0.88	0.80	0.74	0.89	0.81	0.75	0.90	0.82	0.75	0.91	0.83	0.76
	6	0.86	0.79	0.72	0.87	0.79	0.73	0.88	0.80	0.73	0.89	0.81	0.74
	8	0.83	0.76	0.70	0.84	0.76	0.70	0.85	0.77	0.70	0.86	0.78	0.71
	10	0.81	0.74	0.68	0.82	0.74	0.68	0.83	0.75	0.68	0.84	0.76	0.69
PVC cables 0.6/1 to 6/10 kV	1	1.01	1.02	0.99	1.04	1.05	1.00	1.07	1.06	1.01	1.11	1.08	1.01
	2	0.97	0.95	0.89	1.00	0.96	0.90	1.03	0.97	0.91	1.06	0.98	0.92
	3	0.94	0.88	0.82	0.97	0.88	0.82	0.97	0.89	0.83	0.98	0.90	0.84
	4	0.91	0.84	0.78	0.92	0.85	0.79	0.93	0.86	0.79	0.95	0.87	0.80
	5	0.88	0.81	0.75	0.89	0.82	0.76	0.90	0.82	0.76	0.91	0.83	0.77
	6	0.86	0.79	0.73	0.87	0.80	0.74	0.88	0.81	0.74	0.89	0.81	0.75
	8	0.83	0.76	0.70	0.84	0.77	0.71	0.85	0.78	0.71	0.86	0.78	0.72
	10	0.82	0.75	0.69	0.82	0.75	0.69	0.83	0.76	0.69	0.84	0.76	0.70
		Load factor			Load factor			Load factor			Load factor		
		0.85	1.0		0.85	1.0		0.85	1.0		0.85	1.0	
All types	1	0.93	0.87		0.93	0.87		0.94	0.87		0.94	0.87	
	2	0.82	0.75		0.82	0.75		0.82	0.75		0.83	0.75	
	3	0.74	0.67		0.74	0.67		0.74	0.67		0.74	0.67	
	4	0.70	0.64		0.70	0.64		0.70	0.64		0.71	0.64	
	5	0.67	0.60		0.67	0.60		0.67	0.60		0.67	0.60	
	6	0.65	0.59		0.65	0.59		0.65	0.59		0.65	0.59	
	8	0.62	0.56		0.62	0.56		0.62	0.56		0.62	0.56	
	10	0.60	0.54		0.60	0.54		0.61	0.54		0.61	0.54	

Table 4.2.7 Rating factors f_2 for laying in ground.
Three-core[1] cables in three-phase systems.
f_2 shall only be applied together with the rating factor f_1 according to Tables 4.2.3 and 4.2.4

Design	Number of systems	Soil thermal resistivity of the moist area											
		0.7 Km/W			1.0 Km/W			1.5 Km/W			2.5 Km/W		
		Load factor			Load factor			Load factor			Load factor		
		0.5	0.6	0.7	0.5	0.6	0.7	0.5	0.6	0.7	0.5	0.6	0.7
XLPE cables 0.6/1 and 6/10 kV	1	1.02	1.03	0.99	1.06	1.05	1.00	1.09	1.06	1.01	1.11	1.07	1.02
	2	0.95	0.89	0.84	0.98	0.91	0.85	0.99	0.92	0.86	1.01	0.94	0.87
	3	0.86	0.80	0.74	0.89	0.81	0.75	0.90	0.83	0.77	0.92	0.84	0.77
	4	0.82	0.75	0.69	0.84	0.76	0.70	0.85	0.78	0.71	0.86	0.78	0.72
	5	0.78	0.71	0.65	0.80	0.72	0.66	0.81	0.73	0.67	0.82	0.74	0.67
	6	0.75	0.68	0.63	0.77	0.69	0.63	0.78	0.70	0.64	0.79	0.71	0.65
	8	0.71	0.64	0.59	0.72	0.65	0.59	0.73	0.66	0.60	0.74	0.66	0.60
	10	0.68	0.61	0.56	0.69	0.62	0.56	0.70	0.63	0.57	0.71	0.63	0.57
PVC cables 0.6/1 and 3.6/6 kV	1	0.91	0.92	0.94	0.97	0.97	1.00	1.04	1.03	1.01	1.13	1.07	1.02
	2	0.86	0.87	0.85	0.91	0.90	0.86	0.97	0.93	0.87	1.01	0.94	0.88
	3	0.82	0.80	0.75	0.86	0.82	0.76	0.91	0.84	0.77	0.92	0.84	0.78
	4	0.80	0.76	0.70	0.84	0.77	0.71	0.86	0.78	0.72	0.87	0.79	0.73
	5	0.78	0.72	0.66	0.81	0.73	0.67	0.81	0.74	0.68	0.82	0.75	0.68
	6	0.76	0.69	0.64	0.77	0.70	0.64	0.78	0.71	0.65	0.79	0.72	0.65
	8	0.72	0.65	0.59	0.73	0.66	0.60	0.74	0.67	0.61	0.75	0.67	0.61
	10	0.69	0.62	0.57	0.70	0.63	0.57	0.71	0.64	0.58	0.71	0.64	0.58
		Load factor			Load factor			Load factor			Load factor		
		0.85	1.0		0.85	1.0		0.85	1.0		0.85	1.0	
All types	1	0.94	0.89		0.94	0.89		0.94	0.89		0.95	0.89	
	2	0.77	0.72		0.78	0.72		0.78	0.72		0.79	0.72	
	3	0.68	0.62		0.68	0.62		0.69	0.62		0.69	0.62	
	4	0.63	0.57		0.63	0.57		0.63	0.57		0.64	0.57	
	5	0.59	0.53		0.59	0.53		0.59	0.53		0.60	0.53	
	6	0.56	0.51		0.56	0.51		0.57	0.51		0.57	0.51	
	8	0.52	0.47		0.52	0.47		0.52	0.47		0.53	0.47	
	10	0.49	0.44		0.50	0.44		0.50	0.44		0.50	0.44	

[1] In three-phase systems these values also apply to cables with rated voltages of 0.6/1 kV and 4 or 5 conductors

Table 4.2.8 Rating factors f_2 for laying in ground.
Three-core cables in three-phase systems.
f_2 shall only be applied together with the rating factor f_1 according to Tables 4.2.3 and 4.2.4

Design	Number of cables	Soil thermal resistivity of the moist area											
		0.7 Km/W			1.0 Km/W			1.5 Km/W			2.5 Km/W		
		Load factor			Load factor			Load factor			Load factor		
		0.5	0.6	0.7	0.5	0.6	0.7	0.5	0.6	0.7	0.5	0.6	0.7
PVC cables 0.6/1 kV[1], PVC cables 6/10 kV, paper-insulated belted cables 6/10 kV and S.L. cables 12/20 and 18/30 kV	1	0.90	0.91	0.93	0.98	0.99	1.00	1.05	1.04	1.03	1.14	1.09	1.04
	2	0.85	0.85	0.85	0.93	0.92	0.89	0.98	0.95	0.90	1.03	0.96	0.90
	3	0.80	0.79	0.78	0.87	0.86	0.80	0.93	0.86	0.80	0.95	0.87	0.81
	4	0.77	0.77	0.74	0.85	0.81	0.75	0.89	0.82	0.75	0.90	0.82	0.76
	5	0.75	0.75	0.70	0.84	0.77	0.71	0.85	0.77	0.71	0.86	0.78	0.72
	6	0.74	0.73	0.67	0.81	0.74	0.68	0.82	0.74	0.68	0.83	0.75	0.69
	8	0.73	0.69	0.63	0.77	0.70	0.64	0.77	0.70	0.64	0.78	0.71	0.64
	10	0.71	0.66	0.60	0.74	0.67	0.61	0.74	0.67	0.61	0.75	0.67	0.61
		Load factor			Load factor			Load factor			Load factor		
		0.85	1.0		0.85	1.0		0.85	1.0		0.85	1.0	
All types	1	0.96	0.91		0.96	0.91		0.97	0.91		0.97	0.91	
	2	0.81	0.76		0.82	0.76		0.82	0.76		0.82	0.76	
	3	0.72	0.66		0.72	0.66		0.73	0.66		0.73	0.66	
	4	0.67	0.61		0.67	0.61		0.68	0.61		0.68	0.61	
	5	0.63	0.57		0.63	0.57		0.63	0.57		0.64	0.57	
	6	0.60	0.55		0.60	0.55		0.61	0.55		0.61	0.55	
	8	0.56	0.51		0.56	0.51		0.57	0.51		0.57	0.51	
	10	0.53	0.48		0.54	0.48		0.54	0.48		0.54	0.48	

[1] Only for three-core PVC cables in single-phase a.c. operation

Installation in Air

The calculation procedure for the determination of the cross-sectional area for installation in air is given in Table 4.2.9. The "reference operating conditions", shown in this table, in most cases represent the relevant project planning data; the current-carrying capacities I_r (rated currents) for these "reference operating conditions" are given in Clause 5, Item 32.

For "site operating conditions" the rated current shall be multiplied by the relevant rating factors in Tables 4.2.10 to 4.2.12 and according to Figure 4.2.1. Instructions for this procedure are shown in the right column of Table 4.2.9.

The rating factors for *multi-core cables* given in Table 4.2.2 shall be applied to the current-carrying capacities of multi-core cables with three loaded cores. These rating factors have already been considered in the tables of Clause 5, on the assumption that all cores are loaded. However, if not all cores of a multi-core cable are loaded, the current-carrying capacity determined shall be converted to the number of loaded cores by means of the values listed in Table 4.2.2.

Table 4.2.9 Operating conditions for installation in air

Reference operating conditions for the determination of rated currents I_r	*Site operating conditions* and calculation of the current-carrying capacity $I_z = I_r \Pi f$
Type of operation (see Part 1, Clause 18.2.2)	
Continuous operation with the current-carrying capacities according to the tables for installation in air	Current-carrying capacity at intermittent operation according to Figure 4.2.1 and Part 1, Clause 18.6
Installation conditions (see Part 1, Clause 18.2.2)	
Arrangement: 1 multi-core cable 1 single-core d.c. cable 3 single-core cables in three-phase system, lying in a single layer with a clearance equal to the cable diameter d 3 single-core cables in three-phase system, bunched[1]	Rating factors for multi-core cables according to Table 4.2.2, for grouping according to Table 4.2.10 and 4.2.11

Installation in free air, i.e. unhindered heat dissipation is ensured:
 	
for cables spaced from the wall, floor or ceiling ≥ 2 cm,
 	
for cables lying in a single layer with a clearance of at least twice the cable diameter,
 	
for cables lying in multiple layers vertical clearance between
 	
    the cables: minimum twice the cable diameter
    and
    cable layers: minimum 30 cm	

Ambient conditions (see Part 1, Clause 18.2.2)	
Ambient temperature 30 °C	Rating factors
Sufficiently large or ventilated rooms in which the ambient temperature is not noticeably increased by the heat losses from the cables	for deviating ambient temperatures according to Table 4.2.12, for grouping according to Table 4.2.10 and 4.2.11, for the current-carrying capacity when installed in ducts, according to Part 1, Clause 18.5
Protection from direct solar radiation etc.	Current-carrying capacity according to Part 1, Clause 18.4.2
Jointing and earthing of the metal sheaths or screens at both ends (see Part 1, Clause 21)	
System frequency 50 to 60 Hz	

[1] Three cables laid touching in trefoil formation are designated as "bunched"

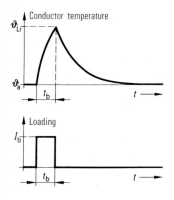

$$f_{KB} = \sqrt{\frac{1}{1-e^{-\frac{t_b}{\tau}}}}$$

without previous loading

$I_0 = 0$

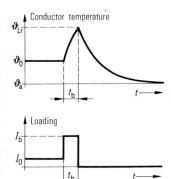

$$f_{KB} = \sqrt{\frac{1-\left(\frac{I_0}{I_b}\right)^2 e^{-\frac{t_b}{\tau}}}{1-e^{-\frac{t_b}{\tau}}}}$$

with previous loading

$I_0 < I_b$

a) Short-time operation (KB [1])

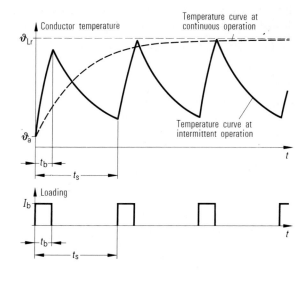

$$f_{AB} = \sqrt{\frac{1-e^{-\frac{t_s}{\tau}}}{1-e^{-\frac{t_b}{\tau}}}} = \sqrt{\frac{1-e^{-\frac{t_s}{\tau}}}{1-e^{-\frac{t_s}{\tau}\frac{ED}{100}}}}$$

$$ED = \frac{t_b}{t_s} 100\%.$$

b) Intermittent operation (AB [2]) without previous loading

ϑ_{Lr} Permissible operating temperature
ϑ_a Initial temperature of the conductor
t_b Load period
t_s Duty cycle time
τ Minimum time value of the cable (see tables in Clause 5)
I_b Load current
I_0 Previous load current
ED   Relative on-time period (%)

---

[1] The index KB is the abbreviation of the German term "Kurzzeitbetrieb"

[2] The index AB is the abbreviation of the German term "Aussetzbetrieb"

Fig. 4.2.1
Determination of the rating factors f_{KB} for short-time operation and f_{AB} for intermittent operation

Table 4.2.10 Rating factors f_H for grouping in air[1].
Single-core cables in three-phase systems

Arrangement of cables	Number of cable trays or cable ladders	Flat formation Clearance = cable diameter d Distance from the wall ≥ 2 cm				Bunched arrangement Clearance = $2\,d$ Distance from the wall ≥ 2 cm			
		Number of cable systems				Number of cable systems			
		1	2	3		1	2	3	
On the floor	–	0.92	0.89	0.88		0.95	0.90	0.88	
On unperforated trays	1	0.92	0.89	0.88		0.95	0.90	0.88	
	2	0.87	0.84	0.83		0.90	0.85	0.83	
	3	0.84	0.82	0.81		0.88	0.83	0.81	
	6	0.82	0.80	0.79		0.86	0.81	0.79	
On cable ladders	1	1.00	0.97	0.96		1.00	0.98	0.96	
	2	0.97	0.94	0.93		1.00	0.95	0.93	
	3	0.96	0.93	0.92		1.00	0.94	0.92	
	6	0.94	0.91	0.90		1.00	0.93	0.90	
On racks or on the wall	–	0.94	0.91	0.89		0.89	0.86	0.84	
Arrangements for which a reduction is not required[1]		For flat formation with an enlarged clearance the increased sheath or screen losses counteract the reduced mutual temperature rise. Therefore, information on reduction-free arrangements cannot be given here.							

[1] If in confined spaces or for large grouping the air temperature is increased by the heat losses from the cables, the rating factor f_ϑ for deviating ambient temperatures given in Table 4.2.12 shall be additionally applied.

Table 4.2.11 Rating factors f_H for grouping in air[1].
Multi-core cables

Arrangement of cables	Number of cable trays or cable ladders	Clearance = cable diameter d Distance from the wall ≥ 2 cm						Touching cables Touching the wall					
		Number of cables						Number of cables					
		1	2	3	6	9		1	2	3	6	9	
On the floor	—	0.95	0.90	0.88	0.85	0.84		0.90	0.84	0.80	0.75	0.73	
On unper-forated trays	1	0.95	0.90	0.88	0.85	0.84		0.95	0.84	0.80	0.75	0.73	
	2	0.90	0.85	0.83	0.81	0.80		0.95	0.80	0.76	0.71	0.69	
	3	0.88	0.83	0.81	0.79	0.78		0.95	0.78	0.74	0.70	0.68	
	6	0.86	0.81	0.79	0.77	0.76		0.95	0.76	0.72	0.68	0.66	
On cable ladders	1	1.00	0.98	0.96	0.93	0.92		0.95	0.84	0.80	0.75	0.73	
	2	1.00	0.95	0.93	0.90	0.89		0.95	0.80	0.76	0.71	0.69	
	3	1.00	0.94	0.92	0.89	0.88		0.95	0.78	0.74	0.70	0.68	
	6	1.00	0.93	0.90	0.87	0.86		0.95	0.76	0.72	0.68	0.66	
On racks or on the wall	—	1.00	0.93	0.90	0.87	0.86		0.95	0.78	0.73	0.68	0.66	
Arrangements for which a reduction is not required[1]		The number of cables in vertical arrangement is optional						The number of cables in horizontal arrangement is optional					

[1] If in confined spaces or for large grouping the air temperature is increased by the heat losses from the cables, the rating factors f_ϑ for deviating ambient temperatures in Table 4.2.12 shall be additionally applied.

Table 4.2.12 Rating factors f_ϑ for ambient temperatures deviating from 30 °C

Design	Permissible operating temperature ϑ_{Lr} °C	Permissible temperature rise K	Ambient temperature								
			10°C	15°C	20°C	25°C	30°C	35°C	40°C	45°C	50°C
			Rating factors								
XLPE cables	90	–	1.15	1.12	1.08	1.04	1.00	0.96	0.91	0.87	0.82
PVC cables	70	–	1.22	1.17	1.12	1.06	1.00	0.94	0.87	0.79	0.71
Paper-insulated belted cables 6/10 kV	65	35	1.00	1.00	1.00	1.00	1.00	0.93	0.85	0.76	0.65
S.L. cables 12/20 kV	65	35	1.00	1.00	1.00	1.00	1.00	0.93	0.85	0.76	0.65
18/30 kV	60	30	1.00	1.00	1.00	1.00	1.00	0.91	0.82	0.71	0.58

Examples for Installation in Air

1st Example: Grouping

According to Figure 4.2.2, 33 multi-core cables are laid on cable ladders in a cable basement. Cable types as well as their load currents I_b are shown in Table 4.2.13. The room is so large that its temperature does not exceed the ambient temperature of 45 °C even if the cables are loaded.

The following condition

$$\frac{I_b}{I_r} \leq \Pi f$$

results from the requirement

$$I_b \leq I_z = I_r \cdot \Pi f.$$

For grouping of up to 5 cables on a total of 7 cable ladders, a reduction factor of $f_H \approx 0.87$ is obtained (Table 4.2.11). The rating factor for deviating ambient temperatures f_ϑ is determined by means of Table 4.2.12.

The results in Table 4.2.13 show that the quotient from loading divided by the current-carrying capacity (I_b/I_r) is always smaller than the product of the rating factors $f_H \cdot f_\vartheta$.

Therefore, the cables can be operated under the operating conditions assumed, and the service life of the cables is not reduced.

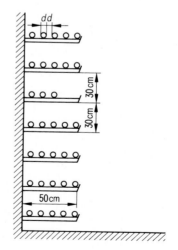

Fig. 4.2.2

Arrangement of cables on cable ladders according to Table 4.2.13

Table 4.2.13
Values for the cables installed according to Figure 4.2.2. For grouping of up to 5 multi-core cables on a total of 7 cable ladders, the rating factor $f_H \approx 0.87$ is to be used (see Table 4.2.11)

Cable type U_0/U	NYFGY 3×120SM 3.6/6 kV	NYCWY 4×240SM 0.6/1 kV	NEKEBA 3×95RM 12/20 kV	NA2XY 4×120SE 0.6/1 kV
Number of cables	13	7	6	7
Loading I_b per cable $m = 1$	180 A	290 A	150 A	200 A
Current-carrying capacity I_z acc. to Table	276 A 5.1.16	429 A 5.1.9	237 A 5.4.2	270 A 5.6.2
Perm. operating temp. ϑ_{Lr} acc. to Table 4.2.12	70 °C	70 °C	65 °C	90 °C
$\dfrac{I_b}{I_r}$	0.65	0.68	0.63	0.74
f_ϑ acc. to Table 4.2.12	0.79	0.79	0.76	0.87
$f_H \cdot f_\vartheta$	0.69	0.69	0.66	0.76

2nd Example: Short-Time Operation and Intermittent Operation

For a four-core PROTODUR cable NAYY 4 × 150 SE 0.6/1 kV, the current-carrying capacity shall be determined for short-time operation with a load period $t_b = 300$ s as well as for intermittent operation with a load period $t_b = 72$ s and a duty cycle $t_s = 360$ s. Moreover, the "reference operating conditions" according to Table 4.2.9 are relevant.

According to Table 5.5.2, for the cable selected the minimum time value $\tau = 1124$ s and the current-carrying capacity $I_r = 238$ A. Thus for short-time operation *without previous loading*

$$f_{KB} = \sqrt{\frac{1}{1 - e^{-\frac{300}{1124}}}} = 2.07$$

and the current-carrying capacity at short-time operation without previous loading

$$I_{KB} = I_r \cdot f_{KB} = 238\text{A} \cdot 2.07 = 493 \text{ A}.$$

With a previous load $I_0 = 150$ A the following equation is obtained for short-time operation according to Figure 4.2.1a:

$$f_{KB} = \sqrt{\frac{1 - \left(\frac{150}{238}\right)^2 \cdot e^{-\frac{300}{1124}}}{1 - e^{-\frac{300}{1124}}}} = 1.72.$$

The current-carrying capacity during short-time operation *with previous loading*

$$I_{KB} = I_r \cdot f_{KB} = 238 \text{ A} \cdot 1.72 = 409 \text{ A}.$$

With the load period $t_b = 72$ s and a duty cycle $t_s = 360$ s the relative on-time period at intermittent operation according to Figure 4.2.1b becomes

$$ED = \frac{72}{360} \cdot 100\% = 20\% .$$

Then

$$f_{AB} = \sqrt{\frac{1 - e^{-\frac{360}{1124}}}{1 - e^{-\frac{360}{1124}} \cdot 0.2}} = 2.10$$

and the current-carrying capacity during intermittent operation

$$I_{AB} = I_r \cdot f_{AB} = 238 \text{ A} \cdot 2.10 = 500 \text{ A}.$$

4.2.3 Determination of the Cross-Sectional Area for Short-Circuit

For the determination of the cross-sectional area for short-circuit, the following condition shall be fulfilled:

$$I_{thz} \geq I_{th}.$$

I_{thz} Thermal short-circuit capacity
I_{th} Thermal equivalent short-circuit current

The short-circuit capacity I_{thz} is determined by multiplying the rated short-circuit current density J_{thr} for cables with copper conductor, Table 4.2.14, and for cables with aluminium conductor, Table 4.2.15, by the respective cross-sectional area of conductor q_n at the short-circuit duration t_k where

$$I_{thz} = q_n \, J_{thr} \sqrt{\frac{t_{kr}}{t_k}} \quad \text{with} \quad t_{kr} = 1 \text{ s}.$$

The short-circuit duration t_k is determined by the characteristic curve of the protective device used or the response time in view of the mechanical delay of the switch and the protective device.

If the cable is operated with a current below its loading before the beginning of the short-circuit, a lower conductor temperature ϑ_a may be assumed for the calculation (see Part 1, Clause 19.3).

The short-circuit capacity of copper wire screens of cables with polymer insulation is specified in Figure 4.2.3.

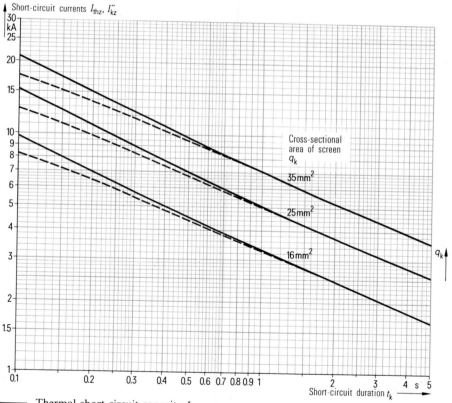

——— Thermal short-circuit capacity I_{thz}
- - - Permissible initial symmetrical short-circuit current $I''_{kz} = I_{thz}/\sqrt{m+1}$

Fig. 4.2.3 Short-circuit capacity of *copper wire screens* of cables with polymer insulation (Details are given in Clause 19.3.1 of Part 1)

Table 4.2.14 Permissible short-circuit temperatures ϑ_e and rated short-time current densities J_{thr} for cables with *copper conductors*

Design, insulation material	Permissible operating temperature ϑ_{Lr} °C	Permissible short-circuit temperature ϑ_e °C	Conductor temperature ϑ_a at the beginning of short-circuit								
			90°C	80°C	70°C	65°C	60°C	50°C	40°C	30°C	20°C
			Rated short-time current density J_{thr} in A/mm² for $t_{kr} = 1$ s								
Soft-soldered connections	–	160	100	107	115	119	122	129	136	143	150
XLPE cables	90	250	143	148	154	157	159	165	170	176	181
PVC cables											
≤ 300 mm²	70	160	–	–	115	119	122	129	136	143	150
> 300 mm²	70	140	–	–	103	107	111	118	126	133	140
Paper-insulated belted cables											
6/10 kV	65	165	–	–	–	121	125	132	138	145	152
S.L. cables											
12/20 kV	65	155	–	–	–	116	119	127	135	141	147
18/30 kV	60	140	–	–	–	–	111	118	126	133	140

Table 4.2.15 Permissible short-circuit temperatures ϑ_e and rated short-time current densities J_{thr} for cables with *aluminium conductors*

Design, insulation material	Permissible operating temperature ϑ_{Lr} °C	Permissible short-circuit temperature ϑ_e °C	Conductor temperature ϑ_a at the beginning of short-circuit								
			90°C	80°C	70°C	65°C	60°C	50°C	40°C	30°C	20°C
			Rated short-time current density J_{thr} in A/mm² for $t_{kr} = 1$ s								
XLPE cables	90	250	94	98	102	104	105	109	113	116	120
PVC cables											
≤ 300 mm²	70	160	–	–	76	78	81	85	90	95	99
> 300 mm²	70	140	–	–	68	71	73	78	83	88	93
Paper-insulated belted cables											
6/10 kV	65	165	–	–	–	80	83	87	92	96	100
S.L. cables											
12/20 kV	65	155	–	–	–	77	79	84	88	93	98
18/30 kV	60	140	–	–	–	–	73	78	83	88	93

Example:
Determination of the Cross-Sectional Area for Short-Circuit

It shall be checked for a given thermal equivalent short-circuit current $I_{th} = 26$ kA and a short-circuit duration $t_k = 0.5$ s whether the conductor cross-sectional area $q_n = 120$ mm^2 of a 10 kV XLPE cable with copper conductor can be regarded as sufficient. The loading before the beginning of short-circuit causes a conductor temperature $\vartheta_a = 60$ °C.

The following result is obtained from Table 4.2.14:

$$J_{thr} = 159 \, A/mm^2.$$

The resulting minimum cross-sectional area is

$$q \geq \frac{I_{th}}{J_{thr}} \cdot \sqrt{\frac{t_k}{t_{kr}}} = \frac{26 \, kA}{159 \, A/mm^2} \cdot \sqrt{\frac{0.5 \, s}{1 \, s}}$$

$$= 115.6 \, mm^2$$

The cable N2XS2Y 1 × 120 RM/16 6/10 kV is therefore sufficient for the short-circuit.

4.2.4 Determination of the Cross-Sectional Area for the Permitted Voltage Drop

Along a cable run the voltage drop ΔU is under load with

direct current

$$\Delta U = 2I_b l \, R',$$

single-phase alternating current

$$\Delta U = 2I_b l \, (R'_w \cos\varphi + X'_L \sin\varphi),$$

three-phase current

$$\Delta U = \sqrt{3} I_b l \, (R'_w \cos\varphi + X'_L \sin\varphi).$$

ΔU Voltage drop
I_b Load current
l Length of cable run
R' d.c. resistance per unit length at operating temperature according to the tables given in Clause 5, Item 23
R'_w Effective resistance per unit length at operating temperature according to the tables given in Clause 5, Item 34
$X'_L = 2\pi \cdot f \cdot L'_b$
L'_b Inductance per unit length of the cable according to the tables given in Clause 5, Item 44
$\cos\varphi$ Power factor of load

The voltage drop Δu related to the nominal voltage of the system U_n is obtained by means of the following equation:

$$\Delta u = \frac{\Delta U}{U_n} \cdot 100\% \, .$$

For conductor cross-sectional areas $q_n \leq 16$ mm^2, the inductance per unit length may be neglected, and for ΔU it is only necessary to take into account the d.c. resistance per unit length at the operating temperature R'.

Example:
Determination of the Cross-Sectional Area for the Permitted Voltage Drop

Along a cable run of $l = 0.15$ km a three-phase effective power of 70 kW shall be transmitted with a power factor $\cos\varphi = 0.9$ and a permissible voltage drop $\Delta u = 2\ \%$. The nominal voltage of the system is 400 V. The conductor cross-sectional area of a four-core PVC cable with copper conductor shall be determined for this purpose.

With the load current

$$I_b = \frac{P_n}{\sqrt{3} \cdot U_n \cdot \cos\varphi}$$

$$= \frac{70\ \text{kW}}{\sqrt{3} \cdot 400\ \text{V} \cdot 0.9} = 112\ \text{A}$$

and the permissible voltage drop

$$\Delta U = \Delta u \cdot U_n = 2\ \% \ \frac{400\ \text{V}}{100\ \%}$$

$$= 0.02 \cdot 400\ \text{V} = 8\ \text{V}$$

the following equation applies to the sum of the relevant impedances:

$$(R'_w \cos\varphi + X'_L \sin\varphi) \leq \frac{\Delta U}{\sqrt{3} \cdot I_b \cdot l} =$$

$$= \frac{8\ \text{V}}{\sqrt{3} \cdot 112\ \text{A} \cdot 0.15\ \text{km}} = 0.275\ \Omega/\text{km}.$$

If the impedances given in Table 5.1.5 are inserted together with $\cos\varphi = 0.9$ and $\sin\varphi = 0.436$ respectively, the following results will be obtained for the conductor cross-sectional areas:

70 mm²: $R'_w = 0.321\ \ \Omega/\text{km}$,
$\qquad\quad L'_b = 0.262\ \ \text{mH/km}$,
$\qquad\quad X'_L = 0.0823\ \Omega\ /\text{km}$,
$\qquad\quad (R'_w \cos\varphi + X'_L \sin\varphi) = 0.325\ \Omega/\text{km}$,

95 mm²: $R'_w = 0.232\ \ \Omega/\text{km}$,
$\qquad\quad L'_b = 0.261\ \ \text{mH/km}$,
$\qquad\quad X'_L = 0.0820\ \Omega\ /\text{km}$,
$\qquad\quad (R'_w \cos\varphi + X'_L \sin\varphi) = 0.245\ \Omega/\text{km}$.

Thus the cable NYY 4 × 95 SM 0.6/1 kV is to be selected for this operating condition assumed.

4.2.5 Protection against Overcurrents

It is possible to heat cables above the permissible operating temperatures by operational overloads as well as by short-circuits.

Overcurrent protective devices, e.g. as given for the low voltage equipment in DIN VDE 0636, DIN VDE 0641 and DIN VDE 0660, protect against overload as well as against short-circuit.

For the scope of the installation rules DIN VDE 0100 Part 430 specifies the conditions for the selection of a overcurrent protective device. The following allocation rules apply:

$$I_b \leq I_n \leq I_z\ ,$$

$$I_2 \leq 1.45\ I_z\ .$$

I_b Load current of a circuit to be expected
I_z Current-carrying capacity of the cable
I_n Rated current of the protective device
I_2 Current causing a tripping or fusing of the protective device under the conditions specified in the installation specifications (tripping or fusing current)

Values for the tripping or fusing currents I_2 are to be taken from the installation specifications of the protective device, e.g. DIN VDE 0636 and 0641.

The application of the allocation rules given does not ensure complete protection in all cases. Owing to the characteristic curves of the protective devices moderate overcurrents can occur for a short time. These overcurrents can cause higher temperatures than permitted especially in the case of small conductor cross-sectional areas. Therefore, according to DIN VDE 0100 Part 430, the circuit shall basically be formed in such a way that small overloads lasting a long time do not occur. If necessary, a lower rating has to be chosen for the overload protection.

4.2.6 Determination of the Cross-Sectional Area of Ship Cables

The current-carrying capacity of ship cables was taken from DIN IEC 92 Part 201. Table 4.2.16 includes details about the "reference operating conditions" which are allocated to the rated values for single-core cables in Table 4.2.17 as well as information on the procedure in case of site operating conditions.

Table 4.2.18 gives rating factors for ambient temperatures deviating from 45 °C. Figure 4.2.5 shows minimum time values τ which are dependent on the overall diameters d of ship cables. These values shall be applied to Figure 4.2.5 at *short-time operation* with $t_b = 30$ min and $t_b = 60$ min or to Figure 4.2.6 at *intermittent operation* with the duty cycle time $t_s = 10$ min and ED = 40 % in order to obtain the corresponding rating factors.

Further details are to be taken from DIN IEC 92 Part 201.

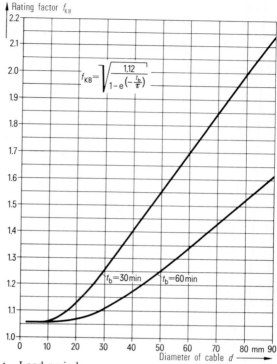

t_b Load period
τ Minimum time value according to Figure 4.2.4

Fig. 4.2.5 Rating factor f_{KB} for short-time operation

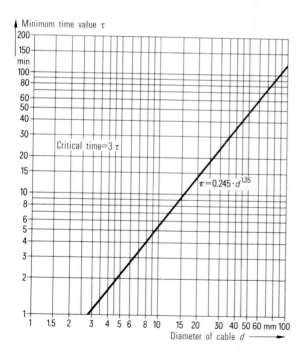

Note:
If a ship cable is loaded with a constant current during a longer period than the critical time 3τ, continuous operation is then assumed.

Fig. 4.2.4 Minimum time values τ of ship cables

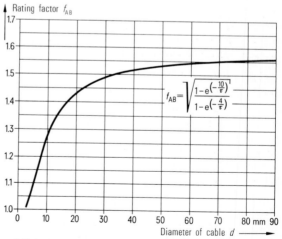

τ Minimum time value according to Figure 4.2.4

Fig. 4.2.6

Rating factor f_{AB} for intermittent operation;
duty cycle $t_s = 10$ min,
relative on-time period (ED) = 40 %

211

Table 4.2.16 Operating conditions for installation in ships according to DIN IEC 92 Part 201

Reference operating conditions for the determination of rated currents I_r	Site operating conditions and calculation of the current-carrying capacity $I_z = I_r \Pi f$
Type of operation	
Continuous operation under constant load longer than the critical time 3τ according to Figure 4.2.4	Ratings factors for short-time operation according to Figure 4.2.5, intermittent operation according to Figure 4.2.6
Installation conditions	
Arrangement: up to 6 single-core cables in three-phase system, bunched, e.g.	Rating factors for two-core cables: 0.85, for three-core and four-core cables: 0.70, for more than 6 single-core or multi-core cables, laid side by side or bunched: 0.85
Installation in free air, in conduits, on unperforated trays or cable ladders	
Ambient conditions	
Ambient temperature 45 °C Sufficiently large or ventilated rooms in which the ambient temperature is not noticeably increased by the heat losses from the cables	Rating factors for deviating ambient temperatures according to Table 4.2.18
Protection from direct solar radiation etc.	

Table 4.2.17 Rated currents I_r at continuous operation for single-core cables with copper conductor according to DIN IEC 92 Part 201 (ambient temperature 45 °C)

Insulation material	PVC for general purposes	Heat-resistant PVC	Butyl rubber	EPR and XLPE	Silicone rubber and mineral insulation
Permissible operating temperature	60 °C	75 °C	80 °C	85 °C	95 °C
Nominal cross-sectional area mm^2	Current-carrying capacity in A				
1	8	13	15	16	20
1.5	12	17	19	20	24
2.5	17	24	26	28	32
4	22	32	35	38	42
6	29	41	45	48	55
10	40	57	63	67	75
16	54	76	84	90	100
25	71	100	110	120	135
35	87	125	140	145	165
50	105	150	165	180	200
70	135	190	215	225	255
95	165	230	260	275	310
120	190	270	300	320	360
150	220	310	340	365	410
185	250	350	390	415	470
240	290	415	460	490	–
300	335	475	530	560	–

Example: Current-Carrying Capacity of Ship Cables

10 three-core cables with a conductor cross-sectional area of 25 mm² each are installed in bunched arrangement on an unperforated tray in a ship. Furthermore 6 two-core cables with a conductor cross-sectional area of 1.5 mm² each are installed in one conduit. The insulation consists of EPR, the ambient temperature is always 50 °C. The current-carrying capacity shall be determined for installation on an unperforated tray as well as for installation in a pipe.

Installation on an unperforated tray

Rated value I_r (Table 4.2.17):	120 A
Rating factors for	
– a three-core cable (Table 4.2.16):	0.70
– grouping, 10 cables (Table 4.2.16):	0.85
– ambient temperature 50 °C (Table 4.2.18):	0.94
Total rating factor Πf :	0.56
Current-carrying capacity $I_z = I_r \cdot \Pi f$:	67 A

Installation in one conduit

Rated value I_r (Table 4.2.17):	20 A
Rating factors for	
– a two-core cable (Table 4.2.16):	0.85
– grouping, 6 cables (Table 4.2.16):	1.00
– ambient temperature 50 °C (Table 4.2.18):	0.94
Total rating factor Πf :	0.80
Current-carrying capacity $I_z = I_r \cdot \Pi f$:	16 A

Table 4.2.18
Rating factors for ambient temperatures deviating from 45 °C

Ambient tempera- ture	Maximum conductor temperature							
	60°C	65°C	70°C	75°C	80°C	85°C	90°C	95°C
35 °C	1.29	1.22	1.18	1.15	1.13	1.12	1.10	1.10
40 °C	1.15	1.12	1.10	1.08	1.07	1.06	1.05	1.05
45 °C	1.00	1.00	1.00	1.00	1.00	1.00	1.00	1.00
50 °C	0.82	0.87	0.89	0.91	0.93	0.94	0.94	0.95
55 °C	–	0.71	0.77	0.82	0.85	0.87	0.88	0.89
60 °C	–	–	0.63	0.71	0.76	0.79	0.82	0.84
65 °C	–	–	–	0.58	0.65	0.71	0.74	0.77
70 °C	–	–	–	–	0.53	0.61	0.67	0.71
75 °C	–	–	–	–	–	0.50	0.58	0.63
80 °C	–	–	–	–	–	–	0.47	0.55
85 °C	–	–	–	–	–	–	–	0.45

5 Project Planning Data of Power Cables

Summary of Tables

Design (with copper conductor)	Type designation code	Rated voltage				
		1 kV	6 kV	10 kV	20 kV	30 kV
		Page				
5.1 PROTODUR cables (PVC insulation)	NYY 1 × ...	216, 218				
	NYY-J 3 × ...	220				
	NYY-O 3 × ...	222				
	NYY 3 × .../..	224				
	NYY 4 × ...	226				
	NYY 5 × ...	228				
	NYY ... × ... (control cable)	230				
	NYCWY 3 × .../..	232				
	NYCWY 4 × .../..	234				
	NYCY ... × ... (control cable)	236				
	NYKY-J 3 × ...	238				
	NYKY 3 × .../..	240				
	NYKY 4 × ...	242				
	NYKY 5 × ...	244				
	NYKY ... × ... (control cable)	246				
	NYFGY 3 × ...		248			
	NYSY 1 × .../...			250, 252		
	NYSEY 3 × .../..			254		
5.2 PROTOTHEN X cables (XLPE insulation)	N2XS2Y 1 × .../..			256, 258	260, 262	264, 266
5.3 SIENOPYR FRNC cables (halogen-free cables with improved performance in the case of fire)	(N)2XH 1 × ...	268				
	(N)2XH-J 3 × ...	270				
	(N)2XH 3 × .../...	270				
	(N)2XH 4 × ...	272				
	(N)2XH 5 × ...	272				
	(N)2XH ... × ... (control cable)	274				
	(N)2XCH 3 × .../...	274				
	(N)2XCH 4 × .../...	276				
	(N)2XCH ... × .../.. (control cable)	276				

Design (with copper conductor)	Type designation code	Rated voltage				
		1 kV	6 kV	10 kV	20 kV	30 kV
		Page				
5.3 SIENOPYR FRNC cables (halogen-free cables with improved performance in the case of fire) (continued)	NHXHX 1 × ... FE	278				
	NHXHX-J 3 × ... FE	280				
	NHXHX3 × .../.. FE	280				
	NHXHX 4 × ... FE	282				
	NHXHX 5 × ... FE	282				
	NHXHX ... × ... FE (control cable)	284				
	NHXCHX 3 × .../.. FE	284				
	NHXCHX 4 × .../.. FE	286				
	NHXCHX ... ×... FE (control cable)	286				
5.4 Cables with paper insulation	NKBA 3 × ...			288		
	NEKEBA 3 × ...				290	292

Design (with aluminium conductor)	Type designation code	Rated voltage				
		1 kV	6 kV	10 kV	20 kV	30 kV
		Page				
5.5 PROTODUR cables (PVC insulation)	NAYY 1 × ...	294, 296				
	NAYY 4 × ...	298				
	NAYCWY 3 × .../..	300				
5.6 PROTOTHEN X cables (XLPE insulation)	NA2XY 1 × ...	302, 304				
	NA2XY 4 × ...	306				
	NA2XS2Y 1 × .../..			308, 310	318, 320	326, 328
	NA2XS(F)2Y 1 × .../..			312, 314	322, 324	330, 332
	NA2XS2Y 3 × .../..			316		
5.7 Cables with paper insulation	NAKBA 3 × ...			334		
	NAEKEBA 3 × ...				336	338

215

Table 5.1.1a
PROTODUR cables (PVC insulation) with copper conductor

Three-phase operation at 50 Hz

Design					4	6	10	16	25
Design	2	Nom. cross-sectional area of conductor		mm^2	4	6	10	16	25
	3	Shape and type of conductor			RE	RE	RE	RE	RM
	9	Thickness of insulation		mm	1.0	1.0	1.0	1.0	1.2
	16	Thickness of outer sheath		mm	1.8	1.8	1.8	1.8	1.8
	17	Overall diameter of cable		mm	9	9	10	11	13
	19	Weight of cable		kg/km	125	155	200	265	395
Mechanical properties	20	Minimum bending radius		mm	135	135	150	165	195
	21	Perm. pulling force with cable grip		N	200	300	500	800	1250
	22	Perm. pulling force with pulling head		N	200	300	500	800	1250
Electrical properties	23	d.c. resistance/unit length at 70 °C		Ω/km	5.52	3.69	2.19	1.38	0.870
	24	d.c. resistance/unit length at 20 °C		Ω/km	4.61	3.08	1.83	1.15	0.727
	31	Reference diameter of cable		mm	7.85	8.36	9.15	10.1	11.9
Laying direct in ground	32	Current-carrying capacity		A	50	62	82	105	136
	34	Eff. a.c. resistance/unit length at 70 °C		Ω/km	5.52	3.69	2.19	1.38	0.870
	38	Ohmic losses per cable		kW/km	13.8	14.2	14.7	15.2	16.1
	39	Fict. thermal resist. of cable for ohmic losses		Km/W	1.193	1.058	0.904	0.772	0.670
	41	Fict. thermal resist. of soil for ohmic losses		Km/W	5.252	5.177	5.069	4.952	4.756
	44	Inductance/unit length per conductor		mH/km	0.459	0.431	0.399	0.371	0.350
Installation in free air	32	Current-carrying capacity		A	36	46	62	83	111
	34	Eff. a.c. resistance/unit length at 70 °C		Ω/km	5.52	3.69	2.19	1.38	0.870
	38	Ohmic losses per cable		kW/km	7.1	7.8	8.4	9.5	10.7
	39	Fict. thermal resist. of cable for ohmic losses		Km/W	1.515	1.356	1.172	1.013	0.889
	43	Thermal resistance of the air		Km/W	4.028	3.818	3.536	3.252	2.840
	44	Inductance/unit length per conductor		mH/km	0.459	0.431	0.399	0.371	0.350
	48	Minimum time value		s	80	110	168	240	328
Short-circuit	49	Rated short-time current of conductor (1s)		kA	0.460	0.690	1.15	1.84	2.87

35 RM	50 RM	70 RM	95 RM	120 RM	150 RM	185 RM	240 RM	300 RM	400 RM	500 RM
1.2	1.4	1.4	1.6	1.6	1.8	2.0	2.2	2.4	2.6	2.8
1.8	1.8	1.8	1.8	1.8	1.8	1.8	1.8	1.9	2.0	2.1
14	16	17	19	21	23	25	28	30	34	38
495	650	850	1150	1400	1700	2050	2650	3300	4150	5250
210	240	255	285	315	345	375	420	450	510	570
1750	2500	3500	4750	6000	7500	9250	12000	15000	20000	25000
1750	2500	3500	4750	6000	7500	9250	12000	15000	20000	25000
0.627	0.463	0.321	0.231	0.183	0.148	0.119	0.0902	0.0719	0.0562	0.0438
0.524	0.387	0.268	0.193	0.153	0.124	0.0991	0.0754	0.0601	0.0470	0.0366
13.0	14.6	16.3	18.3	19.8	21.7	23.7	26.6	29.2	33.0	36.4
163	193	237	284	323	363	411	477	537	612	689
0.627	0.463	0.321	0.232	0.184	0.150	0.121	0.0930	0.0754	0.0607	0.0495
16.7	17.3	18.1	18.7	19.2	19.8	20.4	21.2	21.7	22.7	23.5
0.592	0.551	0.476	0.443	0.402	0.385	0.369	0.342	0.333	0.312	0.299
4.650	4.512	4.380	4.242	4.148	4.039	3.933	3.796	3.684	3.538	3.421
0.333	0.325	0.309	0.302	0.294	0.290	0.287	0.281	0.279	0.275	0.272
136	166	210	259	301	345	398	475	545	636	730
0.627	0.463	0.321	0.232	0.184	0.150	0.121	0.0930	0.0754	0.0607	0.0495
11.6	12.8	14.2	15.6	16.7	17.9	19.1	21.0	22.4	24.6	26.4
0.792	0.742	0.648	0.606	0.553	0.531	0.511	0.475	0.463	0.436	0.419
2.634	2.392	2.175	1.974	1.843	1.706	1.584	1.435	1.326	1.194	1.097
0.333	0.325	0.309	0.302	0.294	0.290	0.287	0.281	0.279	0.275	0.272
428	586	718	869	1027	1221	1396	1649	1957	2555	3030
4.02	5.75	8.05	10.9	13.8	17.2	21.3	27.6	34.5	41.2	51.5

Table 5.1.1b
PROTODUR cables (PVC insulation) with copper conductor

Three-phase operation at 50 Hz

Design					4	6	10	16	25
	2	Nom. cross-sectional area of conductor	mm^2		4	6	10	16	25
	3	Shape and type of conductor			RE	RE	RE	RE	RM
	9	Thickness of insulation	mm		1.0	1.0	1.0	1.0	1.2
	16	Thickness of outer sheath	mm		1.8	1.8	1.8	1.8	1.8
	17	Overall diameter of cable	mm		9	9	10	11	13
	19	Weight of cable	kg/km		125	155	200	265	395
Mechanical properties	20	Minimum bending radius	mm		135	135	150	165	195
	21	Perm. pulling force with cable grip	N		200	300	500	800	1250
	22	Perm. pulling force with pulling head	N		200	300	500	800	1250
Electrical properties	23	d.c. resistance/unit length at 70 °C	Ω/km		5.52	3.69	2.19	1.38	0.870
	24	d.c. resistance/unit length at 20 °C	Ω/km		4.61	3.08	1.83	1.15	0.727
	31	Reference diameter of cable	mm		7.85	8.36	9.15	10.1	11.9
Laying direct in ground	32	Current-carrying capacity	A		59	73	97	125	161
	34	Eff. a.c. resistance/unit length at 70 °C	Ω/km		5.52	3.69	2.19	1.38	0.870
	38	Ohmic losses per cable	kW/km		19.2	19.6	20.6	21.5	22.6
	39	Fict. thermal resist. of cable for ohmic losses	Km/W		1.193	1.058	0.904	0.772	0.670
	41	Fict. thermal resist. of soil for ohmic losses	Km/W		3.428	3.398	3.354	3.305	3.222
	44	Inductance/unit length per conductor	mH/km		0.944	0.904	0.856	0.811	0.761
Installation in free air	32	Current-carrying capacity	A		45	56	77	102	137
	34	Eff. a.c. resistance/unit length at 70 °C	Ω/km		5.52	3.69	2.19	1.38	0.870
	38	Ohmic losses per cable	kW/km		11.2	11.6	13.0	14.3	16.3
	39	Fict. thermal resist. of cable for ohmic losses	Km/W		1.193	1.058	0.904	0.772	0.670
	43	Thermal resistance of the air	Km/W		2.467	2.347	2.186	2.022	1.784
	44	Inductance/unit length per conductor	mH/km		0.623	0.595	0.563	0.535	0.514
	48	Minimum time value	s		51	74	109	159	215
Short-circuit	49	Rated short-time current of conductor (1s)	kA		0.460	0.690	1.15	1.84	2.87

35 RM	50 RM	70 RM	95 RM	120 RM	150 RM	185 RM	240 RM	300 RM	400 RM	500 RM
1.2	1.4	1.4	1.6	1.6	1.8	2.0	2.2	2.4	2.6	2.8
1.8	1.8	1.8	1.8	1.8	1.8	1.8	1.8	1.9	2.0	2.1
14	16	17	19	21	23	25	28	30	34	38
495	650	850	1150	1400	1700	2050	2650	3300	4150	5250
210	240	255	285	315	345	375	420	450	510	570
1750	2500	3500	4750	6000	7500	9250	12000	15000	20000	25000
1750	2500	3500	4750	6000	7500	9250	12000	15000	20000	25000
0.627	0.463	0.321	0.231	0.183	0.148	0.119	0.0902	0.0719	0.0562	0.0438
0.524	0.387	0.268	0.193	0.153	0.124	0.0991	0.0754	0.0601	0.0470	0.0366
13.0	14.6	16.3	18.3	19.8	21.7	23.7	26.6	29.2	33.0	36.4
192	227	278	332	377	423	478	555	626	716	813
0.627	0.463	0.321	0.231	0.184	0.149	0.119	0.0912	0.0732	0.0579	0.0461
23.1	23.9	24.8	25.5	26.1	26.7	27.3	28.1	28.7	29.7	30.5
0.592	0.551	0.476	0.443	0.402	0.385	0.369	0.342	0.333	0.312	0.299
3.176	3.115	3.055	2.991	2.946	2.893	2.841	2.771	2.713	2.635	2.570
0.729	0.702	0.668	0.643	0.621	0.604	0.587	0.564	0.549	0.528	0.512
168	204	258	318	370	424	489	584	672	789	916
0.627	0.463	0.321	0.231	0.184	0.149	0.120	0.0916	0.0736	0.0584	0.0465
17.7	19.3	21.4	23.4	25.2	26.8	28.6	31.2	33.2	36.4	39.0
0.592	0.551	0.476	0.443	0.402	0.385	0.369	0.342	0.333	0.312	0.299
1.662	1.519	1.389	1.268	1.189	1.105	1.030	0.938	0.870	0.788	0.727
0.497	0.489	0.473	0.466	0.458	0.454	0.451	0.445	0.443	0.439	0.436
280	388	475	576	679	808	924	1091	1287	1660	1924
4.02	5.75	8.05	10.9	13.8	17.2	21.3	27.6	34.5	41.2	51.5

Table 5.1.2
PROTODUR cables (PVC insulation) with copper conductor

Single-phase a.c. operation at 50 Hz

				1.5	2.5	4	6	10
Design	2	Nom. cross-sectional area of conductor	mm²	1.5	2.5	4	6	10
	3	Shape and type of conductor		RE	RE	RE	RE	RE
	9	Thickness of insulation	mm	0.8	0.8	1.0	1.0	1.0
	16	Thickness of outer sheath	mm	1.8	1.8	1.8	1.8	1.8
	17	Overall diameter of cable	mm	12	13	16	17	19
	19	Weight of cable	kg/km	220	270	420	510	670
Mechanical properties	20	Minimum bending radius	mm	144	156	192	204	228
	21	Perm. pulling force with cable grip	N	225	375	600	900	1500
	22	Perm. pulling force with pulling head	N	225	375	600	900	1500
Electrical properties	23	d.c. resistance/unit length at 70 °C	Ω/km	14.5	8.87	5.52	3.69	2.19
	24	d.c. resistance/unit length at 20 °C	Ω/km	12.1	7.41	4.61	3.08	1.83
Laying direct in ground	32	Current-carrying capacity	A	31	41	54	68	92
	34	Eff. a.c. resistance/unit length at 70 °C	Ω/km	14.5	8.87	5.52	3.69	2.19
	38	Ohmic losses per cable	kW/km	27.8	29.8	32.2	34.1	37.1
	44	Inductance/unit length per conductor	mH/km	0.343	0.317	0.316	0.298	0.278
Installation in free air	32	Current-carrying capacity	A	22	30	40	50	68
	34	Eff. a.c. resistance/unit length at 70 °C	Ω/km	14.5	8.87	5.52	3.69	2.19
	38	Ohmic losses per cable	kW/km	14.0	16.0	17.7	18.4	20.2
	44	Inductance/unit length per conductor	mH/km	0.343	0.317	0.316	0.298	0.278
	48	Minimum time value	s	30	45	65	93	140
Short-circuit	49	Rated short-time current of conductor (1s)	kA	0.173	0.288	0.460	0.690	1.15

16 RE	25 RM	35 SM	50 SM	70 SM	95 SM	120 SM	150 SM	185 SM	240 SM	300 SM	400 SM
1.0	1.2	1.2	1.4	1.4	1.6	1.6	1.8	2.0	2.2	2.4	2.6
1.8	1.8	1.8	1.8	2.0	2.1	2.2	2.3	2.5	2.7	2.9	3.1
20	25	23	26	30	34	36	40	47	52	57	64
900	1400	1350	1750	2400	3250	4000	4900	6550	8500	10450	13150
240	300	276	312	360	408	432	480	564	624	684	768
2400	3750	5250	7500	10500	14250	18000	22500	27750	36000	45000	60000
2400	3750	5250	7500	10500	14250	18000	22500	27750	36000	45000	60000
1.38	0.870	0.627	0.463	0.321	0.231	0.183	0.148	0.119	0.0902	0.0719	0.0562
1.15	0.727	0.524	0.387	0.268	0.193	0.153	0.124	0.0991	0.0754	0.0601	0.0470
120	156	189	225	276	332	379	426	483	560	633	723
1.38	0.870	0.627	0.463	0.321	0.232	0.184	0.150	0.121	0.0928	0.0752	0.0603
39.6	42.4	44.8	46.9	49.0	51.1	53.0	54.4	56.3	58.2	60.3	63.0
0.262	0.257	0.248	0.247	0.238	0.238	0.233	0.233	0.233	0.231	0.231	0.228
91	122	150	182	230	284	329	377	433	513	588	686
1.38	0.870	0.627	0.463	0.321	0.232	0.184	0.150	0.121	0.0928	0.0752	0.0603
22.8	25.9	28.2	30.7	34.0	37.4	39.9	42.6	45.2	48.8	52.0	56.8
0.262	0.257	0.248	0.247	0.238	0.238	0.233	0.233	0.233	0.231	0.231	0.228
200	271	352	487	598	723	859	1022	1179	1414	1681	2196
1.84	2.87	4.02	5.75	8.05	10.9	13.8	17.2	21.3	27.6	34.5	41.2

Table 5.1.3
PROTODUR cables (PVC insulation) with copper conductor

Three-phase operation at 50 Hz

				1.5	2.5	4	6	10
Design	2	Nom. cross-sectional area of conductor	mm^2	1.5	2.5	4	6	10
	3	Shape and type of conductor		RE	RE	RE	RE	RE
	9	Thickness of insulation	mm	0.8	0.8	1.0	1.0	1.0
	16	Thickness of outer sheath	mm	1.8	1.8	1.8	1.8	1.8
	17	Overall diameter of cable	mm	12	13	16	17	19
	19	Weight of cable	kg/km	220	270	420	510	670
Mechanical properties	20	Minimum bending radius	mm	144	156	192	204	228
	21	Perm. pulling force with cable grip	N	225	375	600	900	1500
	22	Perm. pulling force with pulling head	N	225	375	600	900	1500
Electrical properties	23	d.c. resistance/unit length at 70 °C	Ω/km	14.5	8.87	5.52	3.69	2.19
	24	d.c. resistance/unit length at 20 °C	Ω/km	12.1	7.41	4.61	3.08	1.83
	31	Reference diameter of cable	mm	12.0	12.9	14.8	15.9	17.6
Laying direct in ground	32	Current-carrying capacity	A	25	34	44	55	75
	34	Eff. a.c. resistance/unit length at 70 °C	Ω/km	14.5	8.87	5.52	3.69	2.19
	38	Ohmic losses per cable	kW/km	27.1	30.7	32.0	33.4	36.9
	39	Fict. thermal resist. of cable for ohmic losses	Km/W	1.079	0.973	0.888	0.810	0.717
	41	Fict. thermal resist. of soil for ohmic losses	Km/W	0.706	0.695	0.673	0.662	0.645
	44	Inductance/unit length per conductor	mH/km	0.343	0.317	0.316	0.298	0.278
Installation in free air	32	Current-carrying capacity	A	18.2	24	32	41	56
	34	Eff. a.c. resistance/unit length at 70 °C	Ω/km	14.5	8.87	5.52	3.69	2.19
	38	Ohmic losses per cable	kW/km	14.4	15.3	16.9	18.6	20.6
	39	Fict. thermal resist. of cable for ohmic losses	Km/W	1.079	0.973	0.888	0.810	0.717
	43	Thermal resistance of the air	Km/W	1.697	1.598	1.435	1.352	1.242
	44	Inductance/unit length per conductor	mH/km	0.343	0.317	0.316	0.298	0.278
	48	Minimum time value	s	44	70	101	138	206
Short-circuit	49	Rated short-time current of conductor (1s)	kA	0.173	0.288	0.460	0.690	1.15

16 RE	25 RM	35 SM	50 SM	70 SM	95 SM	120 SM	150 SM	185 SM	240 SM	300 SM	400 SM
1.0	1.2	1.2	1.4	1.4	1.6	1.6	1.8	2.0	2.2	2.4	2.6
1.8	1.8	1.8	1.8	2.0	2.1	2.2	2.3	2.5	2.7	2.9	3.1
20	25	23	26	30	34	36	40	47	52	57	64
900	1400	1350	1750	2400	3250	4000	4900	6550	8500	10450	13150
240	300	276	312	360	408	432	480	564	624	684	768
2400	3750	5250	7500	10500	14250	18000	22500	27750	36000	45000	60000
2400	3750	5250	7500	10500	14250	18000	22500	27750	36000	45000	60000
1.38	0.870	0.627	0.463	0.321	0.231	0.183	0.148	0.119	0.0902	0.0719	0.0562
1.15	0.727	0.524	0.387	0.268	0.193	0.153	0.124	0.0991	0.0754	0.0601	0.0470
19.6	23.5	23.0	25.8	29.6	33.5	36.1	40.4	44.4	50.1	54.9	61.6
98	128	155	184	225	271	309	348	394	458	516	591
1.38	0.870	0.627	0.463	0.321	0.232	0.184	0.150	0.121	0.0928	0.0752	0.0603
39.6	42.8	45.2	47.1	48.8	51.1	52.8	54.5	56.2	58.4	60.1	63.2
0.632	0.556	0.440	0.406	0.379	0.352	0.328	0.322	0.313	0.304	0.293	0.272
0.628	1.498	1.507	1.461	1.407	1.357	1.327	1.283	1.245	1.197	1.161	1.115
0.262	0.257	0.248	0.247	0.238	0.238	0.233	0.233	0.233	0.231	0.231	0.228
74	100	122	149	188	232	268	308	354	419	480	560
1.38	0.870	0.627	0.463	0.321	0.232	0.184	0.150	0.121	0.0928	0.0752	0.0603
22.6	26.1	28.0	30.9	34.1	37.4	39.7	42.7	45.3	48.9	52.0	56.7
0.632	0.556	0.440	0.406	0.379	0.352	0.328	0.322	0.313	0.304	0.293	0.272
1.135	0.978	0.982	0.894	0.798	0.720	0.675	0.616	0.570	0.515	0.477	0.432
0.262	0.257	0.248	0.247	0.238	0.238	0.233	0.233	0.233	0.231	0.231	0.228
302	404	532	727	895	1083	1295	1532	1764	2119	2523	3295
1.84	2.87	4.02	5.75	8.05	10.9	13.8	17.2	21.3	27.6	34.5	41.2

Table 5.1.4
PROTODUR cables (PVC insulation) with copper conductor

Three-phase operation at 50 Hz

				25	35	50	70
Design	2	Nom. cross-sectional area of conductor	mm²	25	35	50	70
	3	Shape and type of conductor		RM	SM	SM	SM
	4	Nom. cross-sectional area of reduced conductor	mm²	16	16	25	35
	5	Shape and type of reduced conductor		RE	RE	RM	SM
	9	Thickness of insulation	mm	1.2	1.2	1.4	1.4
	16	Thickness of outer sheath	mm	1.8	1.8	1.9	2.0
	17	Overall diameter of cable	mm	27	26	31	32
	19	Weight of cable	kg/km	1600	1750	2400	2800
Mechanical properties	20	Minimum bending radius	mm	324	312	372	384
	21	Perm. pulling force with cable grip	N	4550	6050	8750	12250
	22	Perm. pulling force with pulling head	N	4550	6050	8750	12250
Electrical properties	23	d.c. resistance/unit length at 70 °C	Ω/km	0.870	0.627	0.463	0.321
	24	d.c. resistance/unit length at 20 °C	Ω/km	0.727	0.524	0.387	0.268
	31	Reference diameter of cable	mm	23.5	23.0	25.8	29.6
Laying direct in ground	32	Current-carrying capacity	A	128	155	184	225
	34	Eff. a.c. resistance/unit length at 70 °C	Ω/km	0.870	0.627	0.463	0.321
	38	Ohmic losses per cable	kW/km	42.8	45.2	47.1	48.8
	44	Inductance/unit length per conductor	mH/km	0.274	0.261	0.263	0.254
Installation in free air	32	Current-carrying capacity	A	100	122	149	188
	34	Eff. a.c. resistance/unit length at 70 °C	Ω/km	0.870	0.627	0.463	0.321
	38	Ohmic losses per cable	kW/km	26.1	28.0	30.9	34.1
	44	Inductance/unit length per conductor	mH/km	0.274	0.261	0.263	0.254
	48	Minimum time value	s	404	532	727	895
Short-circuit	49	Rated short-time current of conductor (1s)	kA	2.87	4.02	5.75	8.05

95 SM	120 SM	150 SM	185 SM	240 SM	300 SM	400 SM
50 SM	70 SM	70 SM	95 SM	120 SM	150 SM	185 SM
1.6	1.6	1.8	2.0	2.2	2.4	2.6
2.2	2.3	2.4	2.6	2.8	2.9	3.2
37	40	44	50	57	63	71
3800	4750	5700	7650	9850	12100	15350
444	480	528	600	684	756	852
16750	21500	26000	32500	42000	52500	69250
16750	21500	26000	32500	42000	52500	69250
0.231	0.183	0.148	0.119	0.0902	0.0719	0.0562
0.193	0.153	0.124	0.0991	0.0754	0.0601	0.0470
33.5	36.1	40.4	44.4	50.1	54.9	61.6
271	309	348	394	458	518	592
0.232	0.184	0.150	0.120	0.0926	0.0749	0.0600
51.1	52.8	54.5	56.1	58.3	60.3	63.1
0.253	0.250	0.247	0.248	0.245	0.245	0.241
232	269	308	354	419	481	562
0.232	0.184	0.150	0.120	0.0926	0.0749	0.0600
37.4	40.0	42.7	45.3	48.8	52.0	56.9
0.253	0.250	0.247	0.248	0.245	0.245	0.241
1083	1285	1532	1764	2119	2513	3272
10.9	13.8	17.2	21.3	27.6	34.5	41.2

Power Cables

Table 5.1.5
PROTODUR cables (PVC insulation) with copper conductor

Three-phase operation at 50 Hz

				1.5	2.5	4	6	10
Design	2	Nom. cross-sectional area of conductor	mm²	1.5	2.5	4	6	10
	3	Shape and type of conductor		RE	RE	RE	RE	RE
	9	Thickness of insulation	mm	0.8	0.8	1.0	1.0	1.0
	16	Thickness of outer sheath	mm	1.8	1.8	1.8	1.8	1.8
	17	Overall diameter of cable	mm	13	14	17	18	20
	19	Weight of cable	kg/km	250	310	485	600	810
Mechanical properties	20	Minimum bending radius	mm	156	168	204	216	240
	21	Perm. pulling force with cable grip	N	300	500	800	1200	2000
	22	Perm. pulling force with pulling head	N	300	500	800	1200	2000
Electrical properties	23	d.c. resistance/unit length at 70 °C	Ω/km	14.5	8.87	5.52	3.69	2.19
	24	d.c. resistance/unit length at 20 °C	Ω/km	12.1	7.41	4.61	3.08	1.83
	31	Reference diameter of cable	mm	12.8	13.8	15.9	17.1	19.0
Laying direct in ground	32	Current-carrying capacity	A	26	34	44	56	75
	34	Eff. a.c. resistance/unit length at 70 °C	Ω/km	14.5	8.87	5.52	3.69	2.19
	38	Ohmic losses per cable	kW/km	29.4	30.7	32.0	34.7	36.9
	39	Fict. thermal resist. of cable for ohmic losses	Km/W	1.053	0.959	0.877	0.805	0.717
	41	Fict. thermal resist. of soil for ohmic losses	Km/W	0.696	0.684	0.662	0.650	0.633
	44	Inductance/unit length per conductor	mH/km	0.366	0.340	0.339	0.321	0.301
Installation in free air	32	Current-carrying capacity	A	18.6	25	33	42	57
	34	Eff. a.c. resistance/unit length at 70 °C	Ω/km	14.5	8.87	5.52	3.69	2.19
	38	Ohmic losses per cable	kW/km	15.0	16.6	18.0	19.5	21.3
	39	Fict. thermal resist. of cable for ohmic losses	Km/W	1.053	0.959	0.877	0.805	0.717
	43	Thermal resistance of the air	Km/W	1.615	1.519	1.358	1.278	1.171
	44	Inductance/unit length per conductor	mH/km	0.366	0.340	0.339	0.321	0.301
	48	Minimum time value	s	42	65	95	132	199
Short-circuit	49	Rated short-time current of conductor (1s)	kA	0.173	0.288	0.460	0.690	1.15

226

16 RE	25 RM	35 SM	50 SM	70 SM	95 SM	120 SM	150 SM	185 SM	240 SM	300 SM	400 SM
1.0	1.2	1.2	1.4	1.4	1.6	1.6	1.8	2.0	2.2	2.4	2.6
1.8	1.8	1.8	1.9	2.1	2.2	2.4	2.5	2.7	2.9	3.1	3.4
22	28	26	30	33	38	42	46	53	59	65	74
1100	1700	1750	2300	3150	4250	5300	6450	8650	11100	13700	17400
264	336	312	360	396	456	504	552	636	708	780	888
3200	5000	7000	10000	14000	19000	24000	30000	37000	48000	60000	80000
3200	5000	7000	10000	14000	19000	24000	30000	37000	48000	60000	80000
1.38	0.870	0.627	0.463	0.321	0.231	0.183	0.148	0.119	0.0902	0.0719	0.0562
1.15	0.727	0.524	0.387	0.268	0.193	0.153	0.124	0.0991	0.0754	0.0601	0.0470
21.3	25.7	25.8	29.6	33.4	37.9	41.7	45.7	50.7	56.8	62.2	71.0
98	129	156	185	228	274	313	352	398	464	525	602
1.38	0.870	0.627	0.463	0.321	0.232	0.184	0.150	0.120	0.0925	0.0747	0.0597
39.6	43.4	45.8	47.6	50.1	52.2	54.1	55.7	57.2	59.7	61.8	64.9
0.636	0.560	0.446	0.438	0.390	0.363	0.350	0.337	0.336	0.315	0.303	0.289
0.615	1.463	1.461	1.407	1.358	1.308	1.270	1.234	1.192	1.147	1.111	1.058
0.285	0.280	0.271	0.270	0.262	0.261	0.256	0.256	0.256	0.254	0.254	0.251
75	102	126	152	193	238	276	315	361	430	493	575
1.38	0.870	0.627	0.463	0.321	0.232	0.184	0.150	0.120	0.0925	0.0747	0.0597
23.2	27.2	29.9	32.1	35.9	39.4	42.1	44.6	47.0	51.3	54.5	59.2
0.636	0.560	0.446	0.438	0.390	0.363	0.350	0.337	0.336	0.315	0.303	0.289
1.065	0.912	0.898	0.804	0.726	0.653	0.603	0.559	0.513	0.466	0.432	0.386
0.285	0.280	0.271	0.270	0.262	0.261	0.256	0.256	0.256	0.254	0.254	0.251
294	388	498	699	850	1029	1221	1465	1696	2012	2392	3126
1.84	2.87	4.02	5.75	8.05	10.9	13.8	17.2	21.3	27.6	34.5	41.2

Table 5.1.6
PROTODUR cables (PVC insulation) with copper conductor

Three-phase operation at 50 Hz

Design	2	Nom. cross-sectional area of conductor	mm^2	1.5	2.5	4	6
	3	Shape and type of conductor		RE	RE	RE	RE
	9	Thickness of insulation	mm	0.8	0.8	1.0	1.0
	16	Thickness of outer sheath	mm	1.8	1.8	1.8	1.8
	17	Overall diameter of cable	mm	15	16	18	20
	19	Weight of cable	kg/km	345	410	570	700
Mechanical properties	20	Minimum bending radius	mm	180	192	216	240
	21	Perm. pulling force with cable grip	N	375	625	1000	1500
	22	Perm. pulling force with pulling head	N	375	625	1000	1500
Electrical properties	23	d.c. resistance/unit length at 70 °C	Ω/km	14.5	8.87	5.52	3.69
	24	d.c. resistance/unit length at 20 °C	Ω/km	12.1	7.41	4.61	3.08
Laying direct in ground	32	Current-carrying capacity	A	26	34	44	56
	34	Eff. a.c. resistance/unit length at 70 °C	Ω/km	14.5	8.87	5.52	3.69
	38	Ohmic losses per cable	kW/km	29.4	30.7	32.0	34.7
	44	Inductance/unit length per conductor	mH/km	0.375	0.349	0.348	0.330
Installation in free air	32	Current-carrying capacity	A	18.6	25	33	42
	34	Eff. a.c. resistance/unit length at 70 °C	Ω/km	14.5	8.87	5.52	3.69
	38	Ohmic losses per cable	kW/km	15.0	16.6	18.0	19.5
	44	Inductance/unit length per conductor	mH/km	0.375	0.349	0.348	0.330
	48	Minimum time value	s	42	65	95	132
Short-circuit	49	Rated short-time current of conductor (1s)	kA	0.173	0.288	0.460	0.690

10 RE	16 RE	25 RM
1.0	1.0	1.2
1.8	1.8	1.8
22	24	30
960	1300	2100
264	288	360
2500	4000	6250
2500	4000	6250
2.19	1.38	0.870
1.83	1.15	0.727
75	98	129
2.19	1.38	0.870
36.9	39.6	43.4
0.310	0.294	0.289
57	75	102
2.19	1.38	0.870
21.3	23.2	27.2
0.310	0.294	0.289
199	294	388
1.15	1.84	2.87

Table 5.1.7
PROTODUR cables (PVC insulation) with copper conductor

All cores are loaded, operation at 50 Hz

Design				7	10	12	14	19
	1	Number of cores		7	10	12	14	19
	2	Nom. cross-sectional area of conductor	mm^2	1.5	1.5	1.5	1.5	1.5
	3	Shape and type of conductor		RE	RE	RE	RE	RE
	9	Thickness of insulation	mm	0.8	0.8	0.8	0.8	0.8
	16	Thickness of outer sheath	mm	1.8	1.8	1.8	1.8	1.8
	17	Overall diameter of cable	mm	16	19	19	20	22
	19	Weight of cable	kg/km	405	550	600	650	800
Mechanical properties	20	Minimum bending radius	mm	192	228	228	240	264
	21	Perm. pulling force with cable grip	N	525	750	900	1050	1425
Electrical properties	23	d.c. resistance/unit length at 70 °C	Ω/km	14.5	14.5	14.5	14.5	14.5
	24	d.c. resistance/unit length at 20 °C	Ω/km	12.1	12.1	12.1	12.1	12.1
Laying direct in ground	32	Current-carrying capacity	A	15.6	13.0	12.4	11.7	10.4
	34	Eff. a.c. resistance/unit length at 70 °C	Ω/km	14.5	14.5	14.5	14.5	14.5
Installation in free air	32	Current-carrying capacity	A	12.0	10.2	9.7	9.3	8.3
	34	Eff. a.c. resistance/unit length at 70 °C	Ω/km	14.5	14.5	14.5	14.5	14.5
Short-circuit	49	Rated short-time current of conductor (1s)	kA	0.173	0.173	0.173	0.173	0.173

24	30	40	7	10	12	14	19	24	30	40
1.5	1.5	1.5	2.5	2.5	2.5	2.5	2.5	2.5	2.5	2.5
RE	RE	RE	RE	RE	RE	RE	RE	RE	RE	RE
0.8	0.8	0.8	0.8	0.8	0.8	0.8	0.8	0.8	0.8	0.8
1.8	1.8	1.8	1.8	1.8	1.8	1.8	1.8	1.8	1.8	1.9
25	26	29	17	20	21	21	23	27	28	31
980	1150	1450	495	670	740	820	1050	1300	1500	1900
300	312	348	204	240	252	252	276	324	336	372
1800	2250	3000	875	1250	1500	1750	2375	3000	3750	5000
14.5	14.5	14.5	8.87	8.87	8.87	8.87	8.87	8.87	8.87	8.87
12.1	12.1	12.1	7.41	7.41	7.41	7.41	7.41	7.41	7.41	7.41
9.1	8.5	7.8	20	17.0	16.2	15.3	13.6	11.9	11.1	10.2
14.5	14.5	14.5	8.87	8.87	8.87	8.87	8.87	8.87	8.87	8.87
7.4	6.9	6.5	16.3	13.8	13.1	12.5	11.3	10.0	9.4	8.8
14.5	14.5	14.5	8.87	8.87	8.87	8.87	8.87	8.87	8.87	8.87
0.173	0.173	0.173	0.288	0.288	0.288	0.288	0.288	0.288	0.288	0.288

Table 5.1.8
PROTODUR cables (PVC insulation) with copper conductor

Three-phase operation at 50 Hz

Design				1.5	2.5	4	6	10
Design	2	Nom. cross-sectional area of conductor	mm²	1.5	2.5	4	6	10
	3	Shape and type of conductor		RE	RE	RE	RE	RE
	7	Nom. cross-sectional area of concentric cond.	mm²	1.5	2.5	4	6	10
	9	Thickness of insulation	mm	0.8	0.8	1.0	1.0	1.0
	16	Thickness of outer sheath	mm	1.8	1.8	1.8	1.8	1.8
	17	Overall diameter of cable	mm	13	14	17	18	20
	19	Weight of cable	kg/km	245	300	480	590	810
Mechanical properties	20	Minimum bending radius	mm	156	168	204	216	240
	21	Perm. pulling force with cable grip	N	225	375	600	900	1500
	22	Perm. pulling force with pulling head	N	225	375	600	900	1500
Electrical properties	23	d.c. resistance/unit length at 70 °C	Ω/km	14.5	8.87	5.52	3.69	2.19
	24	d.c. resistance/unit length at 20 °C	Ω/km	12.1	7.41	4.61	3.08	1.83
	31	Reference diameter of cable	mm	12.5	13.4	15.3	16.5	18.4
Laying direct in ground	32	Current-carrying capacity	A	26	34	44	56	75
	34	Eff. a.c. resistance/unit length at 70 °C	Ω/km	14.5	8.87	5.52	3.69	2.19
	38	Ohmic losses per cable	kW/km	29.4	30.7	32.0	34.7	37.0
	39	Fict. thermal resist. of cable for ohmic losses	Km/W	1.062	0.960	0.878	0.800	0.706
	41	Fict. thermal resist. of soil for ohmic losses	Km/W	0.700	0.689	0.668	0.656	0.638
	44	Inductance/unit length per conductor	mH/km	0.343	0.317	0.316	0.298	0.278
Installation in free air	32	Current-carrying capacity	A	18.4	24	33	41	57
	34	Eff. a.c. resistance/unit length at 70 °C	Ω/km	14.5	8.87	5.52	3.69	2.19
	38	Ohmic losses per cable	kW/km	14.7	15.3	18.0	18.6	21.3
	39	Fict. thermal resist. of cable for ohmic losses	Km/W	1.062	0.960	0.878	0.800	0.706
	43	Thermal resistance of the air	Km/W	1.645	1.553	1.398	1.313	1.199
	44	Inductance/unit length per conductor	mH/km	0.343	0.317	0.316	0.298	0.278
	48	Minimum time value	s	43	70	95	138	199
Short-circuit	49	Rated short-time current of conductor (1s)	kA	0.173	0.288	0.460	0.690	1.15

Note: Cable type NYCY up to 4 mm² and from 185 mm² upwards

NYCWY 3×…/.. 0.6/1 kV (U_m=1.2 kV)

16	25	35	50	70	95	120	150	185	240	300	400
RE	RM	SM	SM	SM	SM	SM	SM	SM	SM	SM	SM
16	16	16	25	35	50	70	70	95	120	150	185
1.0	1.2	1.2	1.4	1.4	1.6	1.6	1.8	2.0	2.2	2.4	2.6
1.8	1.8	1.8	1.9	2.0	2.2	2.3	2.4	2.6	2.8	3.0	3.3
22	27	26	30	34	39	41	46	51	57	63	70
1100	1600	1700	2250	3100	4100	5050	6100	7500	9650	11900	15000
264	324	312	360	408	468	492	552	612	684	756	840
2400	3750	5250	7500	10500	14250	18000	22500	27750	36000	45000	60000
2400	3750	5250	7500	10500	14250	18000	22500	27750	36000	45000	60000
1.38	0.870	0.627	0.463	0.321	0.231	0.183	0.148	0.119	0.0902	0.0719	0.0562
1.15	0.727	0.524	0.387	0.268	0.193	0.153	0.124	0.0991	0.0754	0.0601	0.0470
20.7	24.6	24.1	27.2	31.0	35.4	38.3	42.6	47.0	53.0	58.1	66.0
99	129	156	184	227	272	310	348	393	453	506	571
1.38	0.871	0.628	0.464	0.323	0.233	0.187	0.152	0.124	0.0969	0.0803	0.0665
40.5	43.5	45.8	47.2	49.9	51.8	53.9	55.4	57.4	59.7	61.7	65.0
0.620	0.548	0.431	0.406	0.371	0.350	0.323	0.317	0.305	0.293	0.278	0.256
0.620	1.480	1.488	1.440	1.388	1.335	1.304	1.262	1.222	1.175	1.138	1.087
0.262	0.257	0.248	0.247	0.238	0.238	0.233	0.233	0.233	0.231	0.231	0.228
75	101	124	151	190	235	272	311	356	419	476	550
1.38	0.871	0.628	0.464	0.323	0.233	0.187	0.152	0.124	0.0969	0.0803	0.0665
23.2	26.6	29.0	31.8	34.9	38.7	41.5	44.2	47.1	51.0	54.6	60.3
0.620	0.548	0.431	0.406	0.371	0.350	0.323	0.317	0.305	0.293	0.278	0.256
1.087	0.943	0.946	0.857	0.768	0.688	0.643	0.589	0.543	0.491	0.454	0.407
0.262	0.257	0.248	0.247	0.238	0.238	0.233	0.233	0.233	0.231	0.231	0.228
294	396	515	708	877	1056	1257	1503	1744	2119	2566	3416
1.84	2.87	4.02	5.75	8.05	10.9	13.8	17.2	21.3	27.6	34.5	41.2

Table 5.1.9
PROTODUR cables (PVC insulation) with copper conductor

Three-phase operation at 50 Hz

Design	2	Nom. cross-sectional area of conductor	mm^2	1.5	2.5	4	6	10
	3	Shape and type of conductor		RE	RE	RE	RE	RE
	7	Nom. cross-sectional area of concentric cond.	mm^2	1.5	2.5	4	6	10
	9	Thickness of insulation	mm	0.8	0.8	1.0	1.0	1.0
	16	Thickness of outer sheath	mm	1.8	1.8	1.8	1.8	1.8
	17	Overall diameter of cable	mm	14	15	18	19	22
	19	Weight of cable	kg/km	275	340	550	680	940
Mechanical properties	20	Minimum bending radius	mm	168	180	216	228	264
	21	Perm. pulling force with cable grip	N	300	500	800	1200	2000
	22	Perm. pulling force with pulling head	N	300	500	800	1200	2000
Electrical properties	23	d.c. resistance/unit length at 70 °C	Ω/km	14.5	8.87	5.52	3.69	2.19
	24	d.c. resistance/unit length at 20 °C	Ω/km	12.1	7.41	4.61	3.08	1.83
	31	Reference diameter of cable	mm	13.3	14.3	16.4	17.7	19.8
Laying direct in ground	32	Current-carrying capacity	A	26	34	45	56	76
	34	Eff. a.c. resistance/unit length at 70 °C	Ω/km	14.5	8.87	5.52	3.69	2.19
	38	Ohmic losses per cable	kW/km	29.4	30.7	33.5	34.7	37.9
	39	Fict. thermal resist. of cable for ohmic losses	Km/W	1.039	0.947	0.869	0.796	0.708
	41	Fict. thermal resist. of soil for ohmic losses	Km/W	0.690	0.678	0.656	0.644	0.627
	44	Inductance/unit length per conductor	mH/km	0.366	0.340	0.339	0.321	0.301
Installation in free air	32	Current-carrying capacity	A	18.8	25	33	42	57
	34	Eff. a.c. resistance/unit length at 70 °C	Ω/km	14.5	8.87	5.52	3.69	2.19
	38	Ohmic losses per cable	kW/km	15.4	16.6	18.0	19.5	21.3
	39	Fict. thermal resist. of cable for ohmic losses	Km/W	1.039	0.947	0.869	0.796	0.708
	43	Thermal resistance of the air	Km/W	1.569	1.478	1.326	1.244	1.134
	44	Inductance/unit length per conductor	mH/km	0.366	0.340	0.339	0.321	0.301
	48	Minimum time value	s	41	65	95	132	199
Short-circuit	49	Rated short-time current of conductor (1s)	kA	0.173	0.288	0.460	0.690	1.15

Note: Cable type NYCY up to 6 mm^2 and from 185 mm^2 upwards

16	25	35	50	70	95	120	150	185	240	300	400
RE	RM	SM	SM	SM	SM	SM	SM	SM	SM	SM	SM
16	16	16	25	35	50	70	70	95	120	150	185
1.0	1.2	1.2	1.4	1.4	1.6	1.6	1.8	2.0	2.2	2.4	2.6
1.8	1.8	1.8	2.0	2.1	2.3	2.4	2.6	2.8	3.0	3.2	3.5
24	30	29	34	38	43	47	51	57	64	70	79
1300	1900	2150	2850	3900	5200	6500	7750	9550	12250	15100	19200
288	360	348	408	456	516	564	612	684	768	840	948
3200	5000	7000	10000	14000	19000	24000	30000	37000	48000	60000	80000
3200	5000	7000	10000	14000	19000	24000	30000	37000	48000	60000	80000
1.38	0.870	0.627	0.463	0.321	0.231	0.183	0.148	0.119	0.0902	0.0719	0.0562
1.15	0.727	0.524	0.387	0.268	0.193	0.153	0.124	0.0991	0.0754	0.0601	0.0470
22.4	26.8	26.9	31.0	34.8	39.8	43.7	47.9	53.3	59.7	65.4	75.2
99	129	157	185	229	275	313	352	396	459	515	583
1.38	0.871	0.628	0.464	0.322	0.233	0.187	0.152	0.124	0.0968	0.0799	0.0660
40.5	43.5	46.4	47.7	50.7	53.0	54.8	56.6	58.2	61.2	63.6	67.3
0.626	0.554	0.438	0.439	0.384	0.360	0.341	0.332	0.328	0.304	0.288	0.267
0.607	1.446	1.445	1.388	1.342	1.289	1.251	1.215	1.172	1.127	1.091	1.035
0.285	0.280	0.271	0.270	0.262	0.261	0.256	0.256	0.256	0.254	0.254	0.251
77	103	128	154	195	241	279	318	363	429	488	565
1.38	0.871	0.628	0.464	0.322	0.233	0.187	0.152	0.124	0.0968	0.0799	0.0660
24.5	27.7	30.9	33.0	36.8	40.7	43.6	46.2	48.9	53.4	57.1	63.2
0.626	0.554	0.438	0.439	0.384	0.360	0.341	0.332	0.328	0.304	0.288	0.267
1.023	0.882	0.868	0.775	0.701	0.627	0.580	0.537	0.492	0.446	0.413	0.366
0.285	0.280	0.271	0.270	0.262	0.261	0.256	0.256	0.256	0.254	0.254	0.251
279	381	483	681	832	1004	1195	1437	1678	2021	2441	3237
1.84	2.87	4.02	5.75	8.05	10.9	13.8	17.2	21.3	27.6	34.5	41.2

Table 5.1.10
PROTODUR cables (PVC insulation) with copper conductor

All cores are loaded, operation at 50 Hz

Design				7	10	12	14	19
	1	Number of cores		7	10	12	14	19
	2	Nom. cross-sectional area of conductor	mm²	1.5	1.5	1.5	1.5	1.5
	3	Shape and type of conductor		RE	RE	RE	RE	RE
	7	Nom. cross-sectional area of concentric cond.	mm²	2.5	2.5	2.5	2.5	4
	9	Thickness of insulation	mm	0.8	0.8	0.8	0.8	0.8
	16	Thickness of outer sheath	mm	1.8	1.8	1.8	1.8	1.8
	17	Overall diameter of cable	mm	17	20	20	21	23
	19	Weight of cable	kg/km	490	590	640	700	860
Mechanical properties	20	Minimum bending radius	mm	204	240	240	252	276
	21	Perm. pulling force with cable grip	N	525	750	900	1050	1425
Electrical properties	23	d.c. resistance/unit length at 70 °C	Ω/km	14.5	14.5	14.5	14.5	14.5
	24	d.c. resistance/unit length at 20 °C	Ω/km	12.1	12.1	12.1	12.1	12.1
Laying direct in ground	32	Current-carrying capacity	A	15.6	13.0	12.4	11.7	10.4
	34	Eff. a.c. resistance/unit length at 70 °C	Ω/km	14.5	14.5	14.5	14.5	14.5
Installation in free air	32	Current-carrying capacity	A	12.0	10.2	9.7	9.3	8.3
	34	Eff. a.c. resistance/unit length at 70 °C	Ω/km	14.5	14.5	14.5	14.5	14.5
Short-circuit	49	Rated short-time current of conductor (1s)	kA	0.173	0.173	0.173	0.173	0.173

24	30	40	7	10	12	14	19	24	30	40
1.5	1.5	1.5	2.5	2.5	2.5	2.5	2.5	2.5	2.5	2.5
RE	RE	RE	RE	RE	RE	RE	RE	RE	RE	RE
6	6	10	2.5	4	4	6	6	10	10	10
0.8	0.8	0.8	0.8	0.8	0.8	0.8	0.8	0.8	0.8	0.8
1.8	1.8	1.8	1.8	1.8	1.8	1.8	1.8	1.8	1.8	1.9
26	27	31	18	21	22	23	25	28	30	33
1100	1250	1550	540	730	800	880	1100	1400	1600	2000
312	324	372	216	252	264	276	300	336	360	396
1800	2250	3000	875	1250	1500	1750	2375	3000	3750	5000
14.5	14.5	14.5	8.87	8.87	8.87	8.87	8.87	8.87	8.87	8.87
12.1	12.1	12.1	7.41	7.41	7.41	7.41	7.41	7.41	7.41	7.41
9.1	8.5	7.8	20	17.0	16.2	15.3	13.6	11.9	11.1	10.2
14.5	14.5	14.5	8.87	8.87	8.87	8.87	8.87	8.87	8.87	8.87
7.4	6.9	6.5	16.3	13.8	13.1	12.5	11.3	10.0	9.4	8.8
14.5	14.5	14.5	8.87	8.87	8.87	8.87	8.87	8.87	8.87	8.87
0.173	0.173	0.173	0.288	0.288	0.288	0.288	0.288	0.288	0.288	0.288

Table 5.1.11
PROTODUR cables (PVC insulation) with copper conductor

Single-phase a.c. operation at 50 Hz

Design				1.5	2.5	4	6	10
Design	2	Nom. cross-sectional area of conductor	mm²	1.5	2.5	4	6	10
	3	Shape and type of conductor		RE	RE	RE	RE	RE
	9	Thickness of insulation	mm	0.8	0.9	1.0	1.0	1.0
	12	Thickness of metal sheath	mm	1.2	1.2	1.2	1.2	1.2
	13	Diameter over metal sheath	mm	10.0	11.5	13.0	14.0	15.5
	16	Thickness of outer sheath	mm	1.4	1.4	1.4	1.4	1.4
	17	Overall diameter of cable	mm	13	15	16	17	19
	19	Weight of cable	kg/km	580	690	835	955	1180
Mechanical properties	20	Minimum bending radius	mm	156	180	192	204	228
	21	Perm. pulling force with cable grip	N	507	675	768	867	1083
	22	Perm. pulling force with pulling head	N	225	375	600	900	1500
Electrical properties	23	d.c. resistance/unit length at 70 °C	Ω/km	14.5	8.87	5.52	3.69	2.19
	24	d.c. resistance/unit length at 20 °C	Ω/km	12.1	7.41	4.61	3.08	1.83
Laying direct in ground	32	Current-carrying capacity	A	33	43	56	71	95
	34	Eff. a.c. resistance/unit length at 70 °C	Ω/km	14.5	8.87	5.52	3.69	2.19
	38	Ohmic losses per cable	kW/km	31.5	32.8	34.6	37.2	39.5
	44	Inductance/unit length per conductor	mH/km	0.343	0.328	0.316	0.298	0.278
Installation in free air	32	Current-carrying capacity	A	24	31	41	52	71
	34	Eff. a.c. resistance/unit length at 70 °C	Ω/km	14.5	8.87	5.52	3.69	2.19
	38	Ohmic losses per cable	kW/km	16.7	17.0	18.5	19.9	22.1
	44	Inductance/unit length per conductor	mH/km	0.343	0.328	0.316	0.298	0.278
	48	Minimum time value	s	25	42	61	86	128
Short-circuit	49	Rated short-time current of conductor (1s)	kA	0.173	0.288	0.460	0.690	1.15

NYKY-J 3×… 0.6/1 kV (U_m=1.2 kV)

16 RE	25 RM	35 SM	50 SM	70 SM	95 SM	120 SM	150 SM	185 SM	240 SM	300 SM	400 SM
1.0	1.2	1.2	1.4	1.4	1.6	1.6	1.8	2.0	2.2	2.4	2.6
1.3	1.3	1.3	1.4	1.5	1.5	1.6	1.7	1.9	2.1	2.2	2.5
18.0	22.5	23.0	26.0	29.0	33.5	36.0	40.0	44.5	50.0	55.0	62.0
1.4	1.4	1.4	1.4	1.4	1.4	1.4	1.4	1.6	1.6	1.8	2.0
21	26	27	30	33	37	40	44	49	54	60	67
1550	2230	2500	3200	4150	5250	6250	7500	9400	11850	14400	18250
252	312	324	360	396	444	480	528	588	648	720	804
1323	2028	2187	2700	3267	4107	4800	5808	7203	8748	10800	13467
2400	3750	5250	7500	10500	14250	18000	22500	27750	36000	45000	60000
1.38	0.870	0.627	0.463	0.321	0.231	0.183	0.148	0.119	0.0902	0.0719	0.0562
1.15	0.727	0.524	0.387	0.268	0.193	0.153	0.124	0.0991	0.0754	0.0601	0.0470
123	159	192	228	282	337	385	434	489	569	640	729
1.38	0.871	0.628	0.464	0.322	0.232	0.185	0.151	0.121	0.0937	0.0762	0.0617
41.6	44.0	46.3	48.2	51.2	52.8	54.8	56.7	58.0	60.7	62.5	65.6
0.262	0.257	0.248	0.247	0.238	0.238	0.233	0.233	0.233	0.231	0.231	0.228
95	127	155	189	239	292	340	391	447	532	605	705
1.38	0.871	0.628	0.464	0.322	0.232	0.185	0.151	0.121	0.0937	0.0762	0.0617
24.8	28.1	30.2	33.1	36.8	39.6	42.7	46.0	48.5	53.0	55.9	61.3
0.262	0.257	0.248	0.247	0.238	0.238	0.233	0.233	0.233	0.231	0.231	0.228
183	250	329	452	554	684	805	951	1106	1314	1588	2079
1.84	2.87	4.02	5.75	8.05	10.9	13.8	17.2	21.3	27.6	34.5	41.2

Table 5.1.12
PROTODUR cables (PVC insulation) with copper conductor

Three-phase operation at 50 Hz

Design				25	35	50	70
	2	Nom. cross-sectional area of conductor	mm^2	25	35	50	70
	3	Shape and type of conductor		RM	SM	SM	SM
	4	Nom. cross-sectional area of reduced conductor	mm^2	16	16	25	35
	5	Shape and type of reduced conductor		RE	RE	RM	SM
	9	Thickness of insulation	mm	1.2	1.2	1.4	1.4
	12	Thickness of metal sheath	mm	1.4	1.4	1.4	1.5
	13	Diameter over metal sheath	mm	24.5	25.0	29.5	32.0
	16	Thickness of outer sheath	mm	1.4	1.4	1.4	1.4
	17	Overall diameter of cable	mm	28	29	33	36
	19	Weight of cable	kg/km	2750	2900	3850	4600
Mechanical properties	20	Minimum bending radius	mm	336	348	396	432
	21	Perm. pulling force with cable grip	N	2352	2523	3267	3888
	22	Perm. pulling force with pulling head	N	4550	6050	8750	12250
Electrical properties	23	d.c. resistance/unit length at 70 °C	Ω/km	0.870	0.627	0.463	0.321
	24	d.c. resistance/unit length at 20 °C	Ω/km	0.727	0.524	0.387	0.268
	31	Reference diameter of cable	mm	25.9	25.2	28.2	31.4
Laying direct in ground	32	Current-carrying capacity	A	130	157	187	230
	34	Eff. a.c. resistance/unit length at 70 °C	Ω/km	0.871	0.628	0.464	0.322
	38	Ohmic losses per cable	kW/km	44.1	46.4	48.6	51.1
	44	Inductance/unit length per conductor	mH/km	0.274	0.261	0.263	0.254
Installation in free air	32	Current-carrying capacity	A	104	127	154	195
	34	Eff. a.c. resistance/unit length at 70 °C	Ω/km	0.871	0.627	0.464	0.322
	38	Ohmic losses per cable	kW/km	28.2	30.4	33.0	36.7
	44	Inductance/unit length per conductor	mH/km	0.274	0.261	0.263	0.254
	48	Minimum time value	s	373	491	681	832
Short-circuit	49	Rated short-time current of conductor (1s)	kA	2.87	4.02	5.75	8.05

95	120	150	185	240	300	400
SM	SM	SM	SM	SM	SM	SM
50	70	70	95	120	150	185
SM	SM	SM	SM	SM	SM	SM
1.6	1.6	1.8	2.0	2.2	2.4	2.6
1.6	1.7	1.8	2.0	2.2	2.4	2.8
36.5	39.5	44.0	48.5	54.5	60.5	69.0
1.4	1.4	1.6	1.6	1.8	2.0	2.0
40	43	49	53	60	66	75
5950	7150	8650	10650	13700	16800	21550
480	516	588	636	720	792	900
4800	5547	7203	8427	10800	13068	16875
16750	21500	26000	32500	42000	52500	69250
0.231	0.183	0.148	0.119	0.0902	0.0719	0.0562
0.193	0.153	0.124	0.0991	0.0754	0.0601	0.0470
35.7	38.3	42.2	47.2	52.5	57.9	65.2
275	314	354	400	465	523	597
0.232	0.185	0.150	0.121	0.0935	0.0759	0.0614
52.7	54.7	56.6	58.1	60.7	62.3	65.7
0.253	0.250	0.247	0.248	0.245	0.246	0.241
239	278	319	366	435	494	577
0.232	0.185	0.150	0.121	0.0935	0.0759	0.0614
39.8	42.8	45.9	48.7	53.1	55.6	61.3
0.253	0.250	0.247	0.248	0.245	0.246	0.241
1020	1203	1428	1650	1966	2382	3104
10.9	13.8	17.2	21.3	27.6	34.5	41.2

Table 5.1.13
PROTODUR cables (PVC insulation) with copper conductor

Three-phase operation at 50 Hz

Design				1.5	2.5	4	6	10
Design	2	Nom. cross-sectional area of conductor	mm²	1.5	2.5	4	6	10
	3	Shape and type of conductor		RE	RE	RE	RE	RE
	9	Thickness of insulation	mm	0.8	0.9	1.0	1.0	1.0
	12	Thickness of metal sheath	mm	1.2	1.2	1.2	1.2	1.2
	13	Diameter over metal sheath	mm	11.0	12.5	15.0	16.0	18.0
	16	Thickness of outer sheath	mm	1.4	1.4	1.4	1.4	1.4
	17	Overall diameter of cable	mm	14	15	18	19	21
	19	Weight of cable	kg/km	640	770	1050	1200	1500
Mechanical properties	20	Minimum bending radius	mm	168	180	216	228	252
	21	Perm. pulling force with cable grip	N	588	675	972	1083	1323
	22	Perm. pulling force with pulling head	N	300	500	800	1200	2000
Electrical properties	23	d.c. resistance/unit length at 70 °C	Ω/km	14.5	8.87	5.52	3.69	2.19
	24	d.c. resistance/unit length at 20 °C	Ω/km	12.1	7.41	4.61	3.08	1.83
	31	Reference diameter of cable	mm	14.0	15.5	17.1	18.3	20.2
Laying direct in ground	32	Current-carrying capacity	A	27	35	46	58	78
	34	Eff. a.c. resistance/unit length at 70 °C	Ω/km	14.5	8.87	5.52	3.69	2.19
	38	Ohmic losses per cable	kW/km	31.7	32.6	35.0	37.2	40.0
	39	Fict. thermal resist. of cable for ohmic losses	Km/W	0.916	0.839	0.773	0.708	0.629
	41	Fict. thermal resist. of soil for ohmic losses	Km/W	0.682	0.666	0.650	0.639	0.624
	44	Inductance/unit length per conductor	mH/km	0.366	0.351	0.339	0.321	0.301
Installation in free air	32	Current-carrying capacity	A	19.5	26	34	44	59
	34	Eff. a.c. resistance/unit length at 70 °C	Ω/km	14.5	8.87	5.52	3.69	2.19
	38	Ohmic losses per cable	kW/km	16.5	18.0	19.1	21.4	22.9
	39	Fict. thermal resist. of cable for ohmic losses	Km/W	0.916	0.839	0.773	0.708	0.629
	43	Thermal resistance of the air	Km/W	1.498	1.381	1.275	1.204	1.109
	44	Inductance/unit length per conductor	mH/km	0.366	0.351	0.339	0.321	0.301
	48	Minimum time value	s	38	60	89	120	186
Short-circuit	49	Rated short-time current of conductor (1s)	kA	0.173	0.288	0.460	0.690	1.15

16 RE	25 RM	35 SM	50 SM	70 SM	95 SM	120 SM	150 SM	185 SM	240 SM	300 SM	400 SM
1.0	1.2	1.2	1.4	1.4	1.6	1.6	1.8	2.0	2.2	2.4	2.6
1.3	1.4	1.4	1.5	1.5	1.6	1.7	1.9	2.1	2.3	2.5	2.8
21.0	26.0	26.0	29.5	33.5	38.0	41.0	46.0	50.5	57.0	62.5	71.0
1.4	1.4	1.4	1.4	1.4	1.4	1.6	1.6	1.8	1.8	2.0	2.2
24	29	30	33	37	42	46	50	55	62	68	77
2000	2850	3050	3900	5000	6500	7850	9700	11950	15200	18650	23800
288	348	360	396	444	504	552	600	660	744	816	924
1728	2523	2700	3267	4107	5292	6348	7500	9075	11532	13872	17787
3200	5000	7000	10000	14000	19000	24000	30000	37000	48000	60000	80000
1.38	0.870	0.627	0.463	0.321	0.231	0.183	0.148	0.119	0.0902	0.0719	0.0562
1.15	0.727	0.524	0.387	0.268	0.193	0.153	0.124	0.0991	0.0754	0.0601	0.0470
23.5	28.1	28.2	31.6	35.6	40.1	43.7	48.5	53.7	60.0	65.8	75.0
100	131	158	188	231	278	318	357	404	470	530	604
1.38	0.871	0.628	0.464	0.322	0.232	0.185	0.150	0.121	0.0936	0.0761	0.0616
41.3	44.8	47.0	49.2	51.5	53.9	56.1	57.5	59.3	62.0	64.1	67.4
0.606	0.539	0.424	0.396	0.364	0.335	0.312	0.313	0.303	0.287	0.276	0.260
0.599	1.427	1.426	1.380	1.333	1.286	1.251	1.210	1.169	1.125	1.088	1.036
0.285	0.280	0.271	0.270	0.262	0.261	0.256	0.256	0.256	0.254	0.254	0.251
78	105	130	158	199	245	285	324	373	442	505	588
1.38	0.871	0.628	0.464	0.322	0.232	0.185	0.150	0.121	0.0936	0.0761	0.0616
25.1	28.8	31.8	34.7	38.2	41.8	45.0	47.4	50.6	54.9	58.2	63.9
0.606	0.539	0.424	0.396	0.364	0.335	0.312	0.313	0.303	0.287	0.276	0.260
0.983	0.848	0.834	0.759	0.687	0.621	0.577	0.530	0.487	0.443	0.410	0.366
0.285	0.280	0.271	0.270	0.262	0.261	0.256	0.256	0.256	0.254	0.254	0.251
272	366	468	647	799	971	1145	1384	1589	1904	2279	2989
1.84	2.87	4.02	5.75	8.05	10.9	13.8	17.2	21.3	27.6	34.5	41.2

Table 5.1.14
PROTODUR cables (PVC insulation) with copper conductor

Three-phase operation at 50 Hz

Design					1.5	2.5	4	6
	2	Nom. cross-sectional area of conductor	mm^2		1.5	2.5	4	6
	3	Shape and type of conductor			RE	RE	RE	RE
	9	Thickness of insulation	mm		0.8	0.9	1.0	1.0
	12	Thickness of metal sheath	mm		1.2	1.2	1.2	1.2
	13	Diameter over metal sheath	mm		13.0	14.0	16.0	17.5
	16	Thickness of outer sheath	mm		1.4	1.4	1.4	1.4
	17	Overall diameter of cable	mm		16	17	19	20
	19	Weight of cable	kg/km		790	950	1150	1350
Mechanical properties	20	Minimum bending radius	mm		192	204	228	240
	21	Perm. pulling force with cable grip	N		768	867	1083	1200
	22	Perm. pulling force with pulling head	N		375	625	1000	1500
Electrical properties	23	d.c. resistance/unit length at 70 °C	Ω/km		14.5	8.87	5.52	3.69
	24	d.c. resistance/unit length at 20 °C	Ω/km		12.1	7.41	4.61	3.08
Laying direct in ground	32	Current-carrying capacity	A		27	35	46	58
	34	Eff. a.c. resistance/unit length at 70 °C	Ω/km		14.5	8.87	5.52	3.69
	38	Ohmic losses per cable	kW/km		31.7	32.6	35.0	37.2
	44	Inductance/unit length per conductor	mH/km		0.375	0.360	0.348	0.330
Installation in free air	32	Current-carrying capacity	A		19.5	26	34	44
	34	Eff. a.c. resistance/unit length at 70 °C	Ω/km		14.5	8.87	5.52	3.69
	38	Ohmic losses per cable	kW/km		16.5	18.0	19.1	21.4
	44	Inductance/unit length per conductor	mH/km		0.375	0.360	0.348	0.330
	48	Minimum time value	s		38	60	89	120
Short-circuit	49	Rated short-time current of conductor (1s)	kA		0.173	0.288	0.460	0.690

| 10 | 16 | 25 |
RE	RE	RM
1.0	1.0	1.2
1.3	1.3	1.5
20.0	23.0	29.5
1.4	1.4	1.4
23	26	33
1800	2300	3400
276	312	396
1587	2028	3267
2500	4000	6250
2.19	1.38	0.870
1.83	1.15	0.727
78	100	131
2.19	1.38	0.871
40.0	41.3	44.8
0.310	0.294	0.289
59	78	105
2.19	1.38	0.871
22.9	25.1	28.8
0.310	0.294	0.289
186	272	366
1.15	1.84	2.87

Table 5.1.15
PROTODUR cables (PVC insulation) with copper conductor

All cores are loaded, operation at 50 Hz

Design					7	10	12	14	19
	1	Number of cores			7	10	12	14	19
	2	Nom. cross-sectional area of conductor	mm²		1.5	1.5	1.5	1.5	1.5
	3	Shape and type of conductor			RE	RE	RE	RE	RE
	9	Thickness of insulation	mm		0.8	0.8	0.8	0.8	0.8
	12	Thickness of metal sheath	mm		1.2	1.2	1.2	1.2	1.3
	13	Diameter over metal sheath	mm		13.5	16.5	17.0	18.0	19.0
	16	Thickness of outer sheath	mm		1.4	1.4	1.4	1.4	1.4
	17	Overall diameter of cable	mm		17	20	20	21	23
	19	Weight of cable	kg/km		900	1200	1250	1300	1650
Mechanical properties	20	Minimum bending radius	mm		204	240	240	252	276
	21	Perm. pulling force with cable grip	N		867	1200	1200	1323	1587
Electrical properties	23	d.c. resistance/unit length at 70 °C	Ω/km		14.5	14.5	14.5	14.5	14.5
	24	d.c. resistance/unit length at 20 °C	Ω/km		12.1	12.1	12.1	12.1	12.1
Laying direct in ground	32	Current-carrying capacity	A		15.6	13.0	12.4	11.7	10.4
	34	Eff. a.c. resistance/unit length at 70 °C	Ω/km		14.5	14.5	14.5	14.5	14.5
Installation in free air	32	Current-carrying capacity	A		12.0	10.2	9.7	9.3	8.3
	34	Eff. a.c. resistance/unit length at 70 °C	Ω/km		14.5	14.5	14.5	14.5	14.5
Short-circuit	49	Rated short-time current of conductor (1s)	kA		0.173	0.173	0.173	0.173	0.173

24	30	40	7	10	12	14	19	24	30	40
1.5	1.5	1.5	2.5	2.5	2.5	2.5	2.5	2.5	2.5	2.5
RE	RE	RE	RE	RE	RE	RE	RE	RE	RE	RE
0.8	0.8	0.8	0.9	0.9	0.9	0.9	0.9	0.9	0.9	0.9
1.4	1.4	1.4	1.2	1.3	1.3	1.3	1.3	1.4	1.4	1.5
24.0	25.0	28.0	15.0	19.5	19.5	21.0	23.0	27.0	29.5	33.0
1.4	1.4	1.4	1.4	1.4	1.4	1.4	1.4	1.4	1.4	1.4
27	28	31	18	22	23	24	26	30	31	36
2000	2300	2600	1100	1500	1550	1800	2150	2450	2950	3700
324	336	372	216	264	276	288	312	360	372	432
2187	2352	2883	972	1452	1587	1728	2028	2700	2883	3888
14.5	14.5	14.5	8.87	8.87	8.87	8.87	8.87	8.87	8.87	8.87
12.1	12.1	12.1	7.41	7.41	7.41	7.41	7.41	7.41	7.41	7.41
9.1	8.5	7.8	20	17.0	16.2	15.3	13.6	11.9	11.1	10.2
14.5	14.5	14.5	8.87	8.87	8.87	8.87	8.87	8.87	8.87	8.87
7.4	6.9	6.4	16.3	13.8	13.1	12.5	11.3	10.0	9.4	8.8
14.5	14.5	14.5	8.87	8.87	8.87	8.87	8.87	8.87	8.87	8.87
0.173	0.173	0.173	0.288	0.288	0.288	0.288	0.288	0.288	0.288	0.288

Table 5.1.16
PROTODUR cables (PVC insulation) with copper conductor

Three-phase operation at 50 Hz

Design				25	35	50	70
Design	2	Nom. cross-sectional area of conductor	mm²	25	35	50	70
	3	Shape and type of conductor		RM	SM	SM	SM
	9	Thickness of insulation	mm	3.4	3.4	3.4	3.4
	14	Thickness of an armour wire	mm	0.8	0.8	0.8	0.8
	16	Thickness of outer sheath	mm	2.0	2.1	2.2	2.3
	17	Overall diameter of cable	mm	37	36	38	41
	19	Weight of cable	kg/km	2750	2500	2950	3750
Mechanical properties	20	Minimum bending radius	mm	555	540	570	615
	21	Perm. pulling force with cable grip	N	12321	11664	12996	15129
	22	Perm. pulling force with pulling head	N	3750	5250	7500	10500
Electrical properties	23	d.c. resistance/unit length at 70 °C	Ω/km	0.870	0.627	0.463	0.321
	24	d.c. resistance/unit length at 20 °C	Ω/km	0.727	0.524	0.387	0.268
	25	Operating capacitance/unit length	μF/km	0.430	0.480	0.550	0.620
	26	Charging current	A/km	0.470	0.520	0.600	0.670
	28	Earth fault current	A/km	0.880	1.01	1.18	1.31
	31	Reference diameter of cable	mm	35.4	33.5	35.7	38.9
Installation in ground	32	Current-carrying capacity	A	126	158	187	229
	33	Permissible transmission power	MVA	1.31	1.64	1.94	2.38
	34	Eff. a.c. resistance/unit length at 70 °C	Ω/km	0.872	0.628	0.464	0.322
	38	Ohmic losses per cable	kW/km	41.5	47.1	48.7	50.7
	39	Fict. thermal resist. of cable for ohmic losses	Km/W	0.678	0.512	0.473	0.424
	41	Fict. thermal resist. of soil for ohmic losses	Km/W	0.534	1.357	1.332	1.298
	44	Inductance/unit length per conductor	mH/km	0.383	0.363	0.346	0.328
Installation in free air	32	Current-carrying capacity	A	105	130	157	197
	33	Permissible transmission power	MVA	1.09	1.35	1.63	2.05
	34	Eff. a.c. resistance/unit length at 70 °C	Ω/km	0.872	0.628	0.464	0.322
	38	Ohmic losses per cable	kW/km	28.8	31.9	34.3	37.5
	39	Fict. thermal resist. of cable for ohmic losses	Km/W	0.678	0.512	0.473	0.424
	43	Thermal resistance of the air	Km/W	0.717	0.736	0.696	0.645
	44	Inductance/unit length per conductor	mH/km	0.383	0.363	0.346	0.328
	48	Minimum time value	s	366	468	655	815
Short-circuit	49	Rated short-time current of conductor (1s)	kA	2.87	4.02	5.75	8.05

95 SM	120 SM	150 SM	185 SM	240 SM	300 SM
3.4	3.4	3.4	3.4	3.4	3.4
0.8	0.8	0.8	0.8	0.8	0.8
2.4	2.6	2.7	2.8	2.9	3.1
44	47	50	53	58	62
4650	5500	6450	7650	9450	11400
660	705	750	795	870	930
17424	19881	22500	25281	30276	34596
14250	18000	22500	27750	36000	45000
0.231	0.183	0.148	0.119	0.0902	0.0719
0.193	0.153	0.124	0.0991	0.0754	0.0601
0.690	0.750	0.820	0.880	0.980	1.07
0.750	0.820	0.890	0.960	1.07	1.16
1.44	1.53	1.67	1.80	1.96	2.12
41.9	45.1	48.1	51.1	55.3	59.3
275	312	351	396	460	517
2.86	3.24	3.65	4.12	4.78	5.37
0.233	0.186	0.152	0.122	0.0948	0.0774
52.9	54.3	56.0	57.5	60.2	62.1
0.388	0.377	0.351	0.328	0.299	0.280
1.268	1.239	1.213	1.189	1.158	1.130
0.316	0.306	0.298	0.290	0.281	0.276
240	276	315	361	426	487
2.49	2.87	3.27	3.75	4.43	5.06
0.233	0.186	0.151	0.122	0.0948	0.0774
40.3	42.5	45.1	47.8	51.6	55.1
0.388	0.377	0.351	0.328	0.299	0.280
0.604	0.568	0.537	0.509	0.475	0.447
0.316	0.306	0.298	0.290	0.281	0.276
1012	1221	1465	1696	2050	2451
10.9	13.8	17.2	21.3	27.6	34.5

Table 5.1.17a
PROTODUR cables (PVC insulation) with copper conductor

Three-phase operation at 50 Hz

Design	2	Nom. cross-sectional area of conductor	mm^2	25	35	50	70
	3	Shape and type of conductor		RM	RM	RM	RM
	6	Nom. cross-sectional area of screen	mm^2	16	16	16	16
	9	Thickness of insulation	mm	4.0	4.0	4.0	4.0
	16	Thickness of outer sheath	mm	2.5	2.5	2.5	2.5
	17	Overall diameter of cable	mm	24	25	27	28
	19	Weight of cable	kg/km	990	1150	1300	1550
Mechanical properties	20	Minimum bending radius	mm	360	375	405	420
	21	Perm. pulling force with cable grip	N	1250	1750	2500	3500
	22	Perm. pulling force with pulling head	N	1250	1750	2500	3500
Electrical properties	23	d.c. resistance/unit length at 70 °C	Ω/km	0.870	0.627	0.463	0.321
	24	d.c. resistance/unit length at 20 °C	Ω/km	0.727	0.524	0.387	0.268
	25	Operating capacitance/unit length	μF/km	0.612	0.676	0.745	0.842
	26	Charging current	A/km	1.11	1.23	1.35	1.53
	28	Earth fault current	A/km	3.33	3.69	4.05	4.59
	29	Dielectric losses per cable	kW/km	0.6	0.7	0.8	0.9
	30	Electric field strength at the conductor	kV/mm	2.10	2.00	2.00	1.90
	31	Reference diameter of cable	mm	22.3	23.4	24.6	26.3
Installation in ground	32	Current-carrying capacity	A	136	162	191	233
	33	Permissible transmission power	MVA	2.36	2.81	3.31	4.04
	34	Eff. a.c. resistance/unit length at 70 °C	Ω/km	0.874	0.631	0.466	0.324
	36	Eff. a.c. resistance/unit length at 20 °C	Ω/km	0.731	0.528	0.390	0.271
	38	Ohmic losses per cable	kW/km	16.2	16.5	17.0	17.6
	39	Fict. thermal resist. of cable for ohmic losses	Km/W	1.197	1.083	0.984	0.870
	40	Fict. thermal resist. of cable for diel. losses	Km/W	0.721	0.659	0.603	0.539
	41	Fict. thermal resist. of soil for ohmic losses	Km/W	4.006	3.949	3.889	3.809
	42	Fict. thermal resist. of soil for diel. losses	Km/W	5.217	5.160	5.100	5.020
	44	Inductance/unit length per conductor	mH/km	0.474	0.450	0.428	0.405
	45	Resistance/unit length in zero-phase seq.system	Ω/km	1.596	1.391	1.251	1.130
	46	Reactance/unit length in zero-phase seq.system	Ω/km	0.556	0.550	0.545	0.540
	47	Reduction factor		0.531	0.534	0.536	0.540
Installation in free air	32	Current-carrying capacity	A	117	141	170	212
	33	Permissible transmission power	MVA	2.03	2.44	2.94	3.67
	34	Eff. a.c. resistance/unit length at 70 °C	Ω/km	0.874	0.631	0.466	0.324
	38	Ohmic losses per cable	kW/km	12.0	12.5	13.5	14.6
	39	Fict. thermal resist. of cable for ohmic losses	Km/W	1.475	1.346	1.232	1.102
	40	Fict. thermal resist. of cable for diel. losses	Km/W	0.893	0.822	0.758	0.685
	43	Thermal resistance of the air	Km/W	1.761	1.684	1.608	1.512
	44	Inductance/unit length per conductor	mH/km	0.474	0.450	0.428	0.405
	48	Minimum time value	s	295	398	559	704
Short-circuit	49	Rated short-time current of conductor (1s)	kA	2.87	4.02	5.75	8.05
	50	Rated short-time current of screen (1s)	kA	3.3	3.3	3.3	3.3

95 RM 16	120 RM 16	150 RM 25	185 RM 25	240 RM 25	300 RM 25	400 RM 35
4.0	4.0	4.0	4.0	4.0	4.0	4.0
2.5	2.5	2.5	2.5	2.5	2.5	2.5
30	31	33	34	37	39	43
1850	2100	2500	2850	3450	4100	5050
450	465	495	510	555	585	645
4750	6000	7500	9250	12000	15000	20000
4750	6000	7500	9250	12000	15000	20000
0.231	0.183	0.148	0.119	0.0902	0.0719	0.0562
0.193	0.153	0.124	0.0991	0.0754	0.0601	0.0470
0.932	1.02	1.10	1.19	1.33	1.44	1.62
1.69	1.84	2.00	2.16	2.42	2.62	2.94
5.07	5.52	6.00	6.48	7.26	7.86	8.82
1.0	1.1	1.2	1.2	1.4	1.5	1.7
1.80	1.80	1.80	1.70	1.70	1.70	1.70
27.9	29.4	30.9	32.5	35.0	37.0	40.2
278	315	351	396	456	511	570
4.82	5.46	6.08	6.86	7.90	8.85	9.87
0.234	0.187	0.154	0.124	0.0965	0.0786	0.0653
0.197	0.157	0.130	0.105	0.0817	0.0668	0.0561
18.1	18.5	19.0	19.5	20.1	20.5	21.2
0.787	0.722	0.663	0.612	0.547	0.503	0.439
0.492	0.454	0.423	0.394	0.356	0.331	0.297
3.739	3.676	3.617	3.557	3.468	3.402	3.303
4.950	4.887	4.828	4.768	4.679	4.613	4.514
0.387	0.373	0.360	0.349	0.335	0.326	0.312
1.052	1.010	0.768	0.742	0.717	0.700	0.536
0.536	0.534	0.286	0.284	0.281	0.280	0.171
0.543	0.546	0.394	0.396	0.399	0.401	0.306
257	297	337	386	455	520	597
4.45	5.14	5.84	6.69	7.88	9.01	10.3
0.235	0.187	0.154	0.124	0.0965	0.0787	0.0654
15.5	16.5	17.5	18.5	20.0	21.3	23.3
1.005	0.929	0.857	0.797	0.719	0.664	0.585
0.630	0.586	0.549	0.515	0.470	0.438	0.398
1.433	1.365	1.304	1.246	1.165	1.107	1.026
0.387	0.373	0.360	0.349	0.335	0.326	0.312
883	1054	1280	1484	1797	2150	2900
10.9	13.8	17.2	21.3	27.6	34.5	41.2
3.3	3.3	5.1	5.1	5.1	5.1	7.1

Table 5.1.17 b
PROTODUR cables (PVC insulation) with copper conductor

Three-phase operation at 50 Hz

Design	2	Nom. cross-sectional area of conductor	mm²	25	35	50	70
	3	Shape and type of conductor		RM	RM	RM	RM
	6	Nom. cross-sectional area of screen	mm²	16	16	16	16
	9	Thickness of insulation	mm	4.0	4.0	4.0	4.0
	16	Thickness of outer sheath	mm	2.5	2.5	2.5	2.5
	17	Overall diameter of cable	mm	24	25	27	28
	19	Weight of cable	kg/km	990	1150	1300	1550
Mechanical properties	20	Minimum bending radius	mm	360	375	405	420
	21	Perm. pulling force with cable grip	N	1250	1750	2500	3500
	22	Perm. pulling force with pulling head	N	1250	1750	2500	3500
Electrical properties	23	d.c. resistance/unit length at 70 °C	Ω/km	0.870	0.627	0.463	0.321
	24	d.c. resistance/unit length at 20 °C	Ω/km	0.727	0.524	0.387	0.268
	25	Operating capacitance/unit length	μF/km	0.612	0.676	0.745	0.842
	26	Charging current	A/km	1.11	1.23	1.35	1.53
	28	Earth fault current	A/km	3.33	3.69	4.05	4.59
	29	Dielectric losses per cable	kW/km	0.6	0.7	0.8	0.9
	30	Electric field strength at the conductor	kV/mm	2.10	2.00	2.00	1.90
	31	Reference diameter of cable	mm	22.3	23.4	24.6	26.3
Installation in ground	32	Current-carrying capacity	A	153	182	215	261
	33	Permissible transmission power	MVA	2.65	3.15	3.72	4.52
	34	Eff. a.c. resistance/unit length at 70 °C	Ω/km	0.890	0.646	0.482	0.339
	36	Eff. a.c. resistance/unit length at 20 °C	Ω/km	0.747	0.543	0.406	0.286
	38	Ohmic losses per cable	kW/km	20.8	21.4	22.3	23.1
	39	Fict. thermal resist. of cable for ohmic losses	Km/W	1.179	1.062	0.959	0.842
	40	Fict. thermal resist. of cable for diel. losses	Km/W	0.721	0.659	0.603	0.539
	41	Fict. thermal resist. of soil for ohmic losses	Km/W	1.151	2.849	2.819	2.778
	42	Fict. thermal resist. of soil for diel. losses	Km/W	1.635	4.060	4.030	3.989
	44	Inductance/unit length per conductor	mH/km	0.777	0.745	0.717	0.684
	45	Resistance/unit length in zero-phase seq.system	Ω/km	1.551	1.347	1.208	1.087
	46	Reactance/unit length in zero-phase seq.system	Ω/km	0.574	0.567	0.562	0.556
	47	Reduction factor		0.577	0.578	0.580	0.582
Installation in free air	32	Current-carrying capacity	A	138	168	202	253
	33	Permissible transmission power	MVA	2.39	2.91	3.50	4.38
	34	Eff. a.c. resistance/unit length at 70 °C	Ω/km	0.884	0.641	0.477	0.334
	35	Mean eff. a.c. resistance/unit length at 70 °C	Ω/km	0.881	0.638	0.474	0.332
	38	Ohmic losses per cable	kW/km	16.8	18.0	19.3	21.2
	39	Fict. thermal resist. of cable for ohmic losses	Km/W	1.186	1.069	0.967	0.850
	40	Fict. thermal resist. of cable for diel. losses	Km/W	0.721	0.659	0.603	0.539
	43	Thermal resistance of the air	Km/W	1.111	1.063	1.016	0.956
	44	Inductance/unit length per conductor	mH/km	0.636	0.612	0.590	0.566
	48	Minimum time value	s	212	280	396	494
Short-circuit	49	Rated short-time current of conductor (1s)	kA	2.87	4.02	5.75	8.05
	50	Rated short-time current of screen (1s)	kA	3.3	3.3	3.3	3.3

95	120	150	185	240	300	400
RM	RM	RM	RM	RM	RM	RM
16	16	25	25	25	25	35
4.0	4.0	4.0	4.0	4.0	4.0	4.0
2.5	2.5	2.5	2.5	2.5	2.5	2.5
30	31	33	34	37	39	43
1850	2100	2500	2850	3450	4100	5050
450	465	495	510	555	585	645
4750	6000	7500	9250	12000	15000	20000
4750	6000	7500	9250	12000	15000	20000
0.231	0.183	0.148	0.119	0.0902	0.0719	0.0562
0.193	0.153	0.124	0.0991	0.0754	0.0601	0.0470
0.932	1.02	1.10	1.19	1.33	1.44	1.62
1.69	1.84	2.00	2.16	2.42	2.62	2.94
5.07	5.52	6.00	6.48	7.26	7.86	8.82
1.0	1.1	1.2	1.2	1.4	1.5	1.7
1.80	1.80	1.80	1.70	1.70	1.70	1.70
27.9	29.4	30.9	32.5	35.0	37.0	40.2
310	349	381	426	485	538	579
5.37	6.04	6.60	7.38	8.40	9.32	10.0
0.248	0.200	0.173	0.142	0.112	0.0936	0.0840
0.210	0.169	0.148	0.122	0.0977	0.0818	0.0748
23.8	24.3	25.1	25.8	26.5	27.1	28.2
0.754	0.687	0.608	0.554	0.486	0.439	0.362
0.492	0.454	0.423	0.394	0.356	0.331	0.297
2.742	2.709	2.677	2.645	2.596	2.559	2.503
3.953	3.920	3.888	3.856	3.807	3.770	3.714
0.657	0.636	0.610	0.593	0.570	0.554	0.524
1.011	0.969	0.751	0.726	0.701	0.685	0.529
0.552	0.549	0.299	0.297	0.294	0.292	0.179
0.584	0.586	0.438	0.439	0.441	0.442	0.348
307	354	397	453	531	602	670
5.32	6.13	6.88	7.85	9.20	10.4	11.6
0.245	0.197	0.169	0.139	0.111	0.0929	0.0847
0.242	0.194	0.165	0.135	0.107	0.0885	0.0784
22.8	24.3	26.0	27.7	30.1	32.1	35.2
0.762	0.694	0.617	0.562	0.491	0.442	0.361
0.492	0.454	0.423	0.394	0.356	0.331	0.297
0.907	0.865	0.825	0.789	0.737	0.701	0.648
0.548	0.534	0.519	0.508	0.494	0.485	0.467
618	742	922	1077	1319	1604	2302
10.9	13.8	17.2	21.3	27.6	34.5	41.2
3.3	3.3	5.1	5.1	5.1	5.1	7.1

Table 5.1.18
PROTODUR cables (PVC insulation) with copper conductor

Three-phase operation at 50 Hz

Design	2	Nom. cross-sectional area of conductor	mm^2	25	35	50	70
	3	Shape and type of conductor		RM	RM	RM	RM
	6	Nom. cross-sectional area of screen	mm^2	16	16	16	16
	9	Thickness of insulation	mm	4.0	4.0	4.0	4.0
	16	Thickness of outer sheath	mm	2.5	2.5	2.5	2.6
	17	Overall diameter of cable	mm	45	47	50	53
	19	Weight of cable	kg/km	3350	3850	4500	5400
Mechanical properties	20	Minimum bending radius	mm	675	705	750	795
	21	Perm. pulling force with cable grip	N	3750	5250	7500	10500
	22	Perm. pulling force with pulling head	N	3750	5250	7500	10500
Electrical properties	23	d.c. resistance/unit length at 70 °C	Ω/km	0.870	0.627	0.463	0.321
	24	d.c. resistance/unit length at 20 °C	Ω/km	0.727	0.524	0.387	0.268
	25	Operating capacitance/unit length	μF/km	0.612	0.676	0.745	0.842
	26	Charging current	A/km	1.11	1.23	1.35	1.53
	28	Earth fault current	A/km	3.33	3.69	4.05	4.59
	29	Dielectric losses per cable	kW/km	1.9	2.1	2.3	2.6
	30	Electric field strength at the conductor	kV/mm	2.10	2.00	2.00	1.90
	31	Reference diameter of cable	mm	45.2	47.5	50.1	54.4
Installation in ground	32	Current-carrying capacity	A	132	159	189	230
	33	Permissible transmission power	MVA	2.29	2.75	3.27	3.98
	34	Eff. a.c. resistance/unit length at 70 °C	Ω/km	0.871	0.628	0.464	0.322
	38	Ohmic losses per cable	kW/km	45.5	47.6	49.7	51.1
	39	Fict. thermal resist. of cable for ohmic losses	Km/W	0.556	0.515	0.479	0.448
	40	Fict. thermal resist. of cable for diel. losses	Km/W	0.369	0.343	0.321	0.304
	41	Fict. thermal resist. of soil for ohmic losses	Km/W	0.495	0.487	1.197	1.164
	42	Fict. thermal resist. of soil for diel. losses	Km/W	0.657	0.649	1.601	1.568
	44	Inductance/unit length per conductor	mH/km	0.404	0.382	0.363	0.342
Installation in free air	32	Current-carrying capacity	A	113	137	165	204
	33	Permissible transmission power	MVA	1.96	2.37	2.86	3.53
	34	Eff. a.c. resistance/unit length at 70 °C	Ω/km	0.871	0.628	0.464	0.322
	38	Ohmic losses per cable	kW/km	33.4	35.3	37.9	40.2
	39	Fict. thermal resist. of cable for ohmic losses	Km/W	0.556	0.515	0.479	0.448
	40	Fict. thermal resist. of cable for diel. losses	Km/W	0.369	0.343	0.321	0.304
	43	Thermal resistance of the air	Km/W	0.581	0.556	0.530	0.493
	44	Inductance/unit length per conductor	mH/km	0.404	0.382	0.363	0.342
	48	Minimum time value	s	316	422	593	760
Short-circuit	49	Rated short-time current of conductor (1s)	kA	2.87	4.02	5.75	8.05
	50	Rated short-time current of screen (1s)	kA	3.3	3.3	3.3	3.3

95 RM 16	120 RM 16	150 RM 25	185 RM 25	240 RM 25	300 RM 25
4.0	4.0	4.0	4.0	4.0	4.0
2.8	2.9	3.0	3.1	3.3	3.4
57	61	64	68	74	79
6550	7600	8900	10300	12700	15000
855	915	960	1020	1110	1185
14250	18000	22500	27750	36000	45000
14250	18000	22500	27750	36000	45000
0.231	0.183	0.148	0.119	0.0902	0.0719
0.193	0.153	0.124	0.0991	0.0754	0.0601
0.932	1.02	1.10	1.19	1.33	1.44
1.69	1.84	2.00	2.16	2.42	2.62
5.07	5.52	6.00	6.48	7.26	7.86
2.9	3.2	3.5	3.7	4.2	4.5
1.80	1.80	1.80	1.70	1.70	1.70
58.1	61.7	65.1	68.8	75.0	79.5
275	312	350	394	455	512
4.76	5.40	6.06	6.82	7.88	8.87
0.232	0.185	0.150	0.121	0.0934	0.0758
52.7	53.9	55.3	56.4	58.0	59.6
0.420	0.401	0.381	0.364	0.348	0.331
0.287	0.276	0.264	0.253	0.245	0.233
1.138	1.114	1.093	1.071	1.036	1.013
1.542	1.518	1.496	1.474	1.440	1.417
0.326	0.315	0.305	0.296	0.284	0.277
248	284	322	368	430	490
4.30	4.92	5.58	6.37	7.45	8.49
0.232	0.185	0.150	0.121	0.0934	0.0758
42.8	44.7	46.8	49.2	51.8	54.6
0.420	0.401	0.381	0.364	0.348	0.331
0.287	0.276	0.264	0.253	0.245	0.233
0.466	0.442	0.421	0.401	0.373	0.354
0.326	0.315	0.305	0.296	0.284	0.277
948	1153	1402	1632	2012	2421
10.9	13.8	17.2	21.3	27.6	34.5
3.3	3.3	5.1	5.1	5.1	5.1

Table 5.2.1a
PROTOTHEN X cables (XLPE insulation) with copper conductor

Three-phase operation at 50 Hz

Design							
	2	Nom. cross-sectional area of conductor	mm²	25	35	50	70
	3	Shape and type of conductor		RM	RM	RM	RM
	5	Nom. cross-sectional area of screen	mm²	16	16	16	16
	9	Thickness of insulation	mm	3.4	3.4	3.4	3.4
	16	Thickness of outer sheath	mm	2.5	2.5	2.5	2.5
	17	Overall diameter of cable	mm	23	24	26	27
	19	Weight of cable	kg/km	720	830	970	1200
Mechanical properties	20	Minimum bending radius	mm	345	360	390	405
	21	Perm. pulling force with cable grip	N	1250	1750	2500	3500
	22	Perm. pulling force with pulling head	N	1250	1750	2500	3500
Electrical properties	23	d.c. resistance/unit length at 90 °C	Ω/km	0.927	0.668	0.493	0.342
	24	d.c. resistance/unit length at 20 °C	Ω/km	0.727	0.524	0.387	0.268
	25	Operating capacitance/unit length	μF/km	0.203	0.225	0.249	0.283
	26	Charging current	A/km	0.370	0.410	0.450	0.510
	28	Earth fault current	A/km	1.11	1.23	1.35	1.53
	30	Electric field strength at the conductor	kV/mm	2.40	2.30	2.20	2.20
	31	Reference diameter of cable	mm	22.9	24.0	25.2	26.9
Installation in ground	32	Current-carrying capacity	A	157	187	221	269
	33	Permissible transmission power	MVA	2.72	3.24	3.83	4.66
	34	Eff. a.c. resistance/unit length at 90 °C	Ω/km	0.931	0.672	0.497	0.345
	36	Eff. a.c. resistance/unit length at 20 °C	Ω/km	0.731	0.528	0.390	0.271
	38	Ohmic losses per cable	kW/km	22.9	23.5	24.3	25.0
	39	Fict. thermal resist. of cable for ohmic losses	Km/W	0.706	0.640	0.581	0.515
	41	Fict. thermal resist. of soil for ohmic losses	Km/W	3.974	3.918	3.860	3.782
	44	Inductance/unit length per conductor	mH/km	0.480	0.455	0.434	0.409
	45	Resistance/unit length in zero-phase seq.system	Ω/km	1.595	1.390	1.251	1.129
	46	Reactance/unit length in zero-phase seq.system	Ω/km	0.555	0.549	0.544	0.539
	47	Reduction factor		0.556	0.559	0.562	0.565
Installation in free air	32	Current-carrying capacity	A	163	198	237	295
	33	Permissible transmission power	MVA	2.82	3.43	4.10	5.11
	34	Eff. a.c. resistance/unit length at 90 °C	Ω/km	0.931	0.672	0.497	0.345
	38	Ohmic losses per cable	kW/km	24.7	26.3	27.9	30.0
	39	Fict. thermal resist. of cable for ohmic losses	Km/W	0.869	0.794	0.726	0.652
	43	Thermal resistance of the air	Km/W	1.559	1.492	1.426	1.343
	44	Inductance/unit length per conductor	mH/km	0.480	0.455	0.433	0.409
	48	Minimum time value	s	214	284	405	512
Short-circuit	49	Rated short-time current of conductor (1s)	kA	3.57	5.00	7.15	10.0
	50	Rated short-time current of screen (1s)	kA	3.3	3.3	3.3	3.3

95 RM 16	120 RM 16	150 RM 25	185 RM 25	240 RM 25	300 RM 25	400 RM 35	500 RM 35
3.4	3.4	3.4	3.4	3.4	3.4	3.4	3.4
2.5	2.5	2.5	2.5	2.5	2.5	2.5	2.5
29	30	32	33	36	39	42	45
1450	1700	2050	2450	3000	3600	4500	5550
435	450	480	495	540	585	630	675
4750	6000	7500	9250	12000	15000	20000	25000
4750	6000	7500	9250	12000	15000	20000	25000
0.246	0.195	0.158	0.126	0.0961	0.0766	0.0599	0.0467
0.193	0.153	0.124	0.0991	0.0754	0.0601	0.0470	0.0366
0.315	0.345	0.374	0.406	0.456	0.495	0.558	0.613
0.570	0.630	0.680	0.740	0.830	0.900	1.01	1.11
1.71	1.89	2.04	2.22	2.49	2.70	3.03	3.33
2.10	2.10	2.00	2.00	2.00	1.90	1.90	1.90
28.5	30.0	31.5	33.1	35.6	37.6	40.8	43.6
321	364	406	457	528	593	665	745
5.56	6.30	7.03	7.92	9.15	10.3	11.5	12.9
0.250	0.199	0.164	0.132	0.102	0.0831	0.0686	0.0563
0.197	0.157	0.129	0.105	0.0814	0.0666	0.0557	0.0462
25.7	26.3	27.0	27.6	28.5	29.2	30.3	31.2
0.466	0.428	0.393	0.363	0.325	0.299	0.262	0.238
3.713	3.652	3.594	3.535	3.448	3.383	3.285	3.206
0.391	0.377	0.364	0.353	0.338	0.329	0.315	0.306
1.051	1.009	0.767	0.742	0.717	0.700	0.536	0.525
0.536	0.533	0.285	0.283	0.281	0.279	0.171	0.169
0.568	0.571	0.416	0.418	0.421	0.423	0.325	0.327
358	413	469	535	633	722	830	949
6.20	7.15	8.12	9.27	11.0	12.5	14.4	16.4
0.250	0.199	0.164	0.132	0.102	0.0831	0.0687	0.0564
32.0	33.9	36.0	37.8	41.0	43.3	47.3	50.8
0.595	0.550	0.508	0.473	0.427	0.396	0.350	0.318
1.274	1.215	1.162	1.111	1.040	0.989	0.918	0.864
0.391	0.377	0.364	0.353	0.338	0.329	0.315	0.306
640	767	930	1087	1307	1570	2111	2524
13.6	17.2	21.4	26.5	34.3	42.9	57.2	71.5
3.3	3.3	5.1	5.1	5.1	5.1	7.1	7.1

Table 5.2.1b
PROTOTHEN X cables (XLPE insulation) with copper conductor

Three-phase operation at 50 Hz

Design				25	35	50	70
Design	2	Nom. cross-sectional area of conductor	mm^2	25	35	50	70
	3	Shape and type of conductor		RM	RM	RM	RM
	6	Nom. cross-sectional area of screen	mm^2	16	16	16	16
	9	Thickness of insulation	mm	3.4	3.4	3.4	3.4
	16	Thickness of outer sheath	mm	2.5	2.5	2.5	2.5
	17	Overall diameter of cable	mm	23	24	26	27
	19	Weight of cable	kg/km	720	830	970	1200
Mechanical properties	20	Minimum bending radius	mm	345	360	390	405
	21	Perm. pulling force with cable grip	N	1250	1750	2500	3500
	22	Perm. pulling force with pulling head	N	1250	1750	2500	3500
Electrical properties	23	d.c. resistance/unit length at 90 °C	Ω/km	0.927	0.668	0.493	0.342
	24	d.c. resistance/unit length at 20 °C	Ω/km	0.727	0.524	0.387	0.268
	25	Operating capacitance/unit length	μF/km	0.203	0.225	0.249	0.283
	26	Charging current	A/km	0.370	0.410	0.450	0.510
	28	Earth fault current	A/km	1.11	1.23	1.35	1.53
	30	Electric field strength at the conductor	kV/mm	2.40	2.30	2.20	2.20
	31	Reference diameter of cable	mm	22.9	24.0	25.2	26.9
Installation in ground	32	Current-carrying capacity	A	179	213	250	303
	33	Permissible transmission power	MVA	3.10	3.69	4.33	5.25
	34	Eff. a.c. resistance/unit length at 90 °C	Ω/km	0.946	0.687	0.511	0.359
	36	Eff. a.c. resistance/unit length at 20 °C	Ω/km	0.746	0.543	0.405	0.285
	38	Ohmic losses per cable	kW/km	30.3	31.2	32.0	32.9
	39	Fict. thermal resist. of cable for ohmic losses	Km/W	0.697	0.629	0.569	0.500
	41	Fict. thermal resist. of soil for ohmic losses	Km/W	2.862	2.834	2.804	2.764
	44	Inductance/unit length per conductor	mH/km	0.779	0.747	0.719	0.685
	45	Resistance/unit length in zero-phase seq.system	Ω/km	1.551	1.347	1.208	1.087
	46	Reactance/unit length in zero-phase seq.system	Ω/km	0.572	0.566	0.561	0.555
	47	Reduction factor		0.600	0.602	0.604	0.606
Installation in free air	32	Current-carrying capacity	A	195	236	284	353
	33	Permissible transmission power	MVA	3.38	4.09	4.92	6.11
	34	Eff. a.c. resistance/unit length at 90 °C	Ω/km	0.941	0.682	0.507	0.355
	35	Mean eff. a.c. resistance/unit length at 90 °C	Ω/km	0.938	0.679	0.504	0.352
	38	Ohmic losses per cable	kW/km	35.7	37.8	40.7	43.9
	39	Fict. thermal resist. of cable for ohmic losses	Km/W	0.700	0.632	0.572	0.504
	43	Thermal resistance of the air	Km/W	0.983	0.941	0.900	0.849
	44	Inductance/unit length per conductor	mH/km	0.642	0.617	0.595	0.570
	48	Minimum time value	s	149	200	282	357
Short-circuit	49	Rated short-time current of conductor (1s)	kA	3.57	5.00	7.15	10.0
	50	Rated short-time current of screen (1s)	kA	3.3	3.3	3.3	3.3

95 RM 16	120 RM 16	150 RM 25	185 RM 25	240 RM 25	300 RM 25	400 RM 35	500 RM 35
3.4	3.4	3.4	3.4	3.4	3.4	3.4	3.4
2.5	2.5	2.5	2.5	2.5	2.5	2.5	2.5
29	30	32	33	36	39	42	45
1450	1700	2050	2450	3000	3600	4500	5550
435	450	480	495	540	585	630	675
4750	6000	7500	9250	12000	15000	20000	25000
4750	6000	7500	9250	12000	15000	20000	25000
0.246	0.195	0.158	0.126	0.0961	0.0766	0.0599	0.0467
0.193	0.153	0.124	0.0991	0.0754	0.0601	0.0470	0.0366
0.315	0.345	0.374	0.406	0.456	0.495	0.558	0.613
0.570	0.630	0.680	0.740	0.830	0.900	1.01	1.11
1.71	1.89	2.04	2.22	2.49	2.70	3.03	3.33
2.10	2.10	2.00	2.00	2.00	1.90	1.90	1.90
28.5	30.0	31.5	33.1	35.6	37.6	40.8	43.6
359	405	443	494	564	627	677	749
6.22	7.01	7.67	8.56	9.77	10.9	11.7	13.0
0.262	0.211	0.181	0.149	0.117	0.0972	0.0864	0.0724
0.209	0.169	0.147	0.121	0.0967	0.0807	0.0735	0.0623
33.8	34.6	35.5	36.3	37.3	38.2	39.6	40.6
0.449	0.410	0.365	0.334	0.294	0.267	0.222	0.198
2.728	2.696	2.665	2.633	2.585	2.548	2.492	2.446
0.659	0.638	0.613	0.595	0.572	0.556	0.526	0.511
1.011	0.969	0.751	0.726	0.701	0.685	0.529	0.519
0.551	0.548	0.298	0.296	0.293	0.291	0.179	0.177
0.608	0.610	0.460	0.461	0.462	0.463	0.366	0.367
429	493	552	627	736	833	926	1044
7.43	8.54	9.56	10.9	12.7	14.4	16.0	18.1
0.259	0.208	0.178	0.147	0.116	0.0968	0.0873	0.0742
0.257	0.206	0.174	0.142	0.112	0.0926	0.0814	0.0684
47.3	50.0	53.0	55.9	60.7	64.3	69.8	74.6
0.453	0.414	0.369	0.337	0.296	0.268	0.221	0.196
0.805	0.769	0.735	0.703	0.658	0.626	0.580	0.546
0.552	0.539	0.523	0.512	0.497	0.489	0.471	0.462
446	539	671	791	967	1179	1696	2085
13.6	17.2	21.4	26.5	34.3	42.9	57.2	71.5
3.3	3.3	5.1	5.1	5.1	5.1	7.1	7.1

Table 5.2.2 a
PROTOTHEN X cables (XLPE insulation) with copper conductor

Three-phase operation at 50 Hz

Design	2	Nom. cross-sectional area of conductor	mm^2	25	35	50	70
	3	Shape and type of conductor		RM	RM	RM	RM
	6	Nom. cross-sectional area of screen	mm^2	16	16	16	16
	9	Thickness of insulation	mm	5.5	5.5	5.5	5.5
	16	Thickness of outer sheath	mm	2.5	2.5	2.5	2.5
	17	Overall diameter of cable	mm	28	29	30	31
	19	Weight of cable	kg/km	870	990	1150	1400
Mechanical properties	20	Minimum bending radius	mm	420	435	450	465
	21	Perm. pulling force with cable grip	N	1250	1750	2500	3500
	22	Perm. pulling force with pulling head	N	1250	1750	2500	3500
Electrical properties	23	d.c. resistance/unit length at 90 °C	Ω/km	0.927	0.668	0.493	0.342
	24	d.c. resistance/unit length at 20 °C	Ω/km	0.727	0.524	0.387	0.268
	25	Operating capacitance/unit length	μF/km	0.145	0.159	0.175	0.196
	26	Charging current	A/km	0.530	0.580	0.630	0.710
	28	Earth fault current	A/km	1.59	1.74	1.89	2.13
	30	Electric field strength at the conductor	kV/mm	3.40	3.30	3.20	3.00
	31	Reference diameter of cable	mm	27.1	28.2	29.4	31.1
Installation in ground	32	Current-carrying capacity	A	159	189	223	272
	33	Permissible transmission power	MVA	5.51	6.55	7.72	9.42
	34	Eff. a.c. resistance/unit length at 90 °C	Ω/km	0.930	0.671	0.497	0.345
	36	Eff. a.c. resistance/unit length at 20 °C	Ω/km	0.730	0.527	0.390	0.271
	38	Ohmic losses per cable	kW/km	23.5	24.0	24.7	25.5
	39	Fict. thermal resist. of cable for ohmic losses	Km/W	0.809	0.737	0.674	0.602
	41	Fict. thermal resist. of soil for ohmic losses	Km/W	3.773	3.726	3.676	3.609
	44	Inductance/unit length per conductor	mH/km	0.514	0.488	0.465	0.438
	45	Resistance/unit length in zero-phase seq.system	Ω/km	1.588	1.383	1.244	1.122
	46	Reactance/unit length in zero-phase seq.system	Ω/km	0.572	0.566	0.560	0.554
	47	Reduction factor		0.558	0.561	0.564	0.567
Installation in free air	32	Current-carrying capacity	A	165	200	239	298
	33	Permissible transmission power	MVA	5.72	6.93	8.28	10.3
	34	Eff. a.c. resistance/unit length at 90 °C	Ω/km	0.930	0.671	0.497	0.345
	38	Ohmic losses per cable	kW/km	25.3	26.9	28.4	30.6
	39	Fict. thermal resist. of cable for ohmic losses	Km/W	0.993	0.911	0.838	0.755
	43	Thermal resistance of the air	Km/W	1.374	1.322	1.270	1.205
	44	Inductance/unit length per conductor	mH/km	0.514	0.488	0.465	0.438
	48	Minimum time value	s	209	278	398	502
Short-circuit	49	Rated short-time current of conductor (1s)	kA	3.57	5.00	7.15	10.0
	50	Rated short-time current of screen (1s)	kA	3.3	3.3	3.3	3.3

95 RM 16	120 RM 16	150 RM 25	185 RM 25	240 RM 25	300 RM 25	400 RM 35	500 RM 35
5.5	5.5	5.5	5.5	5.5	5.5	5.5	5.5
2.5	2.5	2.5	2.5	2.5	2.5	2.5	2.5
33	34	36	38	40	43	46	49
1650	1900	2300	2650	3200	3850	4800	5850
495	510	540	570	600	645	690	735
4750	6000	7500	9250	12000	15000	20000	25000
4750	6000	7500	9250	12000	15000	20000	25000
0.246	0.195	0.158	0.126	0.0961	0.0766	0.0599	0.0467
0.193	0.153	0.124	0.0991	0.0754	0.0601	0.0470	0.0366
0.216	0.235	0.254	0.273	0.304	0.329	0.368	0.402
0.780	0.850	0.920	0.990	1.10	1.19	1.33	1.46
2.34	2.55	2.76	2.97	3.30	3.57	3.99	4.38
2.90	2.80	2.80	2.70	2.60	2.60	2.50	2.50
32.7	34.2	35.7	37.3	39.8	41.8	45.0	47.8
324	368	410	462	534	600	673	755
11.2	12.7	14.2	16.0	18.5	20.8	23.3	26.2
0.249	0.199	0.163	0.132	0.102	0.0827	0.0681	0.0557
0.196	0.156	0.129	0.104	0.0810	0.0662	0.0552	0.0456
26.2	26.9	27.4	28.1	29.0	29.8	30.8	31.8
0.548	0.505	0.465	0.432	0.389	0.358	0.315	0.286
3.549	3.496	3.445	3.392	3.315	3.256	3.168	3.096
0.419	0.403	0.389	0.377	0.361	0.350	0.335	0.326
1.045	1.003	0.765	0.739	0.715	0.698	0.535	0.524
0.549	0.546	0.297	0.294	0.291	0.289	0.179	0.177
0.570	0.572	0.417	0.419	0.422	0.424	0.325	0.328
361	416	471	538	635	724	832	952
12.5	14.4	16.3	18.6	22.0	25.1	28.8	33.0
0.249	0.199	0.163	0.132	0.102	0.0828	0.0683	0.0558
32.5	34.4	36.2	38.1	41.0	43.4	47.3	50.6
0.693	0.643	0.596	0.556	0.505	0.468	0.415	0.379
1.150	1.103	1.059	1.017	0.958	0.915	0.854	0.808
0.419	0.403	0.389	0.377	0.361	0.350	0.335	0.325
630	756	922	1075	1299	1561	2101	2508
13.6	17.2	21.4	26.5	34.3	42.9	57.2	71.5
3.3	3.3	5.1	5.1	5.1	5.1	7.1	7.1

Table 5.2.2b
PROTOTHEN X cables (XLPE insulation) with copper conductor

Three-phase operation at 50 Hz

Design				25	35	50	70
Design	2	Nom. cross-sectional area of conductor	mm²	25	35	50	70
	3	Shape and type of conductor		RM	RM	RM	RM
	6	Nom. cross-sectional area of screen	mm²	16	16	16	16
	9	Thickness of insulation	mm	5.5	5.5	5.5	5.5
	16	Thickness of outer sheath	mm	2.5	2.5	2.5	2.5
	17	Overall diameter of cable	mm	28	29	30	31
	19	Weight of cable	kg/km	870	990	1150	1400
Mechanical properties	20	Minimum bending radius	mm	420	435	450	465
	21	Perm. pulling force with cable grip	N	1250	1750	2500	3500
	22	Perm. pulling force with pulling head	N	1250	1750	2500	3500
Electrical properties	23	d.c. resistance/unit length at 90 °C	Ω/km	0.927	0.668	0.493	0.342
	24	d.c. resistance/unit length at 20 °C	Ω/km	0.727	0.524	0.387	0.268
	25	Operating capacitance/unit length	μF/km	0.145	0.159	0.175	0.196
	26	Charging current	A/km	0.530	0.580	0.630	0.710
	28	Earth fault current	A/km	1.59	1.74	1.89	2.13
	30	Electric field strength at the conductor	kV/mm	3.40	3.30	3.20	3.00
	31	Reference diameter of cable	mm	27.1	28.2	29.4	31.1
Installation in ground	32	Current-carrying capacity	A	179	213	250	304
	33	Permissible transmission power	MVA	6.20	7.38	8.66	10.5
	34	Eff. a.c. resistance/unit length at 90 °C	Ω/km	0.944	0.685	0.509	0.357
	36	Eff. a.c. resistance/unit length at 20 °C	Ω/km	0.744	0.540	0.403	0.283
	38	Ohmic losses per cable	kW/km	30.2	31.1	31.8	33.0
	39	Fict. thermal resist. of cable for ohmic losses	Km/W	0.799	0.725	0.660	0.585
	41	Fict. thermal resist. of soil for ohmic losses	Km/W	2.760	2.735	2.709	2.673
	44	Inductance/unit length per conductor	mH/km	0.789	0.757	0.729	0.695
	45	Resistance/unit length in zero-phase seq.system	Ω/km	1.546	1.342	1.204	1.083
	46	Reactance/unit length in zero-phase seq.system	Ω/km	0.588	0.581	0.575	0.568
	47	Reduction factor		0.598	0.600	0.602	0.604
Installation in free air	32	Current-carrying capacity	A	195	237	284	353
	33	Permissible transmission power	MVA	6.75	8.21	9.84	12.2
	34	Eff. a.c. resistance/unit length at 90 °C	Ω/km	0.941	0.682	0.507	0.355
	35	Mean eff. a.c. resistance/unit length at 90 °C	Ω/km	0.938	0.679	0.504	0.352
	38	Ohmic losses per cable	kW/km	35.7	38.1	40.7	43.9
	39	Fict. thermal resist. of cable for ohmic losses	Km/W	0.801	0.728	0.663	0.588
	43	Thermal resistance of the air	Km/W	0.873	0.841	0.808	0.767
	44	Inductance/unit length per conductor	mH/km	0.675	0.649	0.626	0.599
	48	Minimum time value	s	149	198	282	357
Short-circuit	49	Rated short-time current of conductor (1s)	kA	3.57	5.00	7.15	10.0
	50	Rated short-time current of screen (1s)	kA	3.3	3.3	3.3	3.3

⊚ ⊚ ⊚   N2XS2Y 1×.../.. 12/20 kV (U_m=24 kV)

95	120	150	185	240	300	400	500
RM	RM	RM	RM	RM	RM	RM	RM
16	16	25	25	25	25	35	35
5.5	5.5	5.5	5.5	5.5	5.5	5.5	5.5
2.5	2.5	2.5	2.5	2.5	2.5	2.5	2.5
33	34	36	38	40	43	46	49
1650	1900	2300	2650	3200	3850	4800	5850
495	510	540	570	600	645	690	735
4750	6000	7500	9250	12000	15000	20000	25000
4750	6000	7500	9250	12000	15000	20000	25000
0.246	0.195	0.158	0.126	0.0961	0.0766	0.0599	0.0467
0.193	0.153	0.124	0.0991	0.0754	0.0601	0.0470	0.0366
0.216	0.235	0.254	0.273	0.304	0.329	0.368	0.402
0.780	0.850	0.920	0.990	1.10	1.19	1.33	1.46
2.34	2.55	2.76	2.97	3.30	3.57	3.99	4.38
2.90	2.80	2.80	2.70	2.60	2.60	2.50	2.50
32.7	34.2	35.7	37.3	39.8	41.8	45.0	47.8
361	407	446	498	570	634	687	761
12.5	14.1	15.4	17.3	19.7	22.0	23.8	26.4
0.261	0.209	0.179	0.147	0.116	0.0958	0.0848	0.0709
0.208	0.167	0.145	0.119	0.0951	0.0793	0.0719	0.0608
34.0	34.7	35.6	36.4	37.6	38.5	40.0	41.1
0.529	0.484	0.432	0.396	0.350	0.318	0.264	0.234
2.641	2.611	2.583	2.553	2.509	2.476	2.424	2.381
0.668	0.647	0.622	0.605	0.581	0.565	0.536	0.519
1.007	0.965	0.750	0.724	0.700	0.684	0.529	0.518
0.563	0.560	0.309	0.306	0.303	0.301	0.186	0.184
0.606	0.608	0.458	0.459	0.461	0.462	0.365	0.365
428	492	552	627	734	833	928	1047
14.8	17.0	19.1	21.7	25.4	28.9	32.1	36.3
0.259	0.208	0.178	0.146	0.116	0.0965	0.0870	0.0740
0.256	0.205	0.174	0.142	0.112	0.0924	0.0811	0.0682
47.0	49.7	53.0	55.9	60.2	64.1	69.8	74.8
0.531	0.487	0.434	0.397	0.350	0.316	0.260	0.228
0.732	0.702	0.673	0.646	0.609	0.581	0.541	0.511
0.580	0.565	0.548	0.536	0.521	0.510	0.491	0.481
448	541	671	791	972	1179	1689	2073
13.6	17.2	21.4	26.5	34.3	42.9	57.2	71.5
3.3	3.3	5.1	5.1	5.1	5.1	7.1	7.1

Table 5.2.3 a
PROTOTHEN X cables (XLPE insulation) with copper conductor

Three-phase operation at 50 Hz

Design	2	Nom. cross-sectional area of conductor	mm²	25	35	50	70
	3	Shape and type of conductor		RM	RM	RM	RM
	6	Nom. cross-sectional area of screen	mm²	16	16	16	16
	9	Thickness of insulation	mm	8.0	8.0	8.0	8.0
	16	Thickness of outer sheath	mm	2.5	2.5	2.5	2.5
	17	Overall diameter of cable	mm	33	34	35	36
	19	Weight of cable	kg/km	1100	1250	1400	1600
Mechanical properties	20	Minimum bending radius	mm	495	510	525	540
	21	Perm. pulling force with cable grip	N	1250	1750	2500	3500
	22	Perm. pulling force with pulling head	N	1250	1750	2500	3500
Electrical properties	23	d.c. resistance/unit length at 90 °C	Ω/km	0.927	0.668	0.493	0.342
	24	d.c. resistance/unit length at 20 °C	Ω/km	0.727	0.524	0.387	0.268
	25	Operating capacitance/unit length	μF/km	0.115	0.125	0.136	0.151
	26	Charging current	A/km	0.630	0.680	0.740	0.820
	28	Earth fault current	A/km	1.89	2.04	2.22	2.46
	30	Electric field strength at the conductor	kV/mm	4.10	3.90	3.70	3.50
	31	Reference diameter of cable	mm	32.1	33.2	34.4	36.1
Installation in ground	32	Current-carrying capacity	A	161	192	226	275
	33	Permissible transmission power	MVA	8.37	10.0	11.7	14.3
	34	Eff. a.c. resistance/unit length at 90 °C	Ω/km	0.930	0.671	0.496	0.345
	36	Eff. a.c. resistance/unit length at 20 °C	Ω/km	0.730	0.527	0.390	0.271
	38	Ohmic losses per cable	kW/km	24.1	24.7	25.4	26.1
	39	Fict. thermal resist. of cable for ohmic losses	Km/W	0.910	0.835	0.767	0.690
	41	Fict. thermal resist. of soil for ohmic losses	Km/W	3.571	3.531	3.489	3.431
	44	Inductance/unit length per conductor	mH/km	0.548	0.521	0.496	0.468
	45	Resistance/unit length in zero-phase seq.system	Ω/km	1.580	1.375	1.237	1.116
	46	Reactance/unit length in zero-phase seq.system	Ω/km	0.589	0.582	0.575	0.568
	47	Reduction factor		0.561	0.563	0.566	0.569
Installation in free air	32	Current-carrying capacity	A	167	202	241	300
	33	Permissible transmission power	MVA	8.68	10.5	12.5	15.6
	34	Eff. a.c. resistance/unit length at 90 °C	Ω/km	0.930	0.671	0.497	0.345
	38	Ohmic losses per cable	kW/km	25.9	27.4	28.8	31.0
	39	Fict. thermal resist. of cable for ohmic losses	Km/W	1.114	1.027	0.948	0.860
	43	Thermal resistance of the air	Km/W	1.209	1.169	1.129	1.077
	44	Inductance/unit length per conductor	mH/km	0.548	0.521	0.496	0.468
	48	Minimum time value	s	204	273	391	495
Short-circuit	49	Rated short-time current of conductor (1s)	kA	3.57	5.00	7.15	10.0
	50	Rated short-time current of screen (1s)	kA	3.3	3.3	3.3	3.3

95 RM 16	120 RM 16	150 RM 25	185 RM 25	240 RM 25	300 RM 25	400 RM 35	500 RM 35
8.0	8.0	8.0	8.0	8.0	8.0	8.0	8.0
2.5	2.5	2.5	2.5	2.5	2.5	2.5	2.6
38	39	41	43	45	48	51	54
1900	2200	2550	2950	3550	4200	5150	6250
570	585	615	645	675	720	765	810
4750	6000	7500	9250	12000	15000	20000	25000
4750	6000	7500	9250	12000	15000	20000	25000
0.246	0.195	0.158	0.126	0.0961	0.0766	0.0599	0.0467
0.193	0.153	0.124	0.0991	0.0754	0.0601	0.0470	0.0366
0.165	0.178	0.191	0.205	0.227	0.244	0.271	0.295
0.900	0.970	1.04	1.12	1.24	1.33	1.47	1.61
2.70	2.91	3.12	3.36	3.72	3.99	4.41	4.83
3.30	3.20	3.10	3.00	2.90	2.90	2.80	2.70
37.7	39.2	40.7	42.3	44.8	46.8	50.0	53.0
328	372	415	467	540	607	682	767
17.0	19.3	21.6	24.3	28.1	31.5	35.4	39.9
0.249	0.198	0.163	0.131	0.101	0.0823	0.0677	0.0552
0.196	0.156	0.129	0.104	0.0808	0.0658	0.0548	0.0451
26.8	27.4	28.1	28.7	29.6	30.3	31.5	32.5
0.630	0.584	0.540	0.503	0.454	0.421	0.371	0.339
3.380	3.333	3.288	3.242	3.174	3.122	3.043	2.973
0.447	0.430	0.415	0.402	0.384	0.373	0.357	0.346
1.039	0.997	0.763	0.737	0.712	0.696	0.534	0.523
0.563	0.559	0.309	0.306	0.302	0.299	0.188	0.185
0.572	0.574	0.419	0.421	0.423	0.426	0.327	0.329
363	418	473	539	636	725	833	952
18.9	21.7	24.6	28.0	33.0	37.7	43.3	49.5
0.249	0.198	0.163	0.131	0.102	0.0824	0.0678	0.0554
32.9	34.7	36.5	38.2	41.1	43.3	47.0	50.2
0.792	0.737	0.686	0.642	0.584	0.545	0.483	0.445
1.034	0.996	0.961	0.926	0.877	0.842	0.791	0.749
0.447	0.430	0.415	0.402	0.384	0.373	0.356	0.345
623	749	914	1071	1295	1557	2096	2508
13.6	17.2	21.4	26.5	34.3	42.9	57.2	71.5
3.3	3.3	5.1	5.1	5.1	5.1	7.1	7.1

Table 5.2.3b
PROTOTHEN X cables (XLPE insulation) with copper conductor

Three-phase operation at 50 Hz

Design	2	Nom. cross-sectional area of conductor	mm²	25	35	50	70
	3	Shape and type of conductor		RM	RM	RM	RM
	6	Nom. cross-sectional area of screen	mm²	16	16	16	16
	9	Thickness of insulation	mm	8.0	8.0	8.0	8.0
	16	Thickness of outer sheath	mm	2.5	2.5	2.5	2.5
	17	Overall diameter of cable	mm	33	34	35	36
	19	Weight of cable	kg/km	1100	1250	1400	1600
Mechanical properties	20	Minimum bending radius	mm	495	510	525	540
	21	Perm. pulling force with cable grip	N	1250	1750	2500	3500
	22	Perm. pulling force with pulling head	N	1250	1750	2500	3500
Electrical properties	23	d.c. resistance/unit length at 90 °C	Ω/km	0.927	0.668	0.493	0.342
	24	d.c. resistance/unit length at 20 °C	Ω/km	0.727	0.524	0.387	0.268
	25	Operating capacitance/unit length	μF/km	0.115	0.125	0.136	0.151
	26	Charging current	A/km	0.630	0.680	0.740	0.820
	28	Earth fault current	A/km	1.89	2.04	2.22	2.46
	30	Electric field strength at the conductor	kV/mm	4.10	3.90	3.70	3.50
	31	Reference diameter of cable	mm	32.1	33.2	34.4	36.1
Installation in ground	32	Current-carrying capacity	A	179	214	251	305
	33	Permissible transmission power	MVA	9.30	11.1	13.0	15.8
	34	Eff. a.c. resistance/unit length at 90 °C	Ω/km	0.942	0.683	0.508	0.355
	36	Eff. a.c. resistance/unit length at 20 °C	Ω/km	0.742	0.538	0.401	0.282
	38	Ohmic losses per cable	kW/km	30.2	31.3	32.0	33.1
	39	Fict. thermal resist. of cable for ohmic losses	Km/W	0.900	0.823	0.753	0.672
	41	Fict. thermal resist. of soil for ohmic losses	Km/W	2.653	2.631	2.607	2.575
	44	Inductance/unit length per conductor	mH/km	0.800	0.768	0.740	0.705
	45	Resistance/unit length in zero-phase seq.system	Ω/km	1.541	1.337	1.199	1.078
	46	Reactance/unit length in zero-phase seq.system	Ω/km	0.604	0.596	0.589	0.581
	47	Reduction factor		0.597	0.599	0.600	0.603
Installation in free air	32	Current-carrying capacity	A	195	237	283	352
	33	Permissible transmission power	MVA	10.1	12.3	14.7	18.3
	34	Eff. a.c. resistance/unit length at 90 °C	Ω/km	0.940	0.681	0.506	0.355
	35	Mean eff. a.c. resistance/unit length at 90 °C	Ω/km	0.937	0.678	0.504	0.352
	38	Ohmic losses per cable	kW/km	35.6	38.1	40.3	43.6
	39	Fict. thermal resist. of cable for ohmic losses	Km/W	0.902	0.824	0.754	0.674
	43	Thermal resistance of the air	Km/W	0.775	0.749	0.723	0.690
	44	Inductance/unit length per conductor	mH/km	0.709	0.682	0.657	0.630
	48	Minimum time value	s	149	198	284	360
Short-circuit	49	Rated short-time current of conductor (1s)	kA	3.57	5.00	7.15	10.0
	50	Rated short-time current of screen (1s)	kA	3.3	3.3	3.3	3.3

⊙ ⊙ ⊙   N2XS2Y 1×.../.. 18/30 kV (U_m=36 kV)

95 RM 16	120 RM 16	150 RM 25	185 RM 25	240 RM 25	300 RM 25	400 RM 35	500 RM 35
8.0	8.0	8.0	8.0	8.0	8.0	8.0	8.0
2.5	2.5	2.5	2.5	2.5	2.5	2.5	2.6
38	39	41	43	45	48	51	54
1900	2200	2550	2950	3550	4200	5150	6250
570	585	615	645	675	720	765	810
4750	6000	7500	9250	12000	15000	20000	25000
4750	6000	7500	9250	12000	15000	20000	25000
0.246	0.195	0.158	0.126	0.0961	0.0766	0.0599	0.0467
0.193	0.153	0.124	0.0991	0.0754	0.0601	0.0470	0.0366
0.165	0.178	0.191	0.205	0.227	0.244	0.271	0.295
0.900	0.970	1.04	1.12	1.24	1.33	1.47	1.61
2.70	2.91	3.12	3.36	3.72	3.99	4.41	4.83
3.30	3.20	3.10	3.00	2.90	2.90	2.80	2.70
37.7	39.2	40.7	42.3	44.8	46.8	50.0	53.0
362	410	450	503	576	641	698	774
18.8	21.3	23.4	26.1	29.9	33.3	36.3	40.2
0.259	0.208	0.177	0.145	0.114	0.0944	0.0831	0.0695
0.206	0.166	0.143	0.118	0.0935	0.0779	0.0702	0.0594
34.0	35.0	35.9	36.7	37.9	38.8	40.5	41.6
0.609	0.561	0.503	0.462	0.410	0.374	0.311	0.277
2.546	2.520	2.494	2.467	2.427	2.396	2.348	2.306
0.678	0.656	0.632	0.615	0.591	0.575	0.545	0.529
1.002	0.961	0.748	0.723	0.698	0.682	0.528	0.517
0.576	0.571	0.320	0.317	0.313	0.310	0.195	0.192
0.605	0.607	0.456	0.457	0.459	0.461	0.363	0.364
427	490	550	625	732	829	927	1047
22.2	25.5	28.6	32.5	38.0	43.1	48.2	54.4
0.259	0.208	0.178	0.146	0.116	0.0964	0.0870	0.0739
0.256	0.205	0.174	0.142	0.112	0.0923	0.0810	0.0680
46.7	49.3	52.5	55.4	59.8	63.4	69.6	74.5
0.610	0.561	0.502	0.460	0.406	0.368	0.301	0.265
0.662	0.638	0.614	0.592	0.560	0.537	0.502	0.475
0.609	0.592	0.574	0.561	0.544	0.533	0.512	0.501
450	545	676	797	977	1191	1693	2073
13.6	17.2	21.4	26.5	34.3	42.9	57.2	71.5
3.3	3.3	5.1	5.1	5.1	5.1	7.1	7.1

Table 5.3.1a
SIENOPYR FRNC cables with copper conductor

Three-phase operation at 50 Hz

Design	2	Nom. cross-sectional area of conductor	mm^2	4	6	10	16
	3	Shape and type of conductor		RE	RE	RE	RE
	9	Thickness of insulation	mm	0.7	0.7	0.7	0.7
	16	Thickness of outer sheath	mm	1.4	1.4	1.4	1.4
	17	Overall diameter of cable	mm	10	10	11	11
	19	Weight of cable	kg/km	135	160	210	250
Mechanical properties	20	Minimum bending radius	mm	150	150	165	165
	21	Perm. pulling force with cable grip	N	200	300	500	800
	22	Perm. pulling force with pulling head	N	200	300	500	800
Electrical properties	23	d.c. resistance/unit length at 90 °C	Ω/km	5.88	3.93	2.33	1.47
	24	d.c. resistance/unit length at 20 °C	Ω/km	4.61	3.08	1.83	1.15
Installation in free air	32	Current-carrying capacity	A	44	56	76	101
	34	Eff. a.c. resistance/unit length at 90 °C	Ω/km	5.88	3.93	2.33	1.47
	38	Ohmic losses per cable	kW/km	11.4	12.3	13.5	15.0
	44	Inductance/unit length per conductor	mH/km	0.420	0.394	0.366	0.341
	48	Minimum time value	s	75	104	157	228
Short-circuit	49	Rated short-time current of conductor (1s)	kA	0.572	0.858	1.43	2.29

Table 5.3.1b
SIENOPYR FRNC cables with copper conductor

Three-phase operation at 50 Hz

Design	2	Nom. cross-sectional area of conductor	mm^2	4	6	10	16
	3	Shape and type of conductor		RE	RE	RE	RE
	9	Thickness of insulation	mm	0.7	0.7	0.7	0.7
	16	Thickness of outer sheath	mm	1.4	1.4	1.4	1.4
	17	Overall diameter of cable	mm	10	10	11	11
	19	Weight of cable	kg/km	135	160	210	250
Mechanical properties	20	Minimum bending radius	mm	150	150	165	165
	21	Perm. pulling force with cable grip	N	200	300	500	800
	22	Perm. pulling force with pulling head	N	200	300	500	800
Electrical properties	23	d.c. resistance/unit length at 90 °C	Ω/km	5.88	3.93	2.33	1.47
	24	d.c. resistance/unit length at 20 °C	Ω/km	4.61	3.08	1.83	1.15
Installation in free air	32	Current-carrying capacity	A	55	70	95	126
	34	Eff. a.c. resistance/unit length at 90 °C	Ω/km	5.88	3.93	2.33	1.47
	38	Ohmic losses per cable	kW/km	17.8	19.2	21.1	23.3
	44	Inductance/unit length per conductor	mH/km	0.584	0.558	0.530	0.505
	48	Minimum time value	s	48	67	101	147
Short-circuit	49	Rated short-time current of conductor (1s)	kA	0.572	0.858	1.43	2.29

(N)2XH 1×... 0.6/1 kV (U_m=1.2 kV)

25 RM	35 RM	50 RM	70 RM	95 RM	120 RM	150 RM	185 RM	240 RM	300 RM
0.9	0.9	1.0	1.1	1.1	1.2	1.4	1.6	1.7	1.8
1.4	1.4	1.4	1.4	1.5	1.5	1.6	1.7	1.8	1.8
13	14	16	17	19	21	23	25	28	30
365	470	590	810	1100	1350	1600	2000	2550	3150
195	210	240	255	285	315	345	375	420	450
1250	1750	2500	3500	4750	6000	7500	9250	12000	15000
1250	1750	2500	3500	4750	6000	7500	9250	12000	15000
0.927	0.668	0.493	0.342	0.246	0.195	0.158	0.126	0.0961	0.0766
0.727	0.524	0.387	0.268	0.193	0.153	0.124	0.0991	0.0754	0.0601
137	168	206	262	323	377	434	502	601	691
0.923	0.668	0.494	0.342	0.247	0.196	0.160	0.128	0.0990	0.0802
17.4	18.9	21.0	23.5	25.8	27.9	30.1	32.4	35.8	38.3
0.325	0.311	0.302	0.291	0.284	0.279	0.279	0.278	0.273	0.269
303	395	536	649	786	921	1086	1235	1450	1714
3.57	5.00	7.15	10.0	13.6	17.2	21.4	26.5	34.3	42.9

(N)2XH 1×... 0.6/1 kV (U_m=1.2 kV)

25 RM	35 RM	50 RM	70 RM	95 RM	120 RM	150 RM	185 RM	240 RM	300 RM
0.9	0.9	1.0	1.1	1.1	1.2	1.4	1.6	1.7	1.8
1.4	1.4	1.4	1.4	1.5	1.5	1.6	1.7	1.8	1.8
13	14	16	17	19	21	23	25	28	30
365	470	590	810	1100	1350	1600	2000	2550	3150
195	210	240	255	285	315	345	375	420	450
1250	1750	2500	3500	4750	6000	7500	9250	12000	15000
1250	1750	2500	3500	4750	6000	7500	9250	12000	15000
0.927	0.668	0.493	0.342	0.246	0.195	0.158	0.126	0.0961	0.0766
0.727	0.524	0.387	0.268	0.193	0.153	0.124	0.0991	0.0754	0.0601
170	210	255	325	399	466	536	621	743	856
0.927	0.668	0.494	0.342	0.247	0.196	0.159	0.127	0.0974	0.0783
26.8	29.5	32.1	36.1	39.3	42.5	45.7	49.1	53.8	57.4
0.489	0.475	0.466	0.455	0.448	0.443	0.443	0.442	0.437	0.433
197	253	350	422	515	603	712	807	949	1117
3.57	5.00	7.15	10.0	13.6	17.2	21.4	26.5	34.3	42.9

Table 5.3.2
SIENOPYR FRNC cables with copper conductor

Single-phase a.c. operation at 50 Hz

		Design					
Design	2	Nom. cross-sectional area of conductor	mm²	1.5	2.5	4	6
	3	Shape and type of conductor		RE	RE	RE	RE
	9	Thickness of insulation	mm	0.7	0.7	0.7	0.7
	16	Thickness of outer sheath	mm	1.8	1.8	1.8	1.8
	17	Overall diameter of cable	mm	12	13	14	15
	19	Weight of cable	kg/km	195	240	310	390
Mechanical properties	20	Minimum bending radius	mm	144	156	168	180
	21	Perm. pulling force with cable grip	N	225	375	600	900
	22	Perm. pulling force with pulling head	N	225	375	600	900
Electrical properties	23	d.c. resistance/unit length at 90 °C	Ω/km	15.4	9.45	5.88	3.93
	24	d.c. resistance/unit length at 20 °C	Ω/km	12.1	7.41	4.61	3.08
Installation in free air	32	Current-carrying capacity	A	29	38	50	64
	34	Eff. a.c. resistance/unit length at 90 °C	Ω/km	15.4	9.45	5.88	3.93
	38	Ohmic losses per cable	kW/km	26.0	27.3	29.4	32.2
	44	Inductance/unit length per conductor	mH/km	0.329	0.305	0.285	0.271
	48	Minimum time value	s	24	39	58	91
Short-circuit	49	Rated short-time current of conductor (1s)	kA	0.215	0.358	0.572	0.858

Table 5.3.3
SIENOPYR FRNC cables with copper conductor

Three-phase operation at 50 Hz

		Design					
Design	2	Nom. cross-sectional area of conductor	mm²	25	35	50	70
	3	Shape and type of conductor		RM	RM	SM	SM
	4	Nom. cross-sectional area of reduced conductor	mm²	16	16	25	35
	5	Shape and type of reduced conductor		RE	RE	RM	SM
	9	Thickness of insulation	mm	0.9	0.9	1.0	1.1
	16	Thickness of outer sheath	mm	1.8	1.8	1.8	1.9
	17	Overall diameter of cable	mm	24	26	28	32
	19	Weight of cable	kg/km	1330	1680	2050	2750
Mechanical properties	20	Minimum bending radius	mm	288	312	336	384
	21	Perm. pulling force with cable grip	N	4550	6050	8750	12250
	22	Perm. pulling force with pulling head	N	4550	6050	8750	12250
Electrical properties	23	d.c. resistance/unit length at 90 °C	Ω/km	0.927	0.668	0.493	0.342
	24	d.c. resistance/unit length at 20 °C	Ω/km	0.727	0.524	0.387	0.268
Installation in free air	32	Current-carrying capacity	A	128	158	188	239
	34	Eff. a.c. resistance/unit length at 90 °C	Ω/km	0.927	0.669	0.494	0.342
	38	Ohmic losses per cable	kW/km	45.6	50.1	52.4	58.7
	44	Inductance/unit length per conductor	mH/km	0.258	0.247	0.248	0.243
	48	Minimum time value	s	347	446	643	780
Short-circuit	49	Rated short-time current of conductor (1s)	kA	3.57	5.00	7.15	10.0

(N)2XH-J 3×... 0.6/1 kV (U_m=1.2 kV)

10 RE	16 RE	25 RM	35 RM	50 SM	70 SM	95 SM	120 SM	150 SM	185 SM	240 SM
0.7	0.7	0.9	0.9	1.0	1.1	1.1	1.2	1.4	1.6	1.7
1.8	1.8	1.8	1.8	1.8	1.9	2.0	2.1	2.3	2.4	2.6
17	19	23	25	25	29	32	36	40	44	49
530	740	1140	1490	1700	2400	3150	3900	4850	5950	7750
204	228	276	300	300	348	384	432	480	528	588
1500	2400	3750	5250	7500	10500	14250	18000	22500	27750	36000
1500	2400	3750	5250	7500	10500	14250	18000	22500	27750	36000
2.33	1.47	0.927	0.668	0.493	0.342	0.246	0.195	0.158	0.126	0.0961
1.83	1.15	0.727	0.524	0.387	0.268	0.193	0.153	0.124	0.0991	0.0754
87	116	157	194	230	292	359	418	481	556	659
2.33	1.47	0.927	0.669	0.494	0.342	0.247	0.196	0.160	0.128	0.0988
35.3	39.5	45.7	50.3	52.3	58.4	63.7	68.6	73.9	79.4	85.8
0.255	0.243	0.242	0.234	0.232	0.229	0.224	0.223	0.224	0.225	0.222
120	173	231	296	430	522	637	749	884	1006	1206
1.43	2.29	3.57	5.00	7.15	10.0	13.6	17.2	21.4	26.5	34.3

(N)2XH 3×.../.. 0.6/1 kV (U_m=1.2 kV)

95 SM 50 SM	120 SM 70 SM	150 SM 70 SM	185 SM 95 SM	240 SM 120 SM
1.1	1.2	1.4	1.6	1.7
2.0	2.2	2.3	2.5	2.7
35	39	43	48	54
3650	4650	5550	6950	8950
420	468	516	576	648
16750	21500	26000	32500	42000
16750	21500	26000	32500	42000
0.246	0.195	0.158	0.126	0.0961
0.193	0.153	0.124	0.0991	0.0754
293	342	393	454	539
0.247	0.196	0.160	0.128	0.0985
63.6	68.9	74.0	79.3	85.8
0.239	0.239	0.237	0.239	0.236
956	1119	1324	1510	1802
13.6	17.2	21.4	26.5	34.3

Table 5.3.4
SIENOPYR FRNC cables with copper conductor

Three-phase operation at 50 Hz

Design	2	Nom. cross-sectional area of conductor	mm²	1.5	2.5	4	6
	3	Shape and type of conductor		RE	RE	RE	RE
	9	Thickness of insulation	mm	0.7	0.7	0.7	0.7
	16	Thickness of outer sheath	mm	1.8	1.8	1.8	1.8
	17	Overall diameter of cable	mm	13	14	15	16
	19	Weight of cable	kg/km	225	280	365	470
Mechanical properties	20	Minimum bending radius	mm	156	168	180	192
	21	Perm. pulling force with cable grip	N	300	500	800	1200
	22	Perm. pulling force with pulling head	N	300	500	800	1200
Electrical properties	23	d.c. resistance/unit length at 90 °C	Ω/km	15.4	9.45	5.88	3.93
	24	d.c. resistance/unit length at 20 °C	Ω/km	12.1	7.41	4.61	3.08
Installation in free air	32	Current-carrying capacity	A	24	32	42	53
	34	Eff. a.c. resistance/unit length at 90 °C	Ω/km	15.4	9.45	5.88	3.93
	38	Ohmic losses per cable	kW/km	26.7	29.0	31.1	33.1
	44	Inductance/unit length per conductor	mH/km	0.352	0.328	0.308	0.294
	48	Minimum time value	s	36	55	82	117
Short-circuit	49	Rated short-time current of conductor (1s)	kA	0.215	0.358	0.572	0.858

Table 5.3.5
SIENOPYR FRNC cables with copper conductor

Three-phase operation at 50 Hz

Design	2	Nom. cross-sectional area of conductor	mm²	1.5	2.5	4	6
	3	Shape and type of conductor		RE	RE	RE	RE
	9	Thickness of insulation	mm	0.7	0.7	0.7	0.7
	16	Thickness of outer sheath	mm	1.8	1.8	1.8	1.8
	17	Overall diameter of cable	mm	14	15	16	17
	19	Weight of cable	kg/km	260	330	430	560
Mechanical properties	20	Minimum bending radius	mm	168	180	192	204
	21	Perm. pulling force with cable grip	N	375	625	1000	1500
	22	Perm. pulling force with pulling head	N	375	625	1000	1500
Electrical properties	23	d.c. resistance/unit length at 90 °C	Ω/km	15.4	9.45	5.88	3.93
	24	d.c. resistance/unit length at 20 °C	Ω/km	12.1	7.41	4.61	3.08
Installation in free air	32	Current-carrying capacity	A	24	32	42	53
	34	Eff. a.c. resistance/unit length at 90 °C	Ω/km	15.4	9.45	5.88	3.93
	38	Ohmic losses per cable	kW/km	26.7	29.0	31.1	33.1
	44	Inductance/unit length per conductor	mH/km	0.361	0.337	0.317	0.303
	48	Minimum time value	s	36	55	82	117
Short-circuit	49	Rated short-time current of conductor (1s)	kA	0.215	0.358	0.572	0.858

(N)2XH 4×... 0.6/1 kV (U_m=1.2 kV)

10 RE	16 RE	25 RM	35 RM	50 SM	70 SM	95 SM	120 SM	150 SM	185 SM	240 SM
0.7	0.7	0.9	0.9	1.0	1.1	1.1	1.2	1.4	1.6	1.7
1.8	1.8	1.8	1.8	1.9	2.0	2.1	2.3	2.4	2.6	2.8
18	20	25	27	29	33	37	42	46	51	57
650	920	1430	1880	2200	3100	4100	5300	6300	7800	10200
216	240	300	324	348	396	444	504	552	612	684
2000	3200	5000	7000	10000	14000	19000	24000	30000	37000	48000
2000	3200	5000	7000	10000	14000	19000	24000	30000	37000	48000
2.33	1.47	0.927	0.668	0.493	0.342	0.246	0.195	0.158	0.126	0.0961
1.83	1.15	0.727	0.524	0.387	0.268	0.193	0.153	0.124	0.0991	0.0754
72	96	131	162	195	247	305	355	408	469	558
2.33	1.47	0.927	0.669	0.494	0.342	0.247	0.196	0.159	0.128	0.0984
36.3	40.5	47.7	52.6	56.3	62.7	68.9	74.2	79.7	84.5	91.9
0.278	0.266	0.265	0.258	0.255	0.252	0.247	0.246	0.247	0.248	0.245
175	253	331	424	598	730	882	1039	1229	1415	1682
1.43	2.29	3.57	5.00	7.15	10.0	13.6	17.2	21.4	26.5	34.3

(N)2XH 5×... 0.6/1 kV (U_m=1.2 kV)

10 RE	16 RE
0.7	0.7
1.8	1.8
19	22
785	1120
228	264
2500	4000
2500	4000
2.33	1.47
1.83	1.15
72	96
2.33	1.47
36.3	40.5
0.287	0.275
175	253
1.43	2.29

Table 5.3.6
SIENOPYR FRNC cables with copper conductor

All cores are loaded, operation at 50 Hz

Design							
Design	1	Number of cores		7	10	12	14
	2	Nom. cross-sectional area of conductor	mm^2	1.5	1.5	1.5	1.5
	3	Shape and type of conductor		RE	RE	RE	RE
	9	Thickness of insulation	mm	0.7	0.7	0.7	0.7
	16	Thickness of outer sheath	mm	1.8	1.8	1.8	1.8
	17	Overall diameter of cable	mm	15	17	18	18
	19	Weight of cable	kg/km	310	415	455	505
Mechanical properties	20	Minimum bending radius	mm	180	204	216	216
	21	Perm. pulling force with cable grip	N	525	750	900	1050
Electrical properties	23	d.c. resistance/unit length at 90 °C	Ω/km	15.4	15.4	15.4	15.4
	24	d.c. resistance/unit length at 20 °C	Ω/km	12.1	12.1	12.1	12.1
Installation in free air	32	Current-carrying capacity	A	15.6	13.2	12.6	12.0
	34	Eff. a.c. resistance/unit length at 90 °C	Ω/km	15.4	15.4	15.4	15.4
Short-circuit	49	Rated short-time current of conductor (1s)	kA	0.215	0.215	0.215	0.215

Table 5.3.7
SIENOPYR FRNC cables with copper conductor

Three-phase operation at 50 Hz

Design							
Design	2	Nom. cross-sectional area of conductor	mm^2	1.5	2.5	4	6
	3	Shape and type of conductor		RE	RE	RE	RE
	7	Nom. cross-sectional area of concentric cond.	mm^2	1.5	2.5	4	6
	9	Thickness of insulation	mm	0.7	0.7	0.7	0.7
	16	Thickness of outer sheath	mm	1.8	1.8	1.8	1.8
	17	Overall diameter of cable	mm	13	14	15	16
	19	Weight of cable	kg/km	230	285	365	465
Mechanical properties	20	Minimum bending radius	mm	156	168	180	192
	21	Perm. pulling force with cable grip	N	225	375	600	900
	22	Perm. pulling force with pulling head	N	225	375	600	900
Electrical properties	23	d.c. resistance/unit length at 90 °C	Ω/km	15.4	9.45	5.88	3.93
	24	d.c. resistance/unit length at 20 °C	Ω/km	12.1	7.41	4.61	3.08
Installation in free air	32	Current-carrying capacity	A	24	31	41	53
	34	Eff. a.c. resistance/unit length at 90 °C	Ω/km	15.4	9.45	5.88	3.93
	38	Ohmic losses per cable	kW/km	26.7	27.2	29.6	33.1
	44	Inductance/unit length per conductor	mH/km	0.329	0.305	0.285	0.271
	48	Minimum time value	s	36	59	87	117
Short-circuit	49	Rated short-time current of conductor (1s)	kA	0.215	0.358	0.572	0.858

(N)2XH ... × ... 0.6/1 kV (U_m=1.2 kV)

19	24	30	7	10	12	14	19	24	30
1.5	1.5	1.5	2.5	2.5	2.5	2.5	2.5	2.5	2.5
RE	RE	RE	RE	RE	RE	RE	RE	RE	RE
0.7	0.7	0.7	0.7	0.7	0.7	0.7	0.7	0.7	0.7
1.8	1.8	1.8	1.8	1.8	1.8	1.8	1.8	1.8	1.8
20	23	24	16	19	19	20	22	25	26
620	765	890	395	535	600	670	830	1030	1220
240	276	288	192	228	228	240	264	300	312
1425	1800	2250	875	1250	1500	1750	2375	3000	3750
15.4	15.4	15.4	9.45	9.45	9.45	9.45	9.45	9.45	9.45
12.1	12.1	12.1	7.41	7.41	7.41	7.41	7.41	7.41	7.41
10.8	9.6	9.0	21	17.6	16.8	16.0	14.4	12.8	12.0
15.4	15.4	15.4	9.45	9.45	9.45	9.45	9.45	9.45	9.45
0.215	0.215	0.215	0.358	0.358	0.358	0.358	0.358	0.358	0.358

(N)2XCH 3× ... / .. 0.6/1 kV (U_m=1.2 kV)

10	16	25	35	50	70	95	120	150	185	240
RE	RE	RM	RM	SM	SM	SM	SM	SM	SM	SM
10	16	25	35	50	70	95	120	70	95	120
0.7	0.7	0.9	0.9	1.0	1.1	1.1	1.2	1.4	1.6	1.7
1.8	1.8	1.8	1.8	1.8	1.9	2.0	2.1	2.3	2.4	2.6
18	20	24	27	28	33	36	39	43	48	53
650	920	1420	1840	2150	3050	4100	5050	5500	6900	8900
216	240	288	324	336	396	432	468	516	576	636
1500	2400	3750	5250	7500	10500	14250	18000	22500	27750	36000
1500	2400	3750	5250	7500	10500	14250	18000	22500	27750	36000
2.33	1.47	0.927	0.668	0.493	0.342	0.246	0.195	0.158	0.126	0.0961
1.83	1.15	0.727	0.524	0.387	0.268	0.193	0.153	0.124	0.0991	0.0754
72	96	131	161	192	244	300	348	397	458	539
2.33	1.47	0.928	0.670	0.495	0.344	0.250	0.200	0.162	0.131	0.103
36.3	40.6	47.8	52.1	54.8	61.5	67.5	72.7	76.6	82.8	89.4
0.255	0.243	0.242	0.234	0.232	0.229	0.224	0.223	0.224	0.225	0.222
175	253	331	430	617	748	912	1081	1298	1483	1802
1.43	2.29	3.57	5.00	7.15	10.0	13.6	17.2	21.4	26.5	34.3

Table 5.3.8
SIENOPYR FRNC cables with copper conductor

Three-phase operation at 50 Hz

Design	2	Nom. cross-sectional area of conductor	mm²	1.5	2.5	4	6
	3	Shape and type of conductor		RE	RE	RE	RE
	7	Nom. cross-sectional area of concentric cond.	mm²	1.5	2.5	4	6
	9	Thickness of insulation	mm	0.7	0.7	0.7	0.7
	16	Thickness of outer sheath	mm	1.8	1.8	1.8	1.8
	17	Overall diameter of cable	mm	14	15	16	17
	19	Weight of cable	kg/km	255	320	425	550
Mechanical properties	20	Minimum bending radius	mm	168	180	192	204
	21	Perm. pulling force with cable grip	N	300	500	800	1200
	22	Perm. pulling force with pulling head	N	300	500	800	1200
Electrical properties	23	d.c. resistance/unit length at 90 °C	Ω/km	15.4	9.45	5.88	3.93
	24	d.c. resistance/unit length at 20 °C	Ω/km	12.1	7.41	4.61	3.08
Installation in free air	32	Current-carrying capacity	A	24	32	42	54
	34	Eff. a.c. resistance/unit length at 90 °C	Ω/km	15.4	9.45	5.88	3.93
	38	Ohmic losses per cable	kW/km	26.7	29.0	31.1	34.4
	44	Inductance/unit length per conductor	mH/km	0.352	0.328	0.308	0.294
	48	Minimum time value	s	36	55	82	112
Short-circuit	49	Rated short-time current of conductor (1s)	kA	0.215	0.358	0.572	0.858

Table 5.3.9
SIENOPYR FRNC cables with copper conductor

All cores are loaded, operation at 50 Hz

Design	1	Number of cores		7	10	12	16	21
	2	Nom. cross-sectional area of conductor	mm²	1.5	1.5	1.5	1.5	1.5
	3	Shape and type of conductor		RE	RE	RE	RE	RE
	7	Nom. cross-sectional area of concentric cond.	mm²	2.5	2.5	2.5	4	6
	9	Thickness of insulation	mm	0.7	0.7	0.7	0.7	0.7
	16	Thickness of outer sheath	mm	1.8	1.8	1.8	1.8	1.8
	17	Overall diameter of cable	mm	15	19	19	20	22
	19	Weight of cable	kg/km	360	460	500	620	770
Mechanical properties	20	Minimum bending radius	mm	180	228	228	240	264
	21	Perm. pulling force with cable grip	N	525	750	900	1200	1575
Electrical properties	23	d.c. resistance/unit length at 90 °C	Ω/km	15.4	15.4	15.4	15.4	15.4
	24	d.c. resistance/unit length at 20 °C	Ω/km	12.1	12.1	12.1	12.1	12.1
Installation in free air	32	Current-carrying capacity	A	15.6	13.2	12.6	11.4	10.2
	34	Eff. a.c. resistance/unit length at 90 °C	Ω/km	15.4	15.4	15.4	15.4	15.4
Short-circuit	49	Rated short-time current of conductor (1s)	kA	0.215	0.215	0.215	0.215	0.215

(N)2XCH 4×.../.. 0.6/1 kV (U_m=1.2 kV)

10	16	25	35	50	70	95
RE	RE	RM	RM	SM	SM	SM
10	16	16	16	25	35	50
0.7	0.7	0.9	0.9	1.0	1.1	1.1
1.8	1.8	1.8	1.8	1.9	2.0	2.1
19	21	26	28	30	36	39
770	1110	1620	2070	2450	3450	4600
228	252	312	336	360	432	468
2000	3200	5000	7000	10000	14000	19000
2000	3200	5000	7000	10000	14000	19000
2.33	1.47	0.927	0.668	0.493	0.342	0.246
1.83	1.15	0.727	0.524	0.387	0.268	0.193
74	98	133	164	198	251	309
2.33	1.47	0.928	0.669	0.495	0.343	0.249
38.3	42.3	49.2	54.0	58.2	64.9	71.2
0.278	0.266	0.265	0.258	0.255	0.252	0.247
166	242	321	414	580	707	859
1.43	2.29	3.57	5.00	7.15	10.0	13.6

(N)2XCH ...×.../.. 0.6/1 kV (U_m=1.2 kV)

24	30	7	10	12	16	21	24	30	7	12
1.5	1.5	2.5	2.5	2.5	2.5	2.5	2.5	2.5	4	4
RE	RE	RE	RE	RE	RE	RE	RE	RE	RE	RE
6	10	2.5	4	4	6	10	10	10	4	6
0.7	0.7	0.7	0.7	0.7	0.7	0.7	0.7	0.7	0.7	0.7
1.8	1.8	1.8	1.8	1.8	1.8	1.8	1.8	1.8	1.8	1.8
24	25	17	20	20	22	24	26	27	18	22
850	1020	440	600	660	830	1040	1160	1350	550	910
288	300	204	240	240	264	288	312	324	216	264
1800	2250	875	1250	1500	2000	2625	3000	3750	1400	2400
15.4	15.4	9.45	9.45	9.45	9.45	9.45	9.45	9.45	5.88	5.88
12.1	12.1	7.41	7.41	7.41	7.41	7.41	7.41	7.41	4.61	4.61
9.6	9.0	21	17.6	16.8	15.2	13.6	12.8	12.0	27	22
15.4	15.4	9.45	9.45	9.45	9.45	9.45	9.45	9.45	5.88	5.88
0.215	0.215	0.358	0.358	0.358	0.358	0.358	0.358	0.358	0.572	0.572

Table 5.3.10a
SIENOPYR FRNC cables with copper conductor

Three-phase operation at 50 Hz

Design	2	Nom. cross-sectional area of conductor	mm²	4	6	10	16
	3	Shape and type of conductor		RE	RE	RE	RE
	9	Thickness of insulation	mm	1.0	1.0	1.0	1.0
	16	Thickness of outer sheath	mm	1.4	1.4	1.4	1.4
	17	Overall diameter of cable	mm	8	9	9	10
	19	Weight of cable	kg/km	95	120	165	225
Mechanical properties	20	Minimum bending radius	mm	120	135	135	150
	21	Perm. pulling force with cable grip	N	200	300	500	800
	22	Perm. pulling force with pulling head	N	200	300	500	800
Electrical properties	23	d.c. resistance/unit length at 70 °C	Ω/km	5.52	3.69	2.19	1.38
	24	d.c. resistance/unit length at 20 °C	Ω/km	4.61	3.08	1.83	1.15
Installation in free air	32	Current-carrying capacity	A	36	46	62	83
	34	Eff. a.c. resistance/unit length at 70 °C	Ω/km	5.52	3.69	2.19	1.38
	38	Ohmic losses per cable	kW/km	7.1	7.8	8.4	9.5
	44	Inductance/unit length per conductor	mH/km	0.438	0.411	0.381	0.355
	48	Minimum time value	s	80	110	168	240
Short-circuit	49	Rated short-time current of conductor (1s)	kA	0.460	0.690	1.15	1.84

Table 5.3.10b
SIENOPYR FRNC cables with copper conductor

Three-phase operation at 50 Hz

Design	2	Nom. cross-sectional area of conductor	mm²	4	6	10	16
	3	Shape and type of conductor		RE	RE	RE	RE
	9	Thickness of insulation	mm	1.0	1.0	1.0	1.0
	16	Thickness of outer sheath	mm	1.4	1.4	1.4	1.4
	17	Overall diameter of cable	mm	8	9	9	10
	19	Weight of cable	kg/km	95	120	165	225
Mechanical properties	20	Minimum bending radius	mm	120	135	135	150
	21	Perm. pulling force with cable grip	N	200	300	500	800
	22	Perm. pulling force with pulling head	N	200	300	500	800
Electrical properties	23	d.c. resistance/unit length at 70 °C	Ω/km	5.52	3.69	2.19	1.38
	24	d.c. resistance/unit length at 20 °C	Ω/km	4.61	3.08	1.83	1.15
Installation in free air	32	Current-carrying capacity	A	45	56	77	102
	34	Eff. a.c. resistance/unit length at 70 °C	Ω/km	5.52	3.69	2.19	1.38
	38	Ohmic losses per cable	kW/km	11.2	11.6	13.0	14.3
	44	Inductance/unit length per conductor	mH/km	0.603	0.575	0.545	0.519
	48	Minimum time value	s	51	74	109	159
Short-circuit	49	Rated short-time current of conductor (1s)	kA	0.460	0.690	1.15	1.84

NHXHX 1×…FE 0.6/1 kV (U_m=1.2 kV)

25 RM	35 RM	50 RM	70 RM	95 RM	120 RM	150 RM	185 RM	240 RM	300 RM
1.2	1.2	1.4	1.4	1.6	1.6	1.8	2.0	2.2	2.4
1.4	1.4	1.4	1.4	1.5	1.5	1.6	1.7	1.8	1.9
12	13	15	16	18	19	22	24	27	30
340	440	575	765	1040	1280	1570	1960	2560	3220
180	195	225	240	270	285	330	360	405	450
1250	1750	2500	3500	4750	6000	7500	9250	12000	15000
1250	1750	2500	3500	4750	6000	7500	9250	12000	15000
0.870	0.627	0.463	0.321	0.231	0.183	0.148	0.119	0.0902	0.0719
0.727	0.524	0.387	0.268	0.193	0.153	0.124	0.0991	0.0754	0.0601
111	136	166	210	259	301	345	398	475	545
0.870	0.627	0.463	0.321	0.232	0.184	0.150	0.121	0.0930	0.0754
10.7	11.6	12.8	14.2	15.6	16.7	17.9	19.1	21.0	22.4
0.336	0.321	0.314	0.299	0.296	0.287	0.286	0.285	0.281	0.279
328	428	586	718	869	1027	1221	1396	1649	1957
2.87	4.02	5.75	8.05	10.9	13.8	17.2	21.3	27.6	34.5

NHXHX 1×…FE 0.6/1 kV (U_m=1.2 kV)

25 RM	35 RM	50 RM	70 RM	95 RM	120 RM	150 RM	185 RM	240 RM	300 RM
1.2	1.2	1.4	1.4	1.6	1.6	1.8	2.0	2.2	2.4
1.4	1.4	1.4	1.4	1.5	1.5	1.6	1.7	1.8	1.9
12	13	15	16	18	19	22	24	27	30
340	440	575	765	1040	1280	1570	1960	2560	3220
180	195	225	240	270	285	330	360	405	450
1250	1750	2500	3500	4750	6000	7500	9250	12000	15000
1250	1750	2500	3500	4750	6000	7500	9250	12000	15000
0.870	0.627	0.463	0.321	0.231	0.183	0.148	0.119	0.0902	0.0719
0.727	0.524	0.387	0.268	0.193	0.153	0.124	0.0991	0.0754	0.0601
137	168	204	258	318	370	424	489	584	672
0.870	0.627	0.463	0.321	0.231	0.184	0.149	0.120	0.0916	0.0736
16.3	17.7	19.3	21.4	23.4	25.2	26.8	28.6	31.2	33.2
0.500	0.485	0.478	0.463	0.460	0.451	0.450	0.449	0.445	0.443
215	280	388	475	576	679	808	924	1091	1287
2.87	4.02	5.75	8.05	10.9	13.8	17.2	21.3	27.6	34.5

Table 5.3.11
SIENOPYR FRNC cables with copper conductor

Single-phase a.c. operation at 50 Hz

Design					1.5	2.5	4	6
	2	Nom. cross-sectional area of conductor		mm²	1.5	2.5	4	6
	3	Shape and type of conductor			RE	RE	RE	RE
	9	Thickness of insulation		mm	1.0	1.0	1.0	1.0
	16	Thickness of outer sheath		mm	1.8	1.8	1.8	1.8
	17	Overall diameter of cable		mm	14	15	15	16
	19	Weight of cable		kg/km	275	325	355	440
Mechanical properties	20	Minimum bending radius		mm	168	180	180	192
	21	Perm. pulling force with cable grip		N	225	375	600	900
	22	Perm. pulling force with pulling head		N	225	375	600	900
Electrical properties	23	d.c. resistance/unit length at 70 °C		Ω/km	14.5	8.87	5.52	3.69
	24	d.c. resistance/unit length at 20 °C		Ω/km	12.1	7.41	4.61	3.08
Installation in free air	32	Current-carrying capacity		A	22	30	40	50
	34	Eff. a.c. resistance/unit length at 70 °C		Ω/km	14.5	8.87	5.52	3.69
	38	Ohmic losses per cable		kW/km	14.0	16.0	17.7	18.4
	44	Inductance/unit length per conductor		mH/km	0.368	0.339	0.316	0.298
	48	Minimum time value		s	30	45	65	93
Short-circuit	49	Rated short-time current of conductor (1s)		kA	0.173	0.288	0.460	0.690

Table 5.3.12
SIENOPYR FRNC cables with copper conductor

Three-phase operation at 50 Hz

Design					25	35	50	70
	2	Nom. cross-sectional area of conductor		mm²	25	35	50	70
	3	Shape and type of conductor			RM	RM	RM	RM
	4	Nom. cross-sectional area of reduced conductor		mm²	16	16	25	35
	5	Shape and type of reduced conductor			RE	RE	RM	RM
	9	Thickness of insulation		mm	1.2	1.2	1.4	1.4
	16	Thickness of outer sheath		mm	1.8	1.8	1.9	2.0
	17	Overall diameter of cable		mm	25	27	31	36
	19	Weight of cable		kg/km	1420	1770	2390	3270
Mechanical properties	20	Minimum bending radius		mm	300	324	372	432
	21	Perm. pulling force with cable grip		N	4550	6050	8750	12250
	22	Perm. pulling force with pulling head		N	4550	6050	8750	12250
Electrical properties	23	d.c. resistance/unit length at 70 °C		Ω/km	0.870	0.627	0.463	0.321
	24	d.c. resistance/unit length at 20 °C		Ω/km	0.727	0.524	0.387	0.268
Installation in free air	32	Current-carrying capacity		A	100	122	149	188
	34	Eff. a.c. resistance/unit length at 70 °C		Ω/km	0.870	0.627	0.463	0.321
	38	Ohmic losses per cable		kW/km	26.1	28.0	30.9	34.1
	44	Inductance/unit length per conductor		mH/km	0.274	0.261	0.263	0.254
	48	Minimum time value		s	404	532	727	895
Short-circuit	49	Rated short-time current of conductor (1s)		kA	2.87	4.02	5.75	8.05

NHXHX-J 3×…FE 0.6/1 kV (U_m=1.2 kV)

10 RE	16 RE	25 RM	35 RM	50 RM	70 RM	95 RM	120 RM	150 RM	185 RM	240 RM
1.0	1.0	1.2	1.2	1.4	1.4	1.6	1.6	1.8	2.0	2.2
1.8	1.8	1.8	1.8	1.8	1.9	2.1	2.2	2.3	2.5	2.7
18	20	24	26	30	34	39	42	46	52	59
595	815	1240	1600	2090	2870	3850	4690	5790	7210	9430
216	240	288	312	360	408	468	504	552	624	708
1500	2400	3750	5250	7500	10500	14250	18000	22500	27750	36000
1500	2400	3750	5250	7500	10500	14250	18000	22500	27750	36000
2.19	1.38	0.870	0.627	0.463	0.321	0.231	0.183	0.148	0.119	0.0902
1.83	1.15	0.727	0.524	0.387	0.268	0.193	0.153	0.124	0.0991	0.0754
68	91	122	150	182	230	284	329	377	433	513
2.19	1.38	0.867	0.627	0.463	0.321	0.232	0.184	0.150	0.121	0.0928
20.2	22.8	25.9	28.2	30.7	34.0	37.4	39.9	42.6	45.2	48.8
0.278	0.262	0.257	0.248	0.247	0.238	0.238	0.233	0.233	0.233	0.231
140	200	271	352	487	598	723	859	1022	1179	1414
1.15	1.84	2.87	4.02	5.75	8.05	10.9	13.8	17.2	21.3	27.6

NHXHX 3×…/..FE 0.6/1 kV (U_m=1.2 kV)

95 RM / 50 RM	120 RM / 70 RM	150 RM / 70 RM	185 RM / 95 RM	240 RM / 120 RM
1.6	1.6	1.8	2.0	2.2
2.2	2.3	2.4	2.6	2.8
40	44	48	54	61
4390	5500	6540	8210	10650
480	528	576	648	732
16750	21500	26000	32500	42000
16750	21500	26000	32500	42000
0.231	0.183	0.148	0.119	0.0902
0.193	0.153	0.124	0.0991	0.0754
232	269	308	354	419
0.232	0.184	0.150	0.120	0.0926
37.4	40.0	42.7	45.3	48.8
0.253	0.250	0.247	0.248	0.245
1083	1285	1532	1764	2119
10.9	13.8	17.2	21.3	27.6

Table 5.3.13
SIENOPYR FRNC cables with copper conductor

Three-phase operation at 50 Hz

Design					1.5	2.5	4	6
	2	Nom. cross-sectional area of conductor	mm²		1.5	2.5	4	6
	3	Shape and type of conductor			RE	RE	RE	RE
	9	Thickness of insulation	mm		1.0	1.0	1.0	1.0
	16	Thickness of outer sheath	mm		1.8	1.8	1.8	1.8
	17	Overall diameter of cable	mm		15	16	16	17
	19	Weight of cable	kg/km		320	385	425	535
Mechanical properties	20	Minimum bending radius	mm		180	192	192	204
	21	Perm. pulling force with cable grip	N		300	500	800	1200
	22	Perm. pulling force with pulling head	N		300	500	800	1200
Electrical properties	23	d.c. resistance/unit length at 70 °C	Ω/km		14.5	8.87	5.52	3.69
	24	d.c. resistance/unit length at 20 °C	Ω/km		12.1	7.41	4.61	3.08
Installation in free air	32	Current-carrying capacity	A		18.6	25	33	42
	34	Eff. a.c. resistance/unit length at 70 °C	Ω/km		14.5	8.87	5.52	3.69
	38	Ohmic losses per cable	kW/km		15.0	16.6	18.0	19.5
	44	Inductance/unit length per conductor	mH/km		0.391	0.362	0.339	0.321
	48	Minimum time value	s		42	65	95	132
Short-circuit	49	Rated short-time current of conductor (1s)	kA		0.173	0.288	0.460	0.690

Table 5.3.14
SIENOPYR FRNC cables with copper conductor

Three-phase operation at 50 Hz

Design					1.5	2.5	4	6
	2	Nom. cross-sectional area of conductor	mm²		1.5	2.5	4	6
	3	Shape and type of conductor			RE	RE	RE	RE
	9	Thickness of insulation	mm		1.0	1.0	1.0	1.0
	16	Thickness of outer sheath	mm		1.8	1.8	1.8	1.8
	17	Overall diameter of cable	mm		17	18	18	19
	19	Weight of cable	kg/km		375	455	505	635
Mechanical properties	20	Minimum bending radius	mm		204	216	216	228
	21	Perm. pulling force with cable grip	N		375	625	1000	1500
	22	Perm. pulling force with pulling head	N		375	625	1000	1500
Electrical properties	23	d.c. resistance/unit length at 70 °C	Ω/km		14.5	8.87	5.52	3.69
	24	d.c. resistance/unit length at 20 °C	Ω/km		12.1	7.41	4.61	3.08
Installation in free air	32	Current-carrying capacity	A		18.6	25	33	42
	34	Eff. a.c. resistance/unit length at 70 °C	Ω/km		14.5	8.87	5.52	3.69
	38	Ohmic losses per cable	kW/km		15.0	16.6	18.0	19.5
	44	Inductance/unit length per conductor	mH/km		0.400	0.371	0.348	0.330
	48	Minimum time value	s		42	65	95	132
Short-circuit	49	Rated short-time current of conductor (1s)	kA		0.173	0.288	0.460	0.690

NHXHX 4×…FE 0.6/1 kV (U_m=1.2 kV)

10 RE	16 RE	25 RM	35 RM	50 RM	70 RM	95 RM	120 RM	150 RM	185 RM	240 RM
1.0	1.0	1.2	1.2	1.4	1.4	1.6	1.6	1.8	2.0	2.2
1.8	1.8	1.8	1.8	1.9	2.1	2.2	2.3	2.5	2.7	2.9
19	22	26	29	33	37	42	46	51	58	65
730	1010	1550	2010	2690	3670	4920	6060	7440	9330	12150
228	264	312	348	396	444	504	552	612	696	780
2000	3200	5000	7000	10000	14000	19000	24000	30000	37000	48000
2000	3200	5000	7000	10000	14000	19000	24000	30000	37000	48000
2.19	1.38	0.870	0.627	0.463	0.321	0.231	0.183	0.148	0.119	0.0902
1.83	1.15	0.727	0.524	0.387	0.268	0.193	0.153	0.124	0.0991	0.0754
57	75	102	126	152	193	238	276	315	361	430
2.19	1.38	0.870	0.627	0.463	0.321	0.232	0.184	0.150	0.120	0.0925
21.3	23.2	27.2	29.9	32.1	35.9	39.4	42.1	44.6	47.0	51.3
0.301	0.285	0.280	0.271	0.270	0.262	0.261	0.256	0.256	0.254	0.254
199	294	388	498	699	850	1029	1221	1465	1696	2012
1.15	1.84	2.87	4.02	5.75	8.05	10.9	13.8	17.2	21.3	27.6

NHXHX 5×…FE 0.6/1 kV (U_m=1.2 kV)

10 RE	16 RE
1.0	1.0
1.8	1.8
21	23
885	1230
252	276
2500	4000
2500	4000
2.19	1.38
1.83	1.15
57	75
2.19	1.38
21.3	23.2
0.310	0.294
199	294
1.15	1.84

Table 5.3.15
SIENOPYR FRNC cables with copper conductor

All cores are loaded, operation at 50 Hz

Design								
	1	Number of cores			7	10	12	14
	2	Nom. cross-sectional area of conductor	mm^2		1.5	1.5	1.5	1.5
	3	Shape and type of conductor			RE	RE	RE	RE
	9	Thickness of insulation	mm		1.0	1.0	1.0	1.0
	16	Thickness of outer sheath	mm		1.8	1.8	1.8	1.8
	17	Overall diameter of cable	mm		18	22	22	23
	19	Weight of cable	kg/km		445	610	670	740
Mechanical properties	20	Minimum bending radius	mm		216	264	264	276
	21	Perm. pulling force with cable grip	N		525	750	900	1050
Electrical properties	23	d.c. resistance/unit length at 70 °C	Ω/km		14.5	14.5	14.5	14.5
	24	d.c. resistance/unit length at 20 °C	Ω/km		12.1	12.1	12.1	12.1
Installation in free air	32	Current-carrying capacity	A		12.0	10.2	9.7	9.3
	34	Eff. a.c. resistance/unit length at 70 °C	Ω/km		14.5	14.5	14.5	14.5
Short-circuit	49	Rated short-time current of conductor (1s)	kA		0.173	0.173	0.173	0.173

Table 5.3.16
SIENOPYR FRNC cables with copper conductor

Three-phase operation at 50 Hz

Design								
	2	Nom. cross-sectional area of conductor	mm^2		1.5	2.5	4	6
	3	Shape and type of conductor			RE	RE	RE	RE
	7	Nom. cross-sectional area of concentric cond.	mm^2		1.5	2.5	4	6
	9	Thickness of insulation	mm		1.0	1.0	1.0	1.0
	16	Thickness of outer sheath	mm		1.8	1.8	1.8	1.8
	17	Overall diameter of cable	mm		16	16	16	17
	19	Weight of cable	kg/km		320	380	415	515
Mechanical properties	20	Minimum bending radius	mm		192	192	192	204
	21	Perm. pulling force with cable grip	N		225	375	600	900
	22	Perm. pulling force with pulling head	N		225	375	600	900
Electrical properties	23	d.c. resistance/unit length at 70 °C	Ω/km		14.5	8.87	5.52	3.69
	24	d.c. resistance/unit length at 20 °C	Ω/km		12.1	7.41	4.61	3.08
Installation in free air	32	Current-carrying capacity	A		18.4	24	33	41
	34	Eff. a.c. resistance/unit length at 70 °C	Ω/km		14.5	8.87	5.52	3.69
	38	Ohmic losses per cable	kW/km		14.7	15.3	18.0	18.6
	44	Inductance/unit length per conductor	mH/km		0.368	0.339	0.316	0.298
	48	Minimum time value	s		43	70	95	138
Short-circuit	49	Rated short-time current of conductor (1s)	kA		0.173	0.288	0.460	0.690

NHXHX … × … FE 0.6/1 kV (U_m=1.2 kV)

19	24	30	7	10	12	14	19	24	30
1.5	1.5	1.5	2.5	2.5	2.5	2.5	2.5	2.5	2.5
RE	RE	RE	RE	RE	RE	RE	RE	RE	RE
1.0	1.0	1.0	1.0	1.0	1.0	1.0	1.0	1.0	1.0
1.8	1.8	1.8	1.8	1.8	1.8	1.8	1.8	1.8	1.9
26	29	31	19	23	24	25	27	32	34
915	1140	1330	540	750	825	920	1150	1430	1700
312	348	372	228	276	288	300	324	384	408
1425	1800	2250	875	1250	1500	1750	2375	3000	3750
14.5	14.5	14.5	8.87	8.87	8.87	8.87	8.87	8.87	8.87
12.1	12.1	12.1	7.41	7.41	7.41	7.41	7.41	7.41	7.41
8.3	7.4	6.9	16.3	13.8	13.1	12.5	11.3	10.0	9.4
14.5	14.5	14.5	8.87	8.87	8.87	8.87	8.87	8.87	8.87
0.173	0.173	0.173	0.288	0.288	0.288	0.288	0.288	0.288	0.288

NHXCHX 3×… /..FE 0.6/1 kV (U_m=1.2 kV)

10	16	25	35	50	70	95	120	150	185	240
RE	RE	RM	RM	RM	RM	RM	RM	RM	RM	RM
10	16	25	35	50	70	95	120	70	95	120
1.0	1.0	1.2	1.2	1.4	1.4	1.6	1.6	1.8	2.0	2.2
1.8	1.8	1.8	1.8	1.8	1.9	2.1	2.2	2.3	2.5	2.7
19	21	26	28	32	36	42	45	51	56	63
715	985	1480	1940	2590	3560	4750	5830	7220	8920	11750
228	252	312	336	384	432	504	540	612	672	756
1500	2400	3750	5250	7500	10500	14250	18000	22500	27750	36000
1500	2400	3750	5250	7500	10500	14250	18000	22500	27750	36000
2.19	1.38	0.870	0.627	0.463	0.321	0.231	0.183	0.148	0.119	0.0902
1.83	1.15	0.727	0.524	0.387	0.268	0.193	0.153	0.124	0.0991	0.0754
57	75	101	125	152	191	236	273	311	356	419
2.19	1.38	0.871	0.628	0.465	0.324	0.235	0.188	0.152	0.124	0.0969
21.3	23.2	26.6	29.5	32.2	35.4	39.3	42.1	44.2	47.1	51.0
0.278	0.262	0.257	0.248	0.247	0.238	0.238	0.233	0.233	0.233	0.231
199	294	396	506	699	868	1047	1248	1503	1744	2119
1.15	1.84	2.87	4.02	5.75	8.05	10.9	13.8	17.2	21.3	27.6

Table 5.3.17
SIENOPYR FRNC cables with copper conductor

Three-phase operation at 50 Hz

Design	2	Nom. cross-sectional area of conductor	mm^2	1.5	2.5	4	6
	3	Shape and type of conductor		RE	RE	RE	RE
	7	Nom. cross-sectional area of concentric cond.	mm^2	1.5	2.5	4	6
	9	Thickness of insulation	mm	1.0	1.0	1.0	1.0
	16	Thickness of outer sheath	mm	1.8	1.8	1.8	1.8
	17	Overall diameter of cable	mm	17	17	17	19
	19	Weight of cable	kg/km	365	440	480	605
Mechanical properties	20	Minimum bending radius	mm	204	204	204	228
	21	Perm. pulling force with cable grip	N	300	500	800	1200
	22	Perm. pulling force with pulling head	N	300	500	800	1200
Electrical properties	23	d.c. resistance/unit length at 70 °C	Ω/km	14.5	8.87	5.52	3.69
	24	d.c. resistance/unit length at 20 °C	Ω/km	12.1	7.41	4.61	3.08
Installation in free air	32	Current-carrying capacity	A	18.8	25	33	42
	34	Eff. a.c. resistance/unit length at 70 °C	Ω/km	14.5	8.87	5.52	3.69
	38	Ohmic losses per cable	kW/km	15.4	16.6	18.0	19.5
	44	Inductance/unit length per conductor	mH/km	0.391	0.362	0.339	0.321
	48	Minimum time value	s	41	65	95	132
Short-circuit	49	Rated short-time current of conductor (1s)	kA	0.173	0.288	0.460	0.690

Table 5.3.18
SIENOPYR FRNC cables with copper conductor

All cores are loaded, operation at 50 Hz

Design	1	Number of cores		7	10	12	16	21
	2	Nom. cross-sectional area of conductor	mm^2	1.5	1.5	1.5	1.5	1.5
	3	Shape and type of conductor		RE	RE	RE	RE	RE
	7	Nom. cross-sectional area of concentric cond.	mm^2	2.5	2.5	2.5	2.5	6
	9	Thickness of insulation	mm	1.0	1.0	1.0	1.0	1.0
	16	Thickness of outer sheath	mm	1.8	1.8	1.8	1.8	1.8
	17	Overall diameter of cable	mm	19	23	23	25	28
	19	Weight of cable	kg/km	480	650	785	875	1070
Mechanical properties	20	Minimum bending radius	mm	228	276	276	300	336
	21	Perm. pulling force with cable grip	N	525	750	900	1200	1575
Electrical properties	23	d.c. resistance/unit length at 70 °C	Ω/km	14.5	14.5	14.5	14.5	14.5
	24	d.c. resistance/unit length at 20 °C	Ω/km	12.1	12.1	12.1	12.1	12.1
Installation in free air	32	Current-carrying capacity	A	12.0	10.2	9.7	8.8	7.9
	34	Eff. a.c. resistance/unit length at 70 °C	Ω/km	14.5	14.5	14.5	14.5	14.5
Short-circuit	49	Rated short-time current of conductor (1s)	kA	0.173	0.173	0.173	0.173	0.173

NHXCHX 4×.../..FE 0.6/1 kV (U_m=1.2 kV)

10 RE 10	16 RE 16	25 RM 16	35 RM 16	50 RM 25	70 RM 35	95 RM 50
1.0	1.0	1.2	1.2	1.4	1.4	1.6
1.8	1.8	1.8	1.8	1.9	2.1	2.2
21	23	28	30	35	39	45
845	1180	1700	2170	2950	4000	5410
252	276	336	360	420	468	540
2000	3200	5000	7000	10000	14000	19000
2000	3200	5000	7000	10000	14000	19000
2.19	1.38	0.870	0.627	0.463	0.321	0.231
1.83	1.15	0.727	0.524	0.387	0.268	0.193
57	77	103	128	154	195	241
2.19	1.38	0.871	0.628	0.464	0.322	0.233
21.3	24.5	27.7	30.9	33.0	36.8	40.7
0.301	0.285	0.280	0.271	0.262	0.256	0.256
199	279	381	483	681	832	1004
1.15	1.84	2.87	4.02	5.75	8.05	10.9

NHXCHX ...×.../..FE 0.6/1 kV (U_m=1.2 kV)

24 1.5 RE 6	30 1.5 RE 10	7 2.5 RE 2.5	10 2.5 RE 4	12 2.5 RE 4	16 2.5 RE 6	21 2.5 RE 10	24 2.5 RE 10	30 2.5 RE 10	7 4 RE 4	12 4 RE 6
1.0	1.0	1.0	1.0	1.0	1.0	1.0	1.0	1.0	1.0	1.0
1.8	1.8	1.8	1.8	1.8	1.8	1.8	1.8	1.9	1.8	1.8
30	32	20	24	25	27	30	33	35	20	25
1210	1400	575	800	880	1100	1380	1560	1820	665	1030
360	384	240	288	300	324	360	396	420	240	300
1800	2250	875	1250	1500	2000	2625	3000	3750	1400	2400
14.5	14.5	8.87	8.87	8.87	8.87	8.87	8.87	8.87	5.52	5.52
12.1	12.1	7.41	7.41	7.41	7.41	7.41	7.41	7.41	4.61	4.61
7.4	6.9	16.3	13.8	13.1	11.9	10.6	10.0	9.4	22	17.3
14.5	14.5	8.87	8.87	8.87	8.87	8.87	8.87	8.87	5.52	5.52
0.173	0.173	0.288	0.288	0.288	0.288	0.288	0.288	0.288	0.460	0.460

Table 5.4.1
Cables with paper insulation with copper conductor

Three-phase operation at 50 Hz

Design			mm²	35	50	70	95
Design	2	Nom. cross-sectional area of conductor	mm²	35	50	70	95
	3	Shape and type of conductor		SM	SM	SM	SM
	10	Thickness of insulation conductor-conductor	mm	6.4	6.4	6.4	6.4
	11	Thickness of insulation conductor-metal sheath	mm	3.7	3.7	3.7	3.7
	12	Thickness of metal sheath	mm	1.5	1.5	1.6	1.6
	13	Diameter over metal sheath	mm	30.0	32.0	35.5	39.0
	15	Thickness of an armour tape	mm	0.5	0.5	0.8	0.8
	17	Overall diameter of cable	mm	38	40	45	48
	19	Weight of cable	kg/km	4000	4550	6050	7100
Mechanical properties	20	Minimum bending radius	mm	600	630	720	765
	21	Perm. pulling force with cable grip	N	4332	4800	6075	6912
	22	Perm. pulling force with pulling head	N	5250	7500	10500	14250
Electrical properties	23	d.c. resistance/unit length at 65 °C	Ω/km	0.617	0.455	0.315	0.227
	24	d.c. resistance/unit length at 20 °C	Ω/km	0.524	0.387	0.268	0.193
	25	Operating capacitance/unit length	μF/km	0.280	0.300	0.340	0.380
	26	Charging current	A/km	0.510	0.540	0.620	0.690
	28	Earth fault current	A/km	1.03	1.09	1.20	1.31
	31	Reference diameter of cable	mm	43.7	46.1	49.5	53.9
Installation in ground	32	Current-carrying capacity	A	143	171	212	257
	33	Permissible transmission power	MVA	2.48	2.96	3.67	4.45
	34	Eff. a.c. resistance/unit length at 65 °C	Ω/km	0.619	0.458	0.318	0.230
	38	Ohmic losses per cable	kW/km	37.9	40.1	42.9	45.6
	39	Fict. thermal resist. of cable for ohmic losses	Km/W	0.685	0.635	0.572	0.520
	41	Fict. thermal resist. of soil for ohmic losses	Km/W	0.500	0.492	0.481	0.467
	44	Inductance/unit length per conductor	mH/km	0.365	0.348	0.330	0.317
Installation in free air	32	Current-carrying capacity	A	120	144	181	222
	33	Permissible transmission power	MVA	2.08	2.49	3.14	3.85
	34	Eff. a.c. resistance/unit length at 65 °C	Ω/km	0.619	0.458	0.318	0.230
	38	Ohmic losses per cable	kW/km	26.7	28.5	31.3	34.1
	39	Fict. thermal resist. of cable for ohmic losses	Km/W	0.685	0.635	0.572	0.520
	43	Thermal resistance of the air	Km/W	0.618	0.589	0.552	0.512
	44	Inductance/unit length per conductor	mH/km	0.365	0.348	0.330	0.317
	48	Minimum time value	s	489	693	859	1052
Short-circuit	49	Rated short-time current of conductor (1s)	kA	4.23	6.05	8.47	11.5

120 SM	150 SM	185 SM	240 SM	300 SM
6.4	6.4	6.4	6.4	6.4
3.7	3.7	3.7	3.7	3.7
1.7	1.8	2.0	2.1	2.3
41.0	44.0	47.5	52.0	56.0
0.8	0.8	0.8	0.8	0.8
51	54	58	62	66
8200	9500	11200	13500	16100
795	855	915	975	1050
7803	8748	10092	11532	13068
18000	22500	27750	36000	45000
0.180	0.146	0.117	0.0887	0.0707
0.153	0.124	0.0991	0.0754	0.0601
0.410	0.430	0.460	0.510	0.550
0.740	0.780	0.830	0.930	1.00
1.36	1.41	1.52	1.69	1.90
56.5	59.7	62.9	68.5	72.5
294	332	378	438	494
5.09	5.75	6.55	7.59	8.56
0.184	0.150	0.121	0.0946	0.0776
47.6	49.6	52.0	54.4	56.8
0.488	0.454	0.424	0.399	0.372
0.460	0.451	0.442	0.429	0.420
0.308	0.300	0.293	0.284	0.278
255	291	333	389	444
4.42	5.04	5.77	6.74	7.69
0.184	0.150	0.121	0.0946	0.0775
35.8	38.1	40.4	42.9	45.8
0.488	0.454	0.424	0.399	0.372
0.491	0.467	0.445	0.414	0.393
0.308	0.300	0.293	0.284	0.278
1272	1527	1773	2187	2623
14.5	18.1	22.4	29.0	36.3

Table 5.4.2
Cables with paper insulation with copper conductor

Three-phase operation at 50 Hz

Design					25	35	50	70
	2	Nom. cross-sectional area of conductor		mm^2	25	35	50	70
	3	Shape and type of conductor			RM	RM	RM	RM
	9	Thickness of insulation		mm	5.5	5.5	5.5	5.5
	12	Thickness of metal sheath		mm	1.3	1.3	1.3	1.4
	13	Diameter over metal sheath		mm	19.5	20.5	22.0	24.0
	15	Thickness of an armour tape		mm	0.8	0.8	0.8	0.8
	17	Overall diameter of cable		mm	53	56	58	62
	19	Weight of cable		kg/km	6100	6700	7400	8750
Mechanical properties	20	Minimum bending radius		mm	840	885	915	975
	21	Perm. pulling force with cable grip		N	2809	3136	3364	3844
	22	Perm. pulling force with pulling head		N	3750	5250	7500	10500
Electrical properties	23	d.c. resistance/unit length at 65 °C		Ω/km	0.856	0.617	0.455	0.315
	24	d.c. resistance/unit length at 20 °C		Ω/km	0.727	0.524	0.387	0.268
	25	Operating capacitance/unit length		μF/km	0.199	0.222	0.245	0.279
	26	Charging current		A/km	0.720	0.810	0.890	1.01
	28	Earth fault current		A/km	2.16	2.43	2.67	3.03
	30	Electric field strength at the conductor		kV/mm	3.80	3.60	3.40	3.20
	31	Reference diameter of cable		mm	58.3	60.6	63.9	67.7
Installation in ground	32	Current-carrying capacity		A	125	150	179	221
	33	Permissible transmission power		MVA	4.33	5.20	6.20	7.66
	34	Eff. a.c. resistance/unit length at 65 °C		Ω/km	0.858	0.619	0.458	0.318
	38	Ohmic losses per cable		kW/km	40.2	41.8	44.0	46.6
	39	Fict. thermal resist. of cable for ohmic losses		Km/W	0.671	0.623	0.581	0.536
	41	Fict. thermal resist. of soil for ohmic losses		Km/W	0.455	0.448	0.440	0.431
	44	Inductance/unit length per conductor		mH/km	0.504	0.476	0.454	0.427
Installation in free air	32	Current-carrying capacity		A	108	131	157	195
	33	Permissible transmission power		MVA	3.74	4.54	5.44	6.75
	34	Eff. a.c. resistance/unit length at 65 °C		Ω/km	0.858	0.619	0.458	0.318
	38	Ohmic losses per cable		kW/km	30.0	31.9	33.9	36.3
	39	Fict. thermal resist. of cable for ohmic losses		Km/W	0.671	0.623	0.581	0.536
	43	Thermal resistance of the air		Km/W	0.489	0.472	0.449	0.426
	44	Inductance/unit length per conductor		mH/km	0.504	0.476	0.454	0.427
	48	Minimum time value		s	308	410	583	740
Short-circuit	49	Rated short-time current of conductor (1s)		kA	2.90	4.06	5.80	8.12

95 RM	120 RM	150 RM	185 RM	240 RM	300 RM
5.5	5.5	5.5	5.5	5.5	5.5
1.4	1.4	1.5	1.5	1.5	1.6
25.5	27.0	28.5	30.0	32.5	35.0
1.0	1.0	1.0	1.0	1.0	1.0
68	71	74	78	83	87
10500	11600	13200	14700	17100	19800
1050	1110	1155	1215	1290	1365
4624	5041	5476	6084	6889	7569
14250	18000	22500	27750	36000	45000
0.227	0.180	0.146	0.117	0.0887	0.0707
0.193	0.153	0.124	0.0991	0.0754	0.0601
0.310	0.338	0.367	0.398	0.445	0.483
1.12	1.23	1.33	1.44	1.61	1.75
3.36	3.69	3.99	4.32	4.83	5.25
3.10	3.00	2.90	2.90	2.80	2.70
71.8	76.2	80.1	83.7	89.7	94.2
266	302	341	387	451	509
9.21	10.5	11.8	13.4	15.6	17.6
0.230	0.184	0.150	0.121	0.0941	0.0768
48.9	50.3	52.3	54.4	57.4	59.7
0.498	0.482	0.455	0.432	0.399	0.377
0.422	0.412	0.404	0.397	0.386	0.378
0.409	0.393	0.382	0.370	0.355	0.345
237	271	308	350	412	468
8.21	9.39	10.7	12.1	14.3	16.2
0.230	0.184	0.150	0.121	0.0941	0.0768
38.8	40.5	42.7	44.5	47.9	50.5
0.498	0.482	0.455	0.432	0.399	0.377
0.404	0.384	0.367	0.353	0.331	0.317
0.409	0.393	0.382	0.370	0.355	0.345
923	1127	1363	1605	1950	2361
11.0	13.9	17.4	21.5	27.8	34.8

Table 5.4.3
Cables with paper insulation with copper conductor

Three-phase operation at 50 Hz

Design				35	50	70	95
	2	Nom. cross-sectional area of conductor	mm^2	35	50	70	95
	3	Shape and type of conductor		RM	RM	RM	RM
	9	Thickness of insulation	mm	7.5	7.5	7.5	7.5
	12	Thickness of metal sheath	mm	1.4	1.4	1.4	1.5
	13	Diameter over metal sheath	mm	25.0	26.0	28.0	29.5
	15	Thickness of an armour tape	mm	1.0	1.0	1.0	1.0
	17	Overall diameter of cable	mm	67	69	73	76
	19	Weight of cable	kg/km	8950	9700	10900	12500
Mechanical properties	20	Minimum bending radius	mm	1035	1080	1125	1200
	21	Perm. pulling force with cable grip	N	4489	4761	5329	5776
	22	Perm. pulling force with pulling head	N	5250	7500	10500	14250
Electrical properties	23	d.c. resistance/unit length at 60 °C	Ω/km	0.606	0.448	0.310	0.223
	24	d.c. resistance/unit length at 20 °C	Ω/km	0.524	0.387	0.268	0.193
	25	Operating capacitance/unit length	µF/km	0.170	0.194	0.224	0.248
	26	Charging current	A/km	0.930	1.06	1.22	1.35
	28	Earth fault current	A/km	2.79	3.18	3.66	4.05
	30	Electric field strength at the conductor	kV/mm	4.10	4.00	3.90	3.70
	31	Reference diameter of cable	mm	74.9	75.3	77.7	81.8
Installation in ground	32	Current-carrying capacity	A	141	168	208	250
	33	Permissible transmission power	MVA	7.33	8.73	10.8	13.0
	34	Eff. a.c. resistance/unit length at 60 °C	Ω/km	0.609	0.451	0.313	0.227
	38	Ohmic losses per cable	kW/km	36.3	38.2	40.7	42.6
	39	Fict. thermal resist. of cable for ohmic losses	Km/W	0.678	0.629	0.574	0.535
	41	Fict. thermal resist. of soil for ohmic losses	Km/W	0.415	0.414	0.409	0.401
	44	Inductance/unit length per conductor	mH/km	0.531	0.496	0.462	0.442
Installation in free air	32	Current-carrying capacity	A	123	147	182	220
	33	Permissible transmission power	MVA	6.39	7.64	9.46	11.4
	34	Eff. a.c. resistance/unit length at 60 °C	Ω/km	0.609	0.451	0.313	0.227
	38	Ohmic losses per cable	kW/km	27.7	29.2	31.1	33.0
	39	Fict. thermal resist. of cable for ohmic losses	Km/W	0.678	0.629	0.574	0.535
	43	Thermal resistance of the air	Km/W	0.406	0.402	0.390	0.372
	44	Inductance/unit length per conductor	mH/km	0.531	0.496	0.462	0.442
	48	Minimum time value	s	406	579	741	934
Short-circuit	49	Rated short-time current of conductor (1s)	kA	3.88	5.55	7.77	10.5

120 RM	150 RM	185 RM	240 RM	300 RM
7.5	7.5	7.5	7.5	7.5
1.5	1.5	1.6	1.6	1.7
31.0	32.5	34.5	37.0	39.0
1.0	1.0	1.0	1.0	1.0
79	83	86	91	96
13700	15000	16900	19400	22200
1245	1290	1350	1440	1500
6241	6889	7396	8281	9216
18000	22500	27750	36000	45000
0.177	0.143	0.115	0.0873	0.0695
0.153	0.124	0.0991	0.0754	0.0601
0.269	0.290	0.313	0.348	0.376
1.46	1.58	1.70	1.89	2.05
4.38	4.74	5.10	5.67	6.15
3.60	3.50	3.40	3.30	3.20
85.2	89.1	92.7	98.9	103
286	323	365	426	481
14.9	16.8	19.0	22.1	25.0
0.181	0.148	0.120	0.0932	0.0760
44.5	46.4	47.8	50.7	52.8
0.505	0.477	0.453	0.420	0.396
0.394	0.387	0.381	0.370	0.363
0.424	0.411	0.398	0.382	0.369
253	287	326	383	435
13.1	14.9	16.9	19.9	22.6
0.181	0.148	0.120	0.0932	0.0760
34.8	36.6	38.2	41.0	43.1
0.505	0.477	0.453	0.420	0.396
0.358	0.343	0.331	0.312	0.300
0.424	0.411	0.398	0.382	0.369
1127	1368	1613	1967	2382
13.3	16.6	20.5	26.6	33.3

Table 5.5.1a
PROTODUR cables (PVC insulation) with aluminium conductor

Three-phase operation at 50 Hz

				50	70	95	120
Design	2	Nom. cross-sectional area of conductor	mm²	50	70	95	120
	3	Shape and type of conductor		RM	RM	RM	RM
	9	Thickness of insulation	mm	1.4	1.4	1.6	1.6
	16	Thickness of outer sheath	mm	1.8	1.8	1.8	1.8
	17	Overall diameter of cable	mm	16	17	19	21
	19	Weight of cable	kg/km	355	430	550	650
Mechanical properties	20	Minimum bending radius	mm	240	255	285	315
	21	Perm. pulling force with cable grip	N	1500	2100	2850	3600
	22	Perm. pulling force with pulling head	N	1500	2100	2850	3600
Electrical properties	23	d.c. resistance/unit length at 70 °C	Ω/km	0.770	0.532	0.384	0.304
	24	d.c. resistance/unit length at 20 °C	Ω/km	0.641	0.443	0.320	0.253
	31	Reference diameter of cable	mm	14.6	16.3	18.3	19.8
Laying direct in ground	32	Current-carrying capacity	A	150	184	220	251
	34	Eff. a.c. resistance/unit length at 70 °C	Ω/km	0.770	0.533	0.385	0.305
	38	Ohmic losses per cable	kW/km	17.3	18.0	18.6	19.2
	39	Fict. thermal resist. of cable for ohmic losses	Km/W	0.551	0.476	0.443	0.402
	41	Fict. thermal resist. of soil for ohmic losses	Km/W	4.512	4.380	4.242	4.148
	44	Inductance/unit length per conductor	mH/km	0.325	0.309	0.302	0.294
Installation in free air	32	Current-carrying capacity	A	129	163	201	234
	34	Eff. a.c. resistance/unit length at 70 °C	Ω/km	0.770	0.533	0.385	0.305
	38	Ohmic losses per cable	kW/km	12.8	14.2	15.6	16.7
	39	Fict. thermal resist. of cable for ohmic losses	Km/W	0.742	0.648	0.606	0.553
	43	Thermal resistance of the air	Km/W	2.392	2.175	1.974	1.843
	44	Inductance/unit length per conductor	mH/km	0.325	0.309	0.302	0.294
	48	Minimum time value	s	425	522	632	744
Short-circuit	49	Rated short-time current of conductor (1s)	kA	3.80	5.32	7.22	9.12

150 RM	185 RM	240 RM
1.8	2.0	2.2
1.8	1.8	1.8
23	25	27
770	930	1200
345	375	405
4500	5550	7200
4500	5550	7200
0.247	0.197	0.150
0.206	0.164	0.125
21.7	23.7	26.6
282	320	373
0.248	0.198	0.152
19.8	20.3	21.1
0.385	0.369	0.342
4.039	3.933	3.796
0.290	0.287	0.281
268	310	371
0.248	0.198	0.152
17.8	19.1	20.9
0.531	0.511	0.475
1.706	1.584	1.435
0.290	0.287	0.281
886	1008	1184
11.4	14.1	18.2

Table 5.5.1b
PROTODUR cables (PVC insulation) with aluminium conductor

Three-phase operation at 50 Hz

Design	2	Nom. cross-sectional area of conductor	mm^2	50	70	95	120
	3	Shape and type of conductor		RM	RM	RM	RM
	9	Thickness of insulation	mm	1.4	1.4	1.6	1.6
	16	Thickness of outer sheath	mm	1.8	1.8	1.8	1.8
	17	Overall diameter of cable	mm	16	17	19	21
	19	Weight of cable	kg/km	355	430	550	650
Mechanical properties	20	Minimum bending radius	mm	240	255	285	315
	21	Perm. pulling force with cable grip	N	1500	2100	2850	3600
	22	Perm. pulling force with pulling head	N	1500	2100	2850	3600
Electrical properties	23	d.c. resistance/unit length at 70 °C	Ω/km	0.770	0.532	0.384	0.304
	24	d.c. resistance/unit length at 20 °C	Ω/km	0.641	0.443	0.320	0.253
	31	Reference diameter of cable	mm	14.6	16.3	18.3	19.8
Laying direct in ground	32	Current-carrying capacity	A	176	216	257	293
	34	Eff. a.c. resistance/unit length at 70 °C	Ω/km	0.770	0.532	0.385	0.304
	38	Ohmic losses per cable	kW/km	23.9	24.8	25.4	26.1
	39	Fict. thermal resist. of cable for ohmic losses	Km/W	0.551	0.476	0.443	0.402
	41	Fict. thermal resist. of soil for ohmic losses	Km/W	3.115	3.055	2.991	2.946
	44	Inductance/unit length per conductor	mH/km	0.702	0.668	0.643	0.621
Installation in free air	32	Current-carrying capacity	A	158	201	246	287
	34	Eff. a.c. resistance/unit length at 70 °C	Ω/km	0.770	0.532	0.385	0.304
	38	Ohmic losses per cable	kW/km	19.2	21.5	23.3	25.1
	39	Fict. thermal resist. of cable for ohmic losses	Km/W	0.551	0.476	0.443	0.402
	43	Thermal resistance of the air	Km/W	1.519	1.389	1.268	1.189
	44	Inductance/unit length per conductor	mH/km	0.489	0.473	0.466	0.458
	48	Minimum time value	s	283	343	422	495
Short-circuit	49	Rated short-time current of conductor (1s)	kA	3.80	5.32	7.22	9.12

⊙ ⊙ ⊙   NAYY 1×... 0.6/1 kV (U_m=1.2 kV)

150 RM	185 RM	240 RM
1.8	2.0	2.2
1.8	1.8	1.8
23	25	27
770	930	1200
345	375	405
4500	5550	7200
4500	5550	7200
0.247	0.197	0.150
0.206	0.164	0.125
21.7	23.7	26.6
328	372	432
0.248	0.197	0.151
26.7	27.3	28.1
0.385	0.369	0.342
2.893	2.841	2.771
0.604	0.587	0.564
329	380	455
0.248	0.198	0.151
26.8	28.5	31.3
0.385	0.369	0.342
1.105	1.030	0.938
0.454	0.451	0.445
588	671	787
11.4	14.1	18.2

Table 5.5.2
PROTODUR cables (PVC insulation) with aluminium conductor

Three-phase operation at 50 Hz

Design	2	Nom. cross-sectional area of conductor	mm²	25	35	50	70
	3	Shape and type of conductor		RE	RE	SE	SE
	9	Thickness of insulation	mm	1.2	1.2	1.4	1.4
	16	Thickness of outer sheath	mm	1.8	1.8	1.9	2.1
	17	Overall diameter of cable	mm	26	28	30	34
	19	Weight of cable	kg/km	1050	1250	1350	1850
Mechanical properties	20	Minimum bending radius	mm	312	336	360	408
	21	Perm. pulling force with cable grip	N	3000	4200	6000	8400
	22	Perm. pulling force with pulling head	N	3000	4200	6000	8400
Electrical properties	23	d.c. resistance/unit length at 70 °C	Ω/km	1.44	1.04	0.770	0.532
	24	d.c. resistance/unit length at 20 °C	Ω/km	1.20	0.868	0.641	0.443
	31	Reference diameter of cable	mm	25.0	27.3	28.2	32.4
Laying direct in ground	32	Current-carrying capacity	A	99	119	141	176
	34	Eff. a.c. resistance/unit length at 70 °C	Ω/km	1.44	1.04	0.770	0.533
	38	Ohmic losses per cable	kW/km	42.4	44.3	45.9	49.5
	39	Fict. thermal resist. of cable for ohmic losses	Km/W	0.576	0.528	0.466	0.405
	41	Fict. thermal resist. of soil for ohmic losses	Km/W	1.474	1.439	1.426	1.371
	44	Inductance/unit length per conductor	mH/km	0.283	0.274	0.270	0.262
Installation in free air	32	Current-carrying capacity	A	78	96	115	148
	34	Eff. a.c. resistance/unit length at 70 °C	Ω/km	1.44	1.04	0.770	0.533
	38	Ohmic losses per cable	kW/km	26.3	28.8	30.6	35.0
	39	Fict. thermal resist. of cable for ohmic losses	Km/W	0.576	0.528	0.466	0.405
	43	Thermal resistance of the air	Km/W	0.934	0.867	0.839	0.745
	44	Inductance/unit length per conductor	mH/km	0.283	0.274	0.270	0.262
	48	Minimum time value	s	291	376	535	633
Short-circuit	49	Rated short-time current of conductor (1s)	kA	1.90	2.66	3.80	5.32

95 SE	120 SE	150 SE	185 SE	240 SE
1.6	1.6	1.8	2.0	2.2
2.2	2.4	2.5	2.7	2.9
38	42	46	50	56
2300	2800	3300	4050	5050
456	504	552	600	672
11400	14400	18000	22200	28800
11400	14400	18000	22200	28800
0.384	0.304	0.247	0.197	0.150
0.320	0.253	0.206	0.164	0.125
36.3	39.9	43.3	48.3	53.6
211	240	270	307	357
0.385	0.305	0.248	0.198	0.152
51.4	52.7	54.3	56.0	58.0
0.381	0.369	0.359	0.356	0.337
1.325	1.288	1.255	1.212	1.170
0.261	0.256	0.256	0.256	0.254
181	210	238	275	326
0.385	0.305	0.248	0.198	0.152
37.8	40.3	42.2	44.9	48.3
0.381	0.369	0.359	0.356	0.337
0.678	0.627	0.586	0.536	0.491
0.261	0.256	0.256	0.256	0.254
780	924	1124	1281	1534
7.22	9.12	11.4	14.1	18.2

Table 5.5.3
PROTODUR cables (PVC insulation) with aluminium conductor

Three-phase operation at 50 Hz

Design	2	Nom. cross-sectional area of conductor	mm^2	50	70	95	120
	3	Shape and type of conductor		SE	SE	SE	SE
	7	Nom. cross-sectional area of concentric cond.	mm^2	50	70	95	120
	9	Thickness of insulation	mm	1.4	1.4	1.6	1.6
	16	Thickness of outer sheath	mm	1.9	2.0	2.2	2.3
	17	Overall diameter of cable	mm	29	33	37	40
	19	Weight of cable	kg/km	1400	1950	2500	2900
Mechanical properties	20	Minimum bending radius	mm	348	396	444	480
	21	Perm. pulling force with cable grip	N	4500	6300	8550	10800
	22	Perm. pulling force with pulling head	N	4500	6300	8550	10800
Electrical properties	23	d.c. resistance/unit length at 70 °C	Ω/km	0.770	0.532	0.384	0.304
	24	d.c. resistance/unit length at 20 °C	Ω/km	0.641	0.443	0.320	0.253
	31	Reference diameter of cable	mm	26.2	30.3	34.3	37.3
Laying direct in ground	32	Current-carrying capacity	A	142	175	210	240
	34	Eff. a.c. resistance/unit length at 70 °C	Ω/km	0.771	0.534	0.387	0.307
	38	Ohmic losses per cable	kW/km	46.7	49.1	51.2	53.1
	39	Fict. thermal resist. of cable for ohmic losses	Km/W	0.429	0.386	0.367	0.338
	41	Fict. thermal resist. of soil for ohmic losses	Km/W	1.455	1.397	1.348	1.314
	44	Inductance/unit length per conductor	mH/km	0.247	0.238	0.238	0.233
Installation in free air	32	Current-carrying capacity	A	115	146	179	209
	34	Eff. a.c. resistance/unit length at 70 °C	Ω/km	0.771	0.534	0.387	0.307
	38	Ohmic losses per cable	kW/km	30.6	34.2	37.2	40.3
	39	Fict. thermal resist. of cable for ohmic losses	Km/W	0.429	0.386	0.367	0.338
	43	Thermal resistance of the air	Km/W	0.885	0.784	0.708	0.659
	44	Inductance/unit length per conductor	mH/km	0.247	0.238	0.238	0.233
	48	Minimum time value	s	535	650	797	933
Short-circuit	49	Rated short-time current of conductor (1s)	kA	3.80	5.32	7.22	9.12

150	185
SE	SE
150	185
1.8	2.0
2.4	2.6
44	48
3550	4300
528	576
13500	16650
13500	16650
0.247	0.197
0.206	0.164
40.8	44.9
268	303
0.251	0.202
54.2	55.6
0.339	0.327
1.279	1.241
0.233	0.233
236	272
0.251	0.202
42.0	44.8
0.339	0.327
0.613	0.566
0.233	0.233
1143	1309
11.4	14.1

Table 5.6.1a
PROTOTHEN X cables (XLPE insulation) with aluminium conductor

Three-phase operation at 50 Hz

Design				50	70	95	120
	2	Nom. cross-sectional area of conductor	mm²	50	70	95	120
	3	Shape and type of conductor		RM	RM	RM	RM
	9	Thickness of insulation	mm	1.0	1.1	1.1	1.2
	16	Thickness of outer sheath	mm	1.4	1.4	1.5	1.5
	17	Overall diameter of cable	mm	14	15	17	19
	19	Weight of cable	kg/km	250	325	425	520
Mechanical properties	20	Minimum bending radius	mm	210	225	255	285
	21	Perm. pulling force with cable grip	N	1500	2100	2850	3600
	22	Perm. pulling force with pulling head	N	1500	2100	2850	3600
Electrical properties	23	d.c. resistance/unit length at 90 °C	Ω/km	0.822	0.568	0.410	0.324
	24	d.c. resistance/unit length at 20 °C	Ω/km	0.641	0.443	0.320	0.253
	31	Reference diameter of cable	mm	13.0	14.9	16.7	18.4
Laying direct in ground	32	Current-carrying capacity	A	162	199	238	271
	34	Eff. a.c. resistance/unit length at 90 °C	Ω/km	0.822	0.568	0.411	0.325
	38	Ohmic losses per cable	kW/km	21.6	22.5	23.3	23.9
	39	Fict. thermal resist. of cable for ohmic losses	Km/W	0.354	0.311	0.287	0.264
	41	Fict. thermal resist. of soil for ohmic losses	Km/W	4.650	4.487	4.351	4.236
	44	Inductance/unit length per conductor	mH/km	0.302	0.291	0.284	0.279
Installation in free air	32	Current-carrying capacity	A	159	203	250	293
	34	Eff. a.c. resistance/unit length at 90 °C	Ω/km	0.822	0.568	0.411	0.325
	38	Ohmic losses per cable	kW/km	20.8	23.4	25.7	27.9
	39	Fict. thermal resist. of cable for ohmic losses	Km/W	0.483	0.428	0.397	0.368
	43	Thermal resistance of the air	Km/W	2.391	2.130	1.935	1.782
	44	Inductance/unit length per conductor	mH/km	0.302	0.291	0.284	0.279
	48	Minimum time value	s	393	473	574	667
Short-circuit	49	Rated short-time current of conductor (1s)	kA	4.70	6.58	8.93	11.3

150 RM	185 RM	240 RM
1.4	1.6	1.7
1.6	1.6	1.7
21	23	26
630	780	980
315	345	390
4500	5550	7200
4500	5550	7200
0.264	0.210	0.160
0.206	0.164	0.125
20.5	22.5	25.4
305	346	403
0.265	0.212	0.162
24.7	25.3	26.3
0.260	0.247	0.231
4.107	3.996	3.851
0.279	0.276	0.272
337	392	470
0.265	0.212	0.162
30.1	32.5	35.8
0.363	0.345	0.323
1.629	1.505	1.357
0.279	0.276	0.272
788	886	1037
14.1	17.4	22.6

Table 5.6.1b
PROTOTHEN X cables (XLPE insulation) with aluminium conductor

Three-phase operation at 50 Hz

Design	2	Nom. cross-sectional area of conductor	mm²	50	70	95	120
	3	Shape and type of conductor		RM	RM	RM	RM
	9	Thickness of insulation	mm	1.0	1.1	1.1	1.2
	16	Thickness of outer sheath	mm	1.4	1.4	1.5	1.5
	17	Overall diameter of cable	mm	14	15	17	19
	19	Weight of cable	kg/km	250	325	425	520
Mechanical properties	20	Minimum bending radius	mm	210	225	255	285
	21	Perm. pulling force with cable grip	N	1500	2100	2850	3600
	22	Perm. pulling force with pulling head	N	1500	2100	2850	3600
Electrical properties	23	d.c. resistance/unit length at 90 °C	Ω/km	0.822	0.568	0.410	0.324
	24	d.c. resistance/unit length at 20 °C	Ω/km	0.641	0.443	0.320	0.253
	31	Reference diameter of cable	mm	13.0	14.9	16.7	18.4
Laying direct in ground	32	Current-carrying capacity	A	192	235	280	319
	34	Eff. a.c. resistance/unit length at 90 °C	Ω/km	0.822	0.568	0.410	0.325
	38	Ohmic losses per cable	kW/km	30.3	31.4	32.2	33.0
	39	Fict. thermal resist. of cable for ohmic losses	Km/W	0.354	0.311	0.287	0.264
	41	Fict. thermal resist. of soil for ohmic losses	Km/W	3.176	3.104	3.042	2.988
	44	Inductance/unit length per conductor	mH/km	0.698	0.665	0.639	0.618
Installation in free air	32	Current-carrying capacity	A	198	252	310	362
	34	Eff. a.c. resistance/unit length at 90 °C	Ω/km	0.822	0.568	0.411	0.325
	38	Ohmic losses per cable	kW/km	32.2	36.1	39.5	42.6
	39	Fict. thermal resist. of cable for ohmic losses	Km/W	0.354	0.311	0.287	0.264
	43	Thermal resistance of the air	Km/W	1.509	1.354	1.238	1.145
	44	Inductance/unit length per conductor	mH/km	0.466	0.455	0.448	0.443
	48	Minimum time value	s	254	307	374	437
Short-circuit	49	Rated short-time current of conductor (1s)	kA	4.70	6.58	8.93	11.3

150 RM	185 RM	240 RM
1.4	1.6	1.7
1.6	1.6	1.7
21	23	26
630	780	980
315	345	390
4500	5550	7200
4500	5550	7200
0.264	0.210	0.160
0.206	0.164	0.125
20.5	22.5	25.4
357	404	470
0.264	0.211	0.161
33.7	34.4	35.5
0.260	0.247	0.231
2.927	2.872	2.799
0.601	0.585	0.562
416	482	578
0.265	0.211	0.161
45.8	49.0	53.8
0.260	0.247	0.231
1.052	0.976	0.885
0.443	0.440	0.436
517	586	686
14.1	17.4	22.6

Table 5.6.2
PROTOTHEN X cables (XLPE insulation) with aluminium conductor

Three-phase operation at 50 Hz

Design	2	Nom. cross-sectional area of conductor	mm^2	25	35	50	70
	3	Shape and type of conductor		RE	RE	SE	SE
	9	Thickness of insulation	mm	0.9	0.9	1.0	1.1
	16	Thickness of outer sheath	mm	1.8	1.8	1.9	2.0
	17	Overall diameter of cable	mm	24	26	27	32
	19	Weight of cable	kg/km	820	1050	1100	1500
Mechanical properties	20	Minimum bending radius	mm	288	312	324	384
	21	Perm. pulling force with cable grip	N	3000	4200	6000	8400
	22	Perm. pulling force with pulling head	N	3000	4200	6000	8400
Electrical properties	23	d.c. resistance/unit length at 90 °C	Ω/km	1.54	1.11	0.822	0.568
	24	d.c. resistance/unit length at 20 °C	Ω/km	1.20	0.868	0.641	0.443
	31	Reference diameter of cable	mm	23.5	25.8	25.8	30.7
Laying direct in ground	32	Current-carrying capacity	A	111	133	157	194
	34	Eff. a.c. resistance/unit length at 90 °C	Ω/km	1.54	1.11	0.822	0.568
	38	Ohmic losses per cable	kW/km	56.9	59.1	60.8	64.2
	39	Fict. thermal resist. of cable for ohmic losses	Km/W	0.393	0.358	0.305	0.275
	41	Fict. thermal resist. of soil for ohmic losses	Km/W	1.498	1.461	1.461	1.392
	44	Inductance/unit length per conductor	mH/km	0.267	0.260	0.255	0.252
Installation in free air	32	Current-carrying capacity	A	101	123	147	189
	34	Eff. a.c. resistance/unit length at 90 °C	Ω/km	1.54	1.11	0.822	0.568
	38	Ohmic losses per cable	kW/km	47.1	50.5	53.3	60.9
	39	Fict. thermal resist. of cable for ohmic losses	Km/W	0.393	0.358	0.305	0.275
	43	Thermal resistance of the air	Km/W	0.893	0.826	0.818	0.708
	44	Inductance/unit length per conductor	mH/km	0.267	0.260	0.255	0.252
	48	Minimum time value	s	244	322	460	546
Short-circuit	49	Rated short-time current of conductor (1s)	kA	2.35	3.29	4.70	6.58

95 SE	120 SE	150 SE	185 SE	240 SE
1.1	1.2	1.4	1.6	1.7
2.1	2.3	2.4	2.6	2.8
35	39	42	47	52
1850	2350	2800	3400	4300
420	468	504	564	624
11400	14400	18000	22200	28800
11400	14400	18000	22200	28800
0.410	0.324	0.264	0.210	0.160
0.320	0.253	0.206	0.164	0.125
33.7	37.8	41.2	45.7	51.0
233	266	299	340	395
0.411	0.325	0.265	0.211	0.162
66.9	69.0	71.0	73.3	75.7
0.253	0.250	0.243	0.235	0.228
1.355	1.309	1.275	1.234	1.190
0.247	0.246	0.247	0.248	0.245
232	270	308	356	422
0.411	0.325	0.265	0.211	0.162
66.3	71.1	75.4	80.4	86.4
0.253	0.250	0.243	0.235	0.228
0.654	0.596	0.555	0.510	0.465
0.247	0.246	0.247	0.248	0.245
667	786	943	1074	1287
8.93	11.3	14.1	17.4	22.6

Table 5.6.3 a
PROTOTHEN X cables (XLPE insulation) with aluminium conductor

Three-phase operation at 50 Hz

Design	2	Nom. cross-sectional area of conductor	mm^2	25	35	50	70
	3	Shape and type of conductor		RM	RM	RM	RM
	6	Nom. cross-sectional area of screen	mm^2	16	16	16	16
	9	Thickness of insulation	mm	3.4	3.4	3.4	3.4
	16	Thickness of outer sheath	mm	2.5	2.5	2.5	2.5
	17	Overall diameter of cable	mm	23	24	26	27
	19	Weight of cable	kg/km	570	620	680	770
Mechanical properties	20	Minimum bending radius	mm	345	360	390	405
	21	Perm. pulling force with cable grip	N	750	1050	1500	2100
	22	Perm. pulling force with pulling head	N	750	1050	1500	2100
Electrical properties	23	d.c. resistance/unit length at 90 °C	Ω/km	1.54	1.11	0.822	0.568
	24	d.c. resistance/unit length at 20 °C	Ω/km	1.20	0.868	0.641	0.443
	25	Operating capacitance/unit length	μF/km	0.203	0.225	0.249	0.283
	26	Charging current	A/km	0.370	0.410	0.450	0.510
	28	Earth fault current	A/km	1.11	1.23	1.35	1.53
	30	Electric field strength at the conductor	kV/mm	2.40	2.30	2.20	2.20
	31	Reference diameter of cable	mm	22.9	24.0	25.2	26.9
Installation in ground	32	Current-carrying capacity	A	122	145	171	209
	33	Permissible transmission power	MVA	2.11	2.51	2.96	3.62
	34	Eff. a.c. resistance/unit length at 90 °C	Ω/km	1.54	1.12	0.825	0.571
	36	Eff. a.c. resistance/unit length at 20 °C	Ω/km	1.20	0.872	0.644	0.446
	38	Ohmic losses per cable	kW/km	23.0	23.5	24.1	25.0
	39	Fict. thermal resist. of cable for ohmic losses	Km/W	0.707	0.640	0.582	0.516
	41	Fict. thermal resist. of soil for ohmic losses	Km/W	3.974	3.918	3.860	3.782
	44	Inductance/unit length per conductor	mH/km	0.480	0.455	0.434	0.409
	45	Resistance/unit length in zero-phase seq.system	Ω/km	2.068	1.734	1.505	1.304
	46	Reactance/unit length in zero-phase seq.system	Ω/km	0.555	0.549	0.544	0.539
	47	Reduction factor		0.556	0.559	0.562	0.565
Installation in free air	32	Current-carrying capacity	A	127	153	184	229
	33	Permissible transmission power	MVA	2.20	2.65	3.19	3.97
	34	Eff. a.c. resistance/unit length at 90 °C	Ω/km	1.54	1.12	0.825	0.571
	38	Ohmic losses per cable	kW/km	24.9	26.1	27.9	30.0
	39	Fict. thermal resist. of cable for ohmic losses	Km/W	0.870	0.795	0.728	0.653
	43	Thermal resistance of the air	Km/W	1.559	1.492	1.426	1.343
	44	Inductance/unit length per conductor	mH/km	0.480	0.455	0.433	0.409
	48	Minimum time value	s	154	208	294	372
Short-circuit	49	Rated short-time current of conductor (1s)	kA	2.35	3.29	4.70	6.58
	50	Rated short-time current of screen (1s)	kA	3.3	3.3	3.3	3.3

95 RM 16	120 RM 16	150 RM 25	185 RM 25	240 RM 25	300 RM 25	400 RM 35	500 RM 35
3.4	3.4	3.4	3.4	3.4	3.4	3.4	3.4
2.5	2.5	2.5	2.5	2.5	2.5	2.5	2.5
29	30	32	33	36	39	42	45
880	980	1200	1350	1550	1800	2150	2550
435	450	480	495	540	585	630	675
2850	3600	4500	5550	7200	9000	12000	15000
2850	3600	4500	5550	7200	9000	12000	15000
0.410	0.324	0.264	0.210	0.160	0.128	0.0997	0.0776
0.320	0.253	0.206	0.164	0.125	0.1000	0.0778	0.0605
0.315	0.345	0.374	0.406	0.456	0.495	0.558	0.613
0.570	0.630	0.680	0.740	0.830	0.900	1.01	1.11
1.71	1.89	2.04	2.22	2.49	2.70	3.03	3.33
2.10	2.10	2.00	2.00	2.00	1.90	1.90	1.90
28.5	30.0	31.5	33.1	35.6	37.6	40.8	43.6
249	283	316	358	415	467	531	603
4.31	4.90	5.47	6.20	7.19	8.09	9.20	10.4
0.414	0.328	0.269	0.215	0.166	0.134	0.107	0.0857
0.323	0.256	0.211	0.169	0.130	0.106	0.0853	0.0686
25.6	26.3	26.9	27.6	28.5	29.2	30.2	31.2
0.467	0.430	0.396	0.367	0.329	0.303	0.267	0.244
3.713	3.652	3.594	3.535	3.448	3.383	3.285	3.206
0.391	0.377	0.364	0.353	0.338	0.329	0.315	0.306
1.178	1.109	0.849	0.807	0.766	0.740	0.567	0.549
0.536	0.533	0.285	0.283	0.281	0.279	0.171	0.169
0.568	0.571	0.416	0.418	0.421	0.423	0.325	0.327
278	322	365	419	496	568	662	766
4.82	5.58	6.32	7.26	8.59	9.84	11.5	13.3
0.414	0.328	0.269	0.215	0.166	0.134	0.107	0.0858
32.0	34.0	35.9	37.8	40.8	43.2	47.0	50.3
0.597	0.552	0.512	0.477	0.432	0.401	0.358	0.326
1.274	1.216	1.162	1.111	1.040	0.990	0.919	0.865
0.391	0.377	0.364	0.353	0.338	0.329	0.315	0.306
465	552	672	775	931	1110	1452	1695
8.93	11.3	14.1	17.4	22.6	28.2	37.6	47.0
3.3	3.3	5.1	5.1	5.1	5.1	7.1	7.1

Table 5.6.3 b
PROTOTHEN X cables (XLPE insulation) with aluminium conductor

Three-phase operation at 50 Hz

Design	2	Nom. cross-sectional area of conductor	mm²	25	35	50	70
	3	Shape and type of conductor		RM	RM	RM	RM
	6	Nom. cross-sectional area of screen	mm²	16	16	16	16
	9	Thickness of insulation	mm	3.4	3.4	3.4	3.4
	16	Thickness of outer sheath	mm	2.5	2.5	2.5	2.5
	17	Overall diameter of cable	mm	23	24	26	27
	19	Weight of cable	kg/km	570	620	680	770
Mechanical properties	20	Minimum bending radius	mm	345	360	390	405
	21	Perm. pulling force with cable grip	N	750	1050	1500	2100
	22	Perm. pulling force with pulling head	N	750	1050	1500	2100
Electrical properties	23	d.c. resistance/unit length at 90 °C	Ω/km	1.54	1.11	0.822	0.568
	24	d.c. resistance/unit length at 20 °C	Ω/km	1.20	0.868	0.641	0.443
	25	Operating capacitance/unit length	μF/km	0.203	0.225	0.249	0.283
	26	Charging current	A/km	0.370	0.410	0.450	0.510
	28	Earth fault current	A/km	1.11	1.23	1.35	1.53
	30	Electric field strength at the conductor	kV/mm	2.40	2.30	2.20	2.20
	31	Reference diameter of cable	mm	22.9	24.0	25.2	26.9
Installation in ground	32	Current-carrying capacity	A	139	166	195	237
	33	Permissible transmission power	MVA	2.41	2.88	3.38	4.10
	34	Eff. a.c. resistance/unit length at 90 °C	Ω/km	1.56	1.13	0.840	0.585
	36	Eff. a.c. resistance/unit length at 20 °C	Ω/km	1.22	0.887	0.659	0.460
	38	Ohmic losses per cable	kW/km	30.1	31.2	31.9	32.9
	39	Fict. thermal resist. of cable for ohmic losses	Km/W	0.701	0.634	0.575	0.507
	41	Fict. thermal resist. of soil for ohmic losses	Km/W	2.862	2.834	2.804	2.764
	44	Inductance/unit length per conductor	mH/km	0.779	0.747	0.719	0.685
	45	Resistance/unit length in zero-phase seq.system	Ω/km	2.024	1.691	1.462	1.262
	46	Reactance/unit length in zero-phase seq.system	Ω/km	0.572	0.566	0.561	0.555
	47	Reduction factor		0.600	0.602	0.604	0.606
Installation in free air	32	Current-carrying capacity	A	151	184	220	276
	33	Permissible transmission power	MVA	2.62	3.19	3.81	4.78
	34	Eff. a.c. resistance/unit length at 90 °C	Ω/km	1.55	1.13	0.835	0.581
	35	Mean eff. a.c. resistance/unit length at 90 °C	Ω/km	1.55	1.12	0.833	0.579
	38	Ohmic losses per cable	kW/km	35.3	38.1	40.3	44.1
	39	Fict. thermal resist. of cable for ohmic losses	Km/W	0.703	0.636	0.577	0.509
	43	Thermal resistance of the air	Km/W	0.983	0.942	0.901	0.849
	44	Inductance/unit length per conductor	mH/km	0.642	0.617	0.595	0.570
	48	Minimum time value	s	109	144	205	256
Short-circuit	49	Rated short-time current of conductor (1s)	kA	2.35	3.29	4.70	6.58
	50	Rated short-time current of screen (1s)	kA	3.3	3.3	3.3	3.3

95 RM 16	120 RM 16	150 RM 25	185 RM 25	240 RM 25	300 RM 25	400 RM 35	500 RM 35
3.4	3.4	3.4	3.4	3.4	3.4	3.4	3.4
2.5	2.5	2.5	2.5	2.5	2.5	2.5	2.5
29	30	32	33	36	39	42	45
880	980	1200	1350	1550	1800	2150	2550
435	450	480	495	540	585	630	675
2850	3600	4500	5550	7200	9000	12000	15000
2850	3600	4500	5550	7200	9000	12000	15000
0.410	0.324	0.264	0.210	0.160	0.128	0.0997	0.0776
0.320	0.253	0.206	0.164	0.125	0.1000	0.0778	0.0605
0.315	0.345	0.374	0.406	0.456	0.495	0.558	0.613
0.570	0.630	0.680	0.740	0.830	0.900	1.01	1.11
1.71	1.89	2.04	2.22	2.49	2.70	3.03	3.33
2.10	2.10	2.00	2.00	2.00	1.90	1.90	1.90
28.5	30.0	31.5	33.1	35.6	37.6	40.8	43.6
281	319	351	394	453	506	560	627
4.87	5.53	6.08	6.82	7.85	8.76	9.70	10.9
0.426	0.340	0.287	0.232	0.181	0.148	0.126	0.103
0.336	0.268	0.229	0.186	0.146	0.120	0.104	0.0855
33.7	34.6	35.3	36.0	37.2	38.0	39.4	40.3
0.457	0.419	0.378	0.347	0.309	0.282	0.239	0.215
2.728	2.696	2.665	2.633	2.585	2.548	2.492	2.446
0.659	0.638	0.613	0.595	0.572	0.556	0.526	0.511
1.138	1.069	0.833	0.791	0.751	0.725	0.560	0.543
0.551	0.548	0.298	0.296	0.293	0.291	0.179	0.177
0.608	0.609	0.459	0.460	0.462	0.463	0.365	0.366
335	386	435	497	586	668	760	868
5.80	6.69	7.53	8.61	10.1	11.6	13.2	15.0
0.423	0.337	0.284	0.230	0.180	0.148	0.126	0.104
0.421	0.335	0.280	0.226	0.176	0.144	0.120	0.0985
47.2	49.9	53.0	55.8	60.4	64.1	69.6	74.2
0.460	0.421	0.381	0.350	0.310	0.282	0.238	0.213
0.806	0.770	0.736	0.704	0.660	0.628	0.583	0.549
0.552	0.539	0.523	0.512	0.497	0.489	0.471	0.462
320	384	473	551	667	802	1102	1320
8.93	11.3	14.1	17.4	22.6	28.2	37.6	47.0
3.3	3.3	5.1	5.1	5.1	5.1	7.1	7.1

Table 5.6.4a
PROTOTHEN X cables (XLPE insulation) with aluminium conductor

Three-phase operation at 50 Hz

Design	2	Nom. cross-sectional area of conductor	mm²	25	35	50	70
	3	Shape and type of conductor		RM	RM	RM	RM
	6	Nom. cross-sectional area of screen	mm²	16	16	16	16
	9	Thickness of insulation	mm	3.4	3.4	3.4	3.4
	16	Thickness of outer sheath	mm	2.5	2.5	2.5	2.5
	17	Overall diameter of cable	mm	25	26	28	30
	19	Weight of cable	kg/km	770	830	910	1050
Mechanical properties	20	Minimum bending radius	mm	375	390	420	450
	21	Perm. pulling force with cable grip	N	750	1050	1500	2100
	22	Perm. pulling force with pulling head	N	750	1050	1500	2100
Electrical properties	23	d.c. resistance/unit length at 90 °C	Ω/km	1.54	1.11	0.822	0.568
	24	d.c. resistance/unit length at 20 °C	Ω/km	1.20	0.868	0.641	0.443
	25	Operating capacitance/unit length	μF/km	0.203	0.225	0.249	0.283
	26	Charging current	A/km	0.370	0.410	0.450	0.510
	28	Earth fault current	A/km	1.11	1.23	1.35	1.53
	30	Electric field strength at the conductor	kV/mm	2.40	2.30	2.20	2.20
	31	Reference diameter of cable	mm	22.9	24.0	25.2	26.9
Installation in ground	32	Current-carrying capacity	A	122	145	171	209
	33	Permissible transmission power	MVA	2.11	2.51	2.96	3.62
	34	Eff. a.c. resistance/unit length at 90 °C	Ω/km	1.54	1.12	0.825	0.571
	36	Eff. a.c. resistance/unit length at 20 °C	Ω/km	1.20	0.872	0.644	0.446
	38	Ohmic losses per cable	kW/km	23.0	23.5	24.1	25.0
	39	Fict. thermal resist. of cable for ohmic losses	Km/W	0.707	0.640	0.582	0.516
	41	Fict. thermal resist. of soil for ohmic losses	Km/W	3.974	3.918	3.860	3.782
	44	Inductance/unit length per conductor	mH/km	0.480	0.455	0.434	0.409
	45	Resistance/unit length in zero-phase seq.system	Ω/km	2.068	1.734	1.505	1.304
	46	Reactance/unit length in zero-phase seq.system	Ω/km	0.555	0.549	0.544	0.539
	47	Reduction factor		0.556	0.559	0.562	0.565
Installation in free air	32	Current-carrying capacity	A	127	153	184	229
	33	Permissible transmission power	MVA	2.20	2.65	3.19	3.97
	34	Eff. a.c. resistance/unit length at 90 °C	Ω/km	1.54	1.12	0.825	0.571
	38	Ohmic losses per cable	kW/km	24.9	26.1	27.9	30.0
	39	Fict. thermal resist. of cable for ohmic losses	Km/W	0.870	0.795	0.728	0.653
	43	Thermal resistance of the air	Km/W	1.559	1.492	1.426	1.343
	44	Inductance/unit length per conductor	mH/km	0.480	0.455	0.433	0.409
	48	Minimum time value	s	154	208	294	372
Short-circuit	49	Rated short-time current of conductor (1s)	kA	2.35	3.29	4.70	6.58
	50	Rated short-time current of screen (1s)	kA	3.3	3.3	3.3	3.3

NA2XS(F)2Y 1×.../.. 6/10 kV (U_m=12 kV)

95 RM 16	120 RM 16	150 RM 25	185 RM 25	240 RM 25	300 RM 25	400 RM 35	500 RM 35
3.4	3.4	3.4	3.4	3.4	3.4	3.4	3.4
2.5	2.5	2.5	2.5	2.5	2.5	2.5	2.5
31	32	34	35	38	41	44	47
1150	1250	1450	1600	1850	2150	2600	3000
465	480	510	525	570	615	660	705
2850	3600	4500	5550	7200	9000	12000	15000
2850	3600	4500	5550	7200	9000	12000	15000
0.410	0.324	0.264	0.210	0.160	0.128	0.0997	0.0776
0.320	0.253	0.206	0.164	0.125	0.1000	0.0778	0.0605
0.315	0.345	0.374	0.406	0.456	0.495	0.558	0.613
0.570	0.630	0.680	0.740	0.830	0.900	1.01	1.11
1.71	1.89	2.04	2.22	2.49	2.70	3.03	3.33
2.10	2.10	2.00	2.00	2.00	1.90	1.90	1.90
28.5	30.0	31.5	33.1	35.6	37.6	40.8	43.6
249	283	316	358	415	467	531	603
4.31	4.90	5.47	6.20	7.19	8.09	9.20	10.4
0.414	0.328	0.269	0.215	0.166	0.134	0.107	0.0857
0.323	0.256	0.211	0.169	0.130	0.106	0.0853	0.0686
25.6	26.3	26.9	27.6	28.5	29.2	30.2	31.2
0.467	0.430	0.396	0.367	0.329	0.303	0.267	0.244
3.713	3.652	3.594	3.535	3.448	3.383	3.285	3.206
0.391	0.377	0.364	0.353	0.338	0.329	0.315	0.306
1.178	1.109	0.849	0.807	0.766	0.740	0.567	0.549
0.536	0.533	0.285	0.283	0.281	0.279	0.171	0.169
0.568	0.571	0.416	0.418	0.421	0.423	0.325	0.327
278	322	365	419	496	568	662	766
4.82	5.58	6.32	7.26	8.59	9.84	11.5	13.3
0.414	0.328	0.269	0.215	0.166	0.134	0.107	0.0858
32.0	34.0	35.9	37.8	40.8	43.2	47.0	50.3
0.597	0.552	0.512	0.477	0.432	0.401	0.358	0.326
1.274	1.216	1.162	1.111	1.040	0.990	0.919	0.865
0.391	0.377	0.364	0.353	0.338	0.329	0.315	0.306
465	552	672	775	931	1110	1452	1695
8.93	11.3	14.1	17.4	22.6	28.2	37.6	47.0
3.3	3.3	5.1	5.1	5.1	5.1	7.1	7.1

Table 5.6.4 b
PROTOTHEN X cables (XLPE insulation) with aluminium conductor

Three-phase operation at 50 Hz

Design							
	2	Nom. cross-sectional area of conductor	mm^2	25	35	50	70
	3	Shape and type of conductor		RM	RM	RM	RM
	6	Nom. cross-sectional area of screen	mm^2	16	16	16	16
	9	Thickness of insulation	mm	3.4	3.4	3.4	3.4
	16	Thickness of outer sheath	mm	2.5	2.5	2.5	2.5
	17	Overall diameter of cable	mm	25	26	28	30
	19	Weight of cable	kg/km	770	830	910	1050
Mechanical properties	20	Minimum bending radius	mm	375	390	420	450
	21	Perm. pulling force with cable grip	N	750	1050	1500	2100
	22	Perm. pulling force with pulling head	N	750	1050	1500	2100
Electrical properties	23	d.c. resistance/unit length at 90 °C	Ω/km	1.54	1.11	0.822	0.568
	24	d.c. resistance/unit length at 20 °C	Ω/km	1.20	0.868	0.641	0.443
	25	Operating capacitance/unit length	μF/km	0.203	0.225	0.249	0.283
	26	Charging current	A/km	0.370	0.410	0.450	0.510
	28	Earth fault current	A/km	1.11	1.23	1.35	1.53
	30	Electric field strength at the conductor	kV/mm	2.40	2.30	2.20	2.20
	31	Reference diameter of cable	mm	22.9	24.0	25.2	26.9
Installation in ground	32	Current-carrying capacity	A	139	166	195	237
	33	Permissible transmission power	MVA	2.41	2.88	3.38	4.10
	34	Eff. a.c. resistance/unit length at 90 °C	Ω/km	1.56	1.13	0.840	0.585
	36	Eff. a.c. resistance/unit length at 20 °C	Ω/km	1.22	0.887	0.659	0.460
	38	Ohmic losses per cable	kW/km	30.1	31.2	31.9	32.9
	39	Fict. thermal resist. of cable for ohmic losses	Km/W	0.701	0.634	0.575	0.507
	41	Fict. thermal resist. of soil for ohmic losses	Km/W	2.862	2.834	2.804	2.764
	44	Inductance/unit length per conductor	mH/km	0.779	0.747	0.719	0.685
	45	Resistance/unit length in zero-phase seq.system	Ω/km	2.024	1.691	1.462	1.262
	46	Reactance/unit length in zero-phase seq.system	Ω/km	0.572	0.566	0.561	0.555
	47	Reduction factor		0.600	0.602	0.604	0.606
Installation in free air	32	Current-carrying capacity	A	151	184	220	276
	33	Permissible transmission power	MVA	2.62	3.19	3.81	4.78
	34	Eff. a.c. resistance/unit length at 90 °C	Ω/km	1.55	1.13	0.835	0.581
	35	Mean eff. a.c. resistance/unit length at 90 °C	Ω/km	1.55	1.12	0.833	0.579
	38	Ohmic losses per cable	kW/km	35.3	38.1	40.3	44.1
	39	Fict. thermal resist. of cable for ohmic losses	Km/W	0.703	0.636	0.577	0.509
	43	Thermal resistance of the air	Km/W	0.983	0.942	0.901	0.849
	44	Inductance/unit length per conductor	mH/km	0.642	0.617	0.595	0.570
	48	Minimum time value	s	109	144	205	256
Short-circuit	49	Rated short-time current of conductor (1s)	kA	2.35	3.29	4.70	6.58
	50	Rated short-time current of screen (1s)	kA	3.3	3.3	3.3	3.3

⊙ ⊙ ⊙   NA2XS(F)2Y 1×…/.. 6/10 kV (U_m=12 kV)

95 RM 16	120 RM 16	150 RM 25	185 RM 25	240 RM 25	300 RM 25	400 RM 35	500 RM 35
3.4	3.4	3.4	3.4	3.4	3.4	3.4	3.4
2.5	2.5	2.5	2.5	2.5	2.5	2.5	2.5
31	32	34	35	38	41	44	47
1150	1250	1450	1600	1850	2150	2600	3000
465	480	510	525	570	615	660	705
2850	3600	4500	5550	7200	9000	12000	15000
2850	3600	4500	5550	7200	9000	12000	15000
0.410	0.324	0.264	0.210	0.160	0.128	0.0997	0.0776
0.320	0.253	0.206	0.164	0.125	0.1000	0.0778	0.0605
0.315	0.345	0.374	0.406	0.456	0.495	0.558	0.613
0.570	0.630	0.680	0.740	0.830	0.900	1.01	1.11
1.71	1.89	2.04	2.22	2.49	2.70	3.03	3.33
2.10	2.10	2.00	2.00	2.00	1.90	1.90	1.90
28.5	30.0	31.5	33.1	35.6	37.6	40.8	43.6
281	319	351	394	453	506	560	627
4.87	5.53	6.08	6.82	7.85	8.76	9.70	10.9
0.426	0.340	0.287	0.232	0.181	0.148	0.126	0.103
0.336	0.268	0.229	0.186	0.146	0.120	0.104	0.0855
33.7	34.6	35.3	36.0	37.2	38.0	39.4	40.3
0.457	0.419	0.378	0.347	0.309	0.282	0.239	0.215
2.728	2.696	2.665	2.633	2.585	2.548	2.492	2.446
0.659	0.638	0.613	0.595	0.572	0.556	0.526	0.511
1.138	1.069	0.833	0.791	0.751	0.725	0.560	0.543
0.551	0.548	0.298	0.296	0.293	0.291	0.179	0.177
0.608	0.609	0.459	0.460	0.462	0.463	0.365	0.366
335	386	435	497	586	668	760	868
5.80	6.69	7.53	8.61	10.1	11.6	13.2	15.0
0.423	0.337	0.284	0.230	0.180	0.148	0.126	0.104
0.421	0.335	0.280	0.226	0.176	0.144	0.120	0.0985
47.2	49.9	53.0	55.8	60.4	64.1	69.6	74.2
0.460	0.421	0.381	0.350	0.310	0.282	0.238	0.213
0.806	0.770	0.736	0.704	0.660	0.628	0.583	0.549
0.552	0.539	0.523	0.512	0.497	0.489	0.471	0.462
320	384	473	551	667	802	1102	1320
8.93	11.3	14.1	17.4	22.6	28.2	37.6	47.0
3.3	3.3	5.1	5.1	5.1	5.1	7.1	7.1

Table 5.6.5
PROTOTHEN X cables (XLPE insulation) with aluminium conductor

Three-phase operation at 50 Hz

Design							
Design	2	Nom. cross-sectional area of conductor	mm²	50	70	95	120
	3	Shape and type of conductor		SE	SE	SE	SE
	6	Nom. cross-sectional area of screen	mm²	16	16	16	16
	9	Thickness of insulation	mm	3.4	3.4	3.4	3.4
	16	Thickness of outer sheath	mm	2.5	2.5	2.7	2.8
	17	Overall diameter of cable	mm	42	46	49	51
	19	Weight of cable	kg/km	1500	1800	2150	2450
Mechanical properties	20	Minimum bending radius	mm	630	690	735	765
	21	Perm. pulling force with cable grip	N	4500	6300	8550	10800
	22	Perm. pulling force with pulling head	N	4500	6300	8550	10800
Electrical properties	23	d.c. resistance/unit length at 90 °C	Ω/km	0.822	0.568	0.410	0.324
	24	d.c. resistance/unit length at 20 °C	Ω/km	0.641	0.443	0.320	0.253
	25	Operating capacitance/unit length	μF/km	0.243	0.271	0.309	0.337
	26	Charging current	A/km	0.440	0.490	0.560	0.610
	28	Earth fault current	A/km	1.32	1.47	1.68	1.83
	30	Electric field strength at the conductor	kV/mm	2.20	2.20	2.10	2.10
	31	Reference diameter of cable	mm	40.2	43.8	46.6	49.2
Installation in ground	32	Current-carrying capacity	A	164	201	240	274
	33	Permissible transmission power	MVA	2.84	3.48	4.16	4.75
	34	Eff. a.c. resistance/unit length at 90 °C	Ω/km	0.823	0.569	0.411	0.325
	38	Ohmic losses per cable	kW/km	66.4	69.0	71.1	73.3
	39	Fict. thermal resist. of cable for ohmic losses	Km/W	0.335	0.304	0.285	0.267
	41	Fict. thermal resist. of soil for ohmic losses	Km/W	1.285	1.251	1.226	1.204
	44	Inductance/unit length per conductor	mH/km	0.337	0.318	0.304	0.294
Installation in free air	32	Current-carrying capacity	A	163	205	248	286
	33	Permissible transmission power	MVA	2.82	3.55	4.30	4.95
	34	Eff. a.c. resistance/unit length at 90 °C	Ω/km	0.823	0.569	0.411	0.325
	38	Ohmic losses per cable	kW/km	65.6	71.7	75.9	79.9
	39	Fict. thermal resist. of cable for ohmic losses	Km/W	0.335	0.304	0.285	0.267
	43	Thermal resistance of the air	Km/W	0.578	0.536	0.508	0.484
	44	Inductance/unit length per conductor	mH/km	0.337	0.318	0.304	0.294
	48	Minimum time value	s	374	464	584	700
Short-circuit	49	Rated short-time current of conductor (1s)	kA	4.70	6.58	8.93	11.3
	50	Rated short-time current of screen (1s)	kA	3.3	3.3	3.3	3.3

150	185
SE	SE
25	25
3.4	3.4
2.9	3.0
53	56
2800	3200
795	840
13500	16650
13500	16650
0.264	0.210
0.206	0.164
0.365	0.402
0.660	0.730
1.98	2.19
2.00	2.00
51.4	54.2
306	347
5.30	6.01
0.266	0.212
74.6	76.6
0.254	0.239
1.187	1.166
0.285	0.277
323	372
5.59	6.44
0.266	0.212
83.2	88.1
0.254	0.239
0.466	0.444
0.285	0.277
858	984
14.1	17.4
5.1	5.1

Table 5.6.6 a
PROTOTHEN X cables (XLPE insulation) with aluminium conductor

Three-phase operation at 50 Hz

Design				25	35	50	70
Design	2	Nom. cross-sectional area of conductor	mm²	25	35	50	70
	3	Shape and type of conductor		RM	RM	RM	RM
	6	Nom. cross-sectional area of screen	mm²	16	16	16	16
	9	Thickness of insulation	mm	5.5	5.5	5.5	5.5
	16	Thickness of outer sheath	mm	2.5	2.5	2.5	2.5
	17	Overall diameter of cable	mm	28	29	30	31
	19	Weight of cable	kg/km	720	780	850	950
Mechanical properties	20	Minimum bending radius	mm	420	435	450	465
	21	Perm. pulling force with cable grip	N	750	1050	1500	2100
	22	Perm. pulling force with pulling head	N	750	1050	1500	2100
Electrical properties	23	d.c. resistance/unit length at 90 °C	Ω/km	1.54	1.11	0.822	0.568
	24	d.c. resistance/unit length at 20 °C	Ω/km	1.20	0.868	0.641	0.443
	25	Operating capacitance/unit length	μF/km	0.145	0.159	0.175	0.196
	26	Charging current	A/km	0.530	0.580	0.630	0.710
	28	Earth fault current	A/km	1.59	1.74	1.89	2.13
	30	Electric field strength at the conductor	kV/mm	3.40	3.30	3.20	3.00
	31	Reference diameter of cable	mm	27.1	28.2	29.4	31.1
Installation in ground	32	Current-carrying capacity	A	123	147	173	211
	33	Permissible transmission power	MVA	4.26	5.09	5.99	7.31
	34	Eff. a.c. resistance/unit length at 90 °C	Ω/km	1.54	1.12	0.825	0.571
	36	Eff. a.c. resistance/unit length at 20 °C	Ω/km	1.20	0.871	0.644	0.446
	38	Ohmic losses per cable	kW/km	23.3	24.1	24.7	25.4
	39	Fict. thermal resist. of cable for ohmic losses	Km/W	0.810	0.738	0.675	0.604
	41	Fict. thermal resist. of soil for ohmic losses	Km/W	3.773	3.726	3.676	3.609
	44	Inductance/unit length per conductor	mH/km	0.514	0.488	0.465	0.438
	45	Resistance/unit length in zero-phase seq.system	Ω/km	2.061	1.727	1.498	1.297
	46	Reactance/unit length in zero-phase seq.system	Ω/km	0.572	0.566	0.560	0.554
	47	Reduction factor		0.558	0.561	0.564	0.567
Installation in free air	32	Current-carrying capacity	A	128	155	186	231
	33	Permissible transmission power	MVA	4.43	5.37	6.44	8.00
	34	Eff. a.c. resistance/unit length at 90 °C	Ω/km	1.54	1.12	0.825	0.571
	38	Ohmic losses per cable	kW/km	25.3	26.8	28.5	30.5
	39	Fict. thermal resist. of cable for ohmic losses	Km/W	0.994	0.913	0.839	0.757
	43	Thermal resistance of the air	Km/W	1.374	1.322	1.270	1.205
	44	Inductance/unit length per conductor	mH/km	0.514	0.488	0.465	0.438
	48	Minimum time value	s	152	203	287	365
Short-circuit	49	Rated short-time current of conductor (1s)	kA	2.35	3.29	4.70	6.58
	50	Rated short-time current of screen (1s)	kA	3.3	3.3	3.3	3.3

NA2XS2Y 1×.../.. 12/20 kV (U_m=24 kV)

95 RM 16	120 RM 16	150 RM 25	185 RM 25	240 RM 25	300 RM 25	400 RM 35	500 RM 35
5.5	5.5	5.5	5.5	5.5	5.5	5.5	5.5
2.5	2.5	2.5	2.5	2.5	2.5	2.5	2.5
33	34	36	38	40	43	46	49
1100	1200	1400	1550	1750	2050	2450	2850
495	510	540	570	600	645	690	735
2850	3600	4500	5550	7200	9000	12000	15000
2850	3600	4500	5550	7200	9000	12000	15000
0.410	0.324	0.264	0.210	0.160	0.128	0.0997	0.0776
0.320	0.253	0.206	0.164	0.125	0.1000	0.0778	0.0605
0.216	0.235	0.254	0.273	0.304	0.329	0.368	0.402
0.780	0.850	0.920	0.990	1.10	1.19	1.33	1.46
2.34	2.55	2.76	2.97	3.30	3.57	3.99	4.38
2.90	2.80	2.80	2.70	2.60	2.60	2.50	2.50
32.7	34.2	35.7	37.3	39.8	41.8	45.0	47.8
252	286	320	361	419	472	537	610
8.73	9.91	11.1	12.5	14.5	16.4	18.6	21.1
0.413	0.328	0.269	0.215	0.165	0.133	0.107	0.0852
0.323	0.256	0.211	0.169	0.130	0.105	0.0849	0.0681
26.3	26.8	27.5	28.0	29.0	29.7	30.8	31.7
0.550	0.508	0.469	0.436	0.393	0.363	0.322	0.294
3.549	3.496	3.445	3.392	3.315	3.256	3.168	3.096
0.419	0.403	0.389	0.377	0.361	0.350	0.335	0.326
1.172	1.103	0.847	0.804	0.764	0.738	0.566	0.548
0.549	0.546	0.297	0.294	0.291	0.289	0.179	0.177
0.570	0.572	0.417	0.419	0.422	0.424	0.325	0.327
280	324	366	420	497	568	662	766
9.70	11.2	12.7	14.5	17.2	19.7	22.9	26.5
0.413	0.328	0.269	0.215	0.165	0.133	0.107	0.0853
32.4	34.4	36.0	38.0	40.9	43.1	46.9	50.1
0.695	0.646	0.601	0.561	0.511	0.475	0.425	0.389
1.150	1.103	1.060	1.017	0.959	0.916	0.856	0.809
0.419	0.403	0.389	0.377	0.361	0.350	0.335	0.325
458	546	668	772	928	1110	1452	1695
8.93	11.3	14.1	17.4	22.6	28.2	37.6	47.0
3.3	3.3	5.1	5.1	5.1	5.1	7.1	7.1

Table 5.6.6b
PROTOTHEN X cables (XLPE insulation) with aluminium conductor

Three-phase operation at 50 Hz

Design	2	Nom. cross-sectional area of conductor	mm²	25	35	50	70
	3	Shape and type of conductor		RM	RM	RM	RM
	6	Nom. cross-sectional area of screen	mm²	16	16	16	16
	9	Thickness of insulation	mm	5.5	5.5	5.5	5.5
	16	Thickness of outer sheath	mm	2.5	2.5	2.5	2.5
	17	Overall diameter of cable	mm	28	29	30	31
	19	Weight of cable	kg/km	720	780	850	950
Mechanical properties	20	Minimum bending radius	mm	420	435	450	465
	21	Perm. pulling force with cable grip	N	750	1050	1500	2100
	22	Perm. pulling force with pulling head	N	750	1050	1500	2100
Electrical properties	23	d.c. resistance/unit length at 90 °C	Ω/km	1.54	1.11	0.822	0.568
	24	d.c. resistance/unit length at 20 °C	Ω/km	1.20	0.868	0.641	0.443
	25	Operating capacitance/unit length	μF/km	0.145	0.159	0.175	0.196
	26	Charging current	A/km	0.530	0.580	0.630	0.710
	28	Earth fault current	A/km	1.59	1.74	1.89	2.13
	30	Electric field strength at the conductor	kV/mm	3.40	3.30	3.20	3.00
	31	Reference diameter of cable	mm	27.1	28.2	29.4	31.1
Installation in ground	32	Current-carrying capacity	A	139	166	195	238
	33	Permissible transmission power	MVA	4.82	5.75	6.75	8.24
	34	Eff. a.c. resistance/unit length at 90 °C	Ω/km	1.56	1.13	0.838	0.583
	36	Eff. a.c. resistance/unit length at 20 °C	Ω/km	1.22	0.885	0.657	0.458
	38	Ohmic losses per cable	kW/km	30.1	31.1	31.9	33.0
	39	Fict. thermal resist. of cable for ohmic losses	Km/W	0.804	0.731	0.667	0.593
	41	Fict. thermal resist. of soil for ohmic losses	Km/W	2.760	2.735	2.709	2.673
	44	Inductance/unit length per conductor	mH/km	0.789	0.757	0.729	0.695
	45	Resistance/unit length in zero-phase seq.system	Ω/km	2.019	1.686	1.458	1.258
	46	Reactance/unit length in zero-phase seq.system	Ω/km	0.588	0.581	0.575	0.568
	47	Reduction factor		0.598	0.600	0.602	0.604
Installation in free air	32	Current-carrying capacity	A	152	184	221	275
	33	Permissible transmission power	MVA	5.27	6.37	7.66	9.53
	34	Eff. a.c. resistance/unit length at 90 °C	Ω/km	1.55	1.13	0.835	0.581
	35	Mean eff. a.c. resistance/unit length at 90 °C	Ω/km	1.55	1.12	0.832	0.578
	38	Ohmic losses per cable	kW/km	35.8	38.0	40.7	43.7
	39	Fict. thermal resist. of cable for ohmic losses	Km/W	0.805	0.733	0.668	0.595
	43	Thermal resistance of the air	Km/W	0.874	0.841	0.809	0.767
	44	Inductance/unit length per conductor	mH/km	0.675	0.649	0.626	0.599
	48	Minimum time value	s	108	144	204	258
Short-circuit	49	Rated short-time current of conductor (1s)	kA	2.35	3.29	4.70	6.58
	50	Rated short-time current of screen (1s)	kA	3.3	3.3	3.3	3.3

95 RM 16	120 RM 16	150 RM 25	185 RM 25	240 RM 25	300 RM 25	400 RM 35	500 RM 35
5.5	5.5	5.5	5.5	5.5	5.5	5.5	5.5
2.5	2.5	2.5	2.5	2.5	2.5	2.5	2.5
33	34	36	38	40	43	46	49
1100	1200	1400	1550	1750	2050	2450	2850
495	510	540	570	600	645	690	735
2850	3600	4500	5550	7200	9000	12000	15000
2850	3600	4500	5550	7200	9000	12000	15000
0.410	0.324	0.264	0.210	0.160	0.128	0.0997	0.0776
0.320	0.253	0.206	0.164	0.125	0.1000	0.0778	0.0605
0.216	0.235	0.254	0.273	0.304	0.329	0.368	0.402
0.780	0.850	0.920	0.990	1.10	1.19	1.33	1.46
2.34	2.55	2.76	2.97	3.30	3.57	3.99	4.38
2.90	2.80	2.80	2.70	2.60	2.60	2.50	2.50
32.7	34.2	35.7	37.3	39.8	41.8	45.0	47.8
282	320	353	397	456	510	565	635
9.77	11.1	12.2	13.8	15.8	17.7	19.6	22.0
0.425	0.338	0.285	0.230	0.180	0.147	0.124	0.101
0.334	0.267	0.227	0.184	0.144	0.119	0.102	0.0840
33.8	34.7	35.5	36.3	37.3	38.2	39.6	40.8
0.538	0.494	0.448	0.413	0.368	0.337	0.287	0.256
2.641	2.611	2.583	2.553	2.509	2.476	2.424	2.381
0.668	0.647	0.622	0.605	0.581	0.565	0.536	0.519
1.134	1.065	0.832	0.789	0.749	0.724	0.560	0.542
0.563	0.560	0.309	0.306	0.303	0.301	0.186	0.184
0.606	0.608	0.457	0.459	0.460	0.461	0.364	0.365
334	385	434	496	584	665	758	868
11.6	13.3	15.0	17.2	20.2	23.0	26.3	30.1
0.423	0.337	0.284	0.230	0.180	0.148	0.126	0.104
0.420	0.335	0.280	0.226	0.176	0.143	0.120	0.0983
46.9	49.6	52.7	55.5	59.9	63.5	69.2	74.1
0.540	0.496	0.449	0.414	0.368	0.335	0.283	0.251
0.733	0.703	0.675	0.648	0.611	0.584	0.544	0.514
0.580	0.565	0.548	0.536	0.521	0.510	0.491	0.481
322	386	475	553	672	810	1108	1320
8.93	11.3	14.1	17.4	22.6	28.2	37.6	47.0
3.3	3.3	5.1	5.1	5.1	5.1	7.1	7.1

Table 5.6.7 a
PROTOTHEN X cables (XLPE insulation) with aluminium conductor

Three-phase operation at 50 Hz

Design				25	35	50	70
Design	2	Nom. cross-sectional area of conductor	mm²	25	35	50	70
	3	Shape and type of conductor		RM	RM	RM	RM
	6	Nom. cross-sectional area of screen	mm²	16	16	16	16
	9	Thickness of insulation	mm	5.5	5.5	5.5	5.5
	16	Thickness of outer sheath	mm	2.5	2.5	2.5	2.5
	17	Overall diameter of cable	mm	30	31	32	33
	19	Weight of cable	kg/km	960	1050	1150	1250
Mechanical properties	20	Minimum bending radius	mm	450	465	480	495
	21	Perm. pulling force with cable grip	N	750	1050	1500	2100
	22	Perm. pulling force with pulling head	N	750	1050	1500	2100
Electrical properties	23	d.c. resistance/unit length at 90 °C	Ω/km	1.54	1.11	0.822	0.568
	24	d.c. resistance/unit length at 20 °C	Ω/km	1.20	0.868	0.641	0.443
	25	Operating capacitance/unit length	μF/km	0.145	0.159	0.175	0.196
	26	Charging current	A/km	0.530	0.580	0.630	0.710
	28	Earth fault current	A/km	1.59	1.74	1.89	2.13
	30	Electric field strength at the conductor	kV/mm	3.40	3.30	3.20	3.00
	31	Reference diameter of cable	mm	27.1	28.2	29.4	31.1
Installation in ground	32	Current-carrying capacity	A	123	147	173	211
	33	Permissible transmission power	MVA	4.26	5.09	5.99	7.31
	34	Eff. a.c. resistance/unit length at 90 °C	Ω/km	1.54	1.12	0.825	0.571
	36	Eff. a.c. resistance/unit length at 20 °C	Ω/km	1.20	0.871	0.644	0.446
	38	Ohmic losses per cable	kW/km	23.3	24.1	24.7	25.4
	39	Fict. thermal resist. of cable for ohmic losses	Km/W	0.810	0.738	0.675	0.604
	41	Fict. thermal resist. of soil for ohmic losses	Km/W	3.773	3.726	3.676	3.609
	44	Inductance/unit length per conductor	mH/km	0.514	0.488	0.465	0.438
	45	Resistance/unit length in zero-phase seq.system	Ω/km	2.061	1.727	1.498	1.297
	46	Reactance/unit length in zero-phase seq.system	Ω/km	0.572	0.566	0.560	0.554
	47	Reduction factor		0.558	0.561	0.564	0.567
Installation in free air	32	Current-carrying capacity	A	128	155	186	231
	33	Permissible transmission power	MVA	4.43	5.37	6.44	8.00
	34	Eff. a.c. resistance/unit length at 90 °C	Ω/km	1.54	1.12	0.825	0.571
	38	Ohmic losses per cable	kW/km	25.3	26.8	28.5	30.5
	39	Fict. thermal resist. of cable for ohmic losses	Km/W	0.994	0.913	0.839	0.757
	43	Thermal resistance of the air	Km/W	1.374	1.322	1.270	1.205
	44	Inductance/unit length per conductor	mH/km	0.514	0.488	0.465	0.438
	48	Minimum time value	s	152	203	287	365
Short-circuit	49	Rated short-time current of conductor (1s)	kA	2.35	3.29	4.70	6.58
	50	Rated short-time current of screen (1s)	kA	3.3	3.3	3.3	3.3

NA2XS(F)2Y 1×.../.. 12/20 kV (U_m=24 kV)

95 RM 16	120 RM 16	150 RM 25	185 RM 25	240 RM 25	300 RM 25	400 RM 35	500 RM 35
5.5	5.5	5.5	5.5	5.5	5.5	5.5	5.5
2.5	2.5	2.5	2.5	2.5	2.5	2.5	2.5
35	36	38	40	42	45	48	51
1400	1500	1700	1900	2150	2450	2500	3350
525	540	570	600	630	675	720	765
2850	3600	4500	5550	7200	9000	12000	15000
2850	3600	4500	5550	7200	9000	12000	15000
0.410	0.324	0.264	0.210	0.160	0.128	0.0997	0.0776
0.320	0.253	0.206	0.164	0.125	0.1000	0.0778	0.0605
0.216	0.235	0.254	0.273	0.304	0.329	0.368	0.402
0.780	0.850	0.920	0.990	1.10	1.19	1.33	1.46
2.34	2.55	2.76	2.97	3.30	3.57	3.99	4.38
2.90	2.80	2.80	2.70	2.60	2.60	2.50	2.50
32.7	34.2	35.7	37.3	39.8	41.8	45.0	47.8
252	286	320	361	419	472	537	610
8.73	9.91	11.1	12.5	14.5	16.4	18.6	21.1
0.413	0.328	0.269	0.215	0.165	0.133	0.107	0.0852
0.323	0.256	0.211	0.169	0.130	0.105	0.0849	0.0681
26.3	26.8	27.5	28.0	29.0	29.7	30.8	31.7
0.550	0.508	0.469	0.436	0.393	0.363	0.322	0.294
3.549	3.496	3.445	3.392	3.315	3.256	3.168	3.096
0.419	0.403	0.389	0.377	0.361	0.350	0.335	0.326
1.172	1.103	0.847	0.804	0.764	0.738	0.566	0.548
0.549	0.546	0.297	0.294	0.291	0.289	0.179	0.177
0.570	0.572	0.417	0.419	0.422	0.424	0.325	0.327
280	324	366	420	497	568	662	766
9.70	11.2	12.7	14.5	17.2	19.7	22.9	26.5
0.413	0.328	0.269	0.215	0.165	0.133	0.107	0.0853
32.4	34.4	36.0	38.0	40.9	43.1	46.9	50.1
0.695	0.646	0.601	0.561	0.511	0.475	0.425	0.389
1.150	1.103	1.060	1.017	0.959	0.916	0.856	0.809
0.419	0.403	0.389	0.377	0.361	0.350	0.335	0.325
458	546	668	772	928	1110	1452	1695
8.93	11.3	14.1	17.4	22.6	28.2	37.6	47.0
3.3	3.3	5.1	5.1	5.1	5.1	7.1	7.1

Table 5.6.7b
PROTOTHEN X cables (XLPE insulation) with aluminium conductor

Three-phase operation at 50 Hz

Design	2	Nom. cross-sectional area of conductor	mm²	25	35	50	70
	3	Shape and type of conductor		RM	RM	RM	RM
	6	Nom. cross-sectional area of screen	mm²	16	16	16	16
	9	Thickness of insulation	mm	5.5	5.5	5.5	5.5
	16	Thickness of outer sheath	mm	2.5	2.5	2.5	2.5
	17	Overall diameter of cable	mm	30	31	32	33
	19	Weight of cable	kg/km	960	1050	1150	1250
Mechanical properties	20	Minimum bending radius	mm	450	465	480	495
	21	Perm. pulling force with cable grip	N	750	1050	1500	2100
	22	Perm. pulling force with pulling head	N	750	1050	1500	2100
Electrical properties	23	d.c. resistance/unit length at 90 °C	Ω/km	1.54	1.11	0.822	0.568
	24	d.c. resistance/unit length at 20 °C	Ω/km	1.20	0.868	0.641	0.443
	25	Operating capacitance/unit length	μF/km	0.145	0.159	0.175	0.196
	26	Charging current	A/km	0.530	0.580	0.630	0.710
	28	Earth fault current	A/km	1.59	1.74	1.89	2.13
	30	Electric field strength at the conductor	kV/mm	3.40	3.30	3.20	3.00
	31	Reference diameter of cable	mm	27.1	28.2	29.4	31.1
Installation in ground	32	Current-carrying capacity	A	139	166	195	238
	33	Permissible transmission power	MVA	4.82	5.75	6.75	8.24
	34	Eff. a.c. resistance/unit length at 90 °C	Ω/km	1.56	1.13	0.838	0.583
	36	Eff. a.c. resistance/unit length at 20 °C	Ω/km	1.22	0.885	0.657	0.458
	38	Ohmic losses per cable	kW/km	30.1	31.1	31.9	33.0
	39	Fict. thermal resist. of cable for ohmic losses	Km/W	0.804	0.731	0.667	0.593
	41	Fict. thermal resist. of soil for ohmic losses	Km/W	2.760	2.735	2.709	2.673
	44	Inductance/unit length per conductor	mH/km	0.789	0.757	0.729	0.695
	45	Resistance/unit length in zero-phase seq.system	Ω/km	2.019	1.686	1.458	1.258
	46	Reactance/unit length in zero-phase seq.system	Ω/km	0.588	0.581	0.575	0.568
	47	Reduction factor		0.598	0.600	0.602	0.604
Installation in free air	32	Current-carrying capacity	A	152	184	221	275
	33	Permissible transmission power	MVA	5.27	6.37	7.66	9.53
	34	Eff. a.c. resistance/unit length at 90 °C	Ω/km	1.55	1.13	0.835	0.581
	35	Mean eff. a.c. resistance/unit length at 90 °C	Ω/km	1.55	1.12	0.832	0.578
	38	Ohmic losses per cable	kW/km	35.8	38.0	40.7	43.7
	39	Fict. thermal resist. of cable for ohmic losses	Km/W	0.805	0.733	0.668	0.595
	43	Thermal resistance of the air	Km/W	0.874	0.841	0.809	0.767
	44	Inductance/unit length per conductor	mH/km	0.675	0.649	0.626	0.599
	48	Minimum time value	s	108	144	204	258
Short-circuit	49	Rated short-time current of conductor (1s)	kA	2.35	3.29	4.70	6.58
	50	Rated short-time current of screen (1s)	kA	3.3	3.3	3.3	3.3

95 RM 16	120 RM 16	150 RM 25	185 RM 25	240 RM 25	300 RM 25	400 RM 35	500 RM 35
5.5	5.5	5.5	5.5	5.5	5.5	5.5	5.5
2.5	2.5	2.5	2.5	2.5	2.5	2.5	2.5
35	36	38	40	42	45	48	51
1400	1500	1700	1900	2150	2450	2500	3350
525	540	570	600	630	675	720	765
2850	3600	4500	5550	7200	9000	12000	15000
2850	3600	4500	5550	7200	9000	12000	15000
0.410	0.324	0.264	0.210	0.160	0.128	0.0997	0.0776
0.320	0.253	0.206	0.164	0.125	0.1000	0.0778	0.0605
0.216	0.235	0.254	0.273	0.304	0.329	0.368	0.402
0.780	0.850	0.920	0.990	1.10	1.19	1.33	1.46
2.34	2.55	2.76	2.97	3.30	3.57	3.99	4.38
2.90	2.80	2.80	2.70	2.60	2.60	2.50	2.50
32.7	34.2	35.7	37.3	39.8	41.8	45.0	47.8
282	320	353	397	456	510	565	635
9.77	11.1	12.2	13.8	15.8	17.7	19.6	22.0
0.425	0.338	0.285	0.230	0.180	0.147	0.124	0.101
0.334	0.267	0.227	0.184	0.144	0.119	0.102	0.0840
33.8	34.7	35.5	36.3	37.3	38.2	39.6	40.8
0.538	0.494	0.448	0.413	0.368	0.337	0.287	0.256
2.641	2.611	2.583	2.553	2.509	2.476	2.424	2.381
0.668	0.647	0.622	0.605	0.581	0.565	0.536	0.519
1.134	1.065	0.832	0.789	0.749	0.724	0.560	0.542
0.563	0.560	0.309	0.306	0.303	0.301	0.186	0.184
0.606	0.608	0.457	0.459	0.460	0.461	0.364	0.365
334	385	434	496	584	665	758	868
11.6	13.3	15.0	17.2	20.2	23.0	26.3	30.1
0.423	0.337	0.284	0.230	0.180	0.148	0.126	0.104
0.420	0.335	0.280	0.226	0.176	0.143	0.120	0.0983
46.9	49.6	52.7	55.5	59.9	63.5	69.2	74.1
0.540	0.496	0.449	0.414	0.368	0.335	0.283	0.251
0.733	0.703	0.675	0.648	0.611	0.584	0.544	0.514
0.580	0.565	0.548	0.536	0.521	0.510	0.491	0.481
322	386	475	553	672	810	1108	1320
8.93	11.3	14.1	17.4	22.6	28.2	37.6	47.0
3.3	3.3	5.1	5.1	5.1	5.1	7.1	7.1

Table 5.6.8 a
PROTOTHEN X cables (XLPE insulation) with aluminium conductor

Three-phase operation at 50 Hz

Design	2	Nom. cross-sectional area of conductor	mm²	25	35	50	70
	3	Shape and type of conductor		RM	RM	RM	RM
	6	Nom. cross-sectional area of screen	mm²	16	16	16	16
	9	Thickness of insulation	mm	8.0	8.0	8.0	8.0
	16	Thickness of outer sheath	mm	2.5	2.5	2.5	2.5
	17	Overall diameter of cable	mm	33	34	35	36
	19	Weight of cable	kg/km	930	1000	1100	1200
Mechanical properties	20	Minimum bending radius	mm	495	510	525	540
	21	Perm. pulling force with cable grip	N	750	1050	1500	2100
	22	Perm. pulling force with pulling head	N	750	1050	1500	2100
Electrical properties	23	d.c. resistance/unit length at 90 °C	Ω/km	1.54	1.11	0.822	0.568
	24	d.c. resistance/unit length at 20 °C	Ω/km	1.20	0.868	0.641	0.443
	25	Operating capacitance/unit length	μF/km	0.115	0.125	0.136	0.151
	26	Charging current	A/km	0.630	0.680	0.740	0.820
	28	Earth fault current	A/km	1.89	2.04	2.22	2.46
	30	Electric field strength at the conductor	kV/mm	4.10	3.90	3.70	3.50
	31	Reference diameter of cable	mm	32.1	33.2	34.4	36.1
Installation in ground	32	Current-carrying capacity	A	125	149	175	214
	33	Permissible transmission power	MVA	6.50	7.74	9.09	11.1
	34	Eff. a.c. resistance/unit length at 90 °C	Ω/km	1.54	1.12	0.825	0.571
	36	Eff. a.c. resistance/unit length at 20 °C	Ω/km	1.20	0.871	0.644	0.446
	38	Ohmic losses per cable	kW/km	24.1	24.8	25.3	26.1
	39	Fict. thermal resist. of cable for ohmic losses	Km/W	0.911	0.836	0.769	0.692
	41	Fict. thermal resist. of soil for ohmic losses	Km/W	3.571	3.531	3.489	3.431
	44	Inductance/unit length per conductor	mH/km	0.548	0.521	0.496	0.468
	45	Resistance/unit length in zero-phase seq.system	Ω/km	2.053	1.719	1.491	1.291
	46	Reactance/unit length in zero-phase seq.system	Ω/km	0.589	0.582	0.575	0.568
	47	Reduction factor		0.561	0.563	0.566	0.569
Installation in free air	32	Current-carrying capacity	A	129	156	187	233
	33	Permissible transmission power	MVA	6.70	8.11	9.72	12.1
	34	Eff. a.c. resistance/unit length at 90 °C	Ω/km	1.54	1.12	0.825	0.571
	38	Ohmic losses per cable	kW/km	25.7	27.2	28.8	31.0
	39	Fict. thermal resist. of cable for ohmic losses	Km/W	1.115	1.029	0.950	0.862
	43	Thermal resistance of the air	Km/W	1.209	1.169	1.129	1.078
	44	Inductance/unit length per conductor	mH/km	0.548	0.521	0.496	0.468
	48	Minimum time value	s	149	200	284	359
Short-circuit	49	Rated short-time current of conductor (1s)	kA	2.35	3.29	4.70	6.58
	50	Rated short-time current of screen (1s)	kA	3.3	3.3	3.3	3.3

95 RM 16	120 RM 16	150 RM 25	185 RM 25	240 RM 25	300 RM 25	400 RM 35	500 RM 35
8.0	8.0	8.0	8.0	8.0	8.0	8.0	8.0
2.5	2.5	2.5	2.5	2.5	2.5	2.5	2.6
38	39	41	43	45	48	51	54
1350	1450	1650	1850	2050	2350	2800	3250
570	585	615	645	675	720	765	810
2850	3600	4500	5550	7200	9000	12000	15000
2850	3600	4500	5550	7200	9000	12000	15000
0.410	0.324	0.264	0.210	0.160	0.128	0.0997	0.0776
0.320	0.253	0.206	0.164	0.125	0.1000	0.0778	0.0605
0.165	0.178	0.191	0.205	0.227	0.244	0.271	0.295
0.900	0.970	1.04	1.12	1.24	1.33	1.47	1.61
2.70	2.91	3.12	3.36	3.72	3.99	4.41	4.83
3.30	3.20	3.10	3.00	2.90	2.90	2.80	2.70
37.7	39.2	40.7	42.3	44.8	46.8	50.0	53.0
255	289	323	365	423	477	543	617
13.3	15.0	16.8	19.0	22.0	24.8	28.2	32.1
0.413	0.327	0.269	0.215	0.165	0.133	0.106	0.0849
0.323	0.256	0.210	0.169	0.130	0.105	0.0846	0.0678
26.9	27.3	28.0	28.6	29.5	30.3	31.4	32.3
0.632	0.587	0.545	0.508	0.460	0.428	0.380	0.349
3.380	3.333	3.288	3.242	3.174	3.122	3.043	2.973
0.447	0.430	0.415	0.402	0.384	0.373	0.357	0.346
1.166	1.097	0.845	0.802	0.762	0.736	0.565	0.547
0.563	0.559	0.309	0.306	0.302	0.299	0.188	0.185
0.572	0.574	0.419	0.420	0.423	0.425	0.326	0.328
282	325	367	421	497	568	661	764
14.7	16.9	19.1	21.9	25.8	29.5	34.3	39.7
0.413	0.327	0.269	0.215	0.165	0.133	0.107	0.0851
32.9	34.6	36.2	38.1	40.8	43.0	46.6	49.7
0.794	0.741	0.692	0.648	0.592	0.553	0.496	0.458
1.034	0.996	0.961	0.927	0.878	0.843	0.792	0.751
0.447	0.430	0.415	0.402	0.384	0.373	0.356	0.345
451	542	665	768	928	1110	1457	1704
8.93	11.3	14.1	17.4	22.6	28.2	37.6	47.0
3.3	3.3	5.1	5.1	5.1	5.1	7.1	7.1

Table 5.6.8 b
PROTOTHEN X cables (XLPE insulation) with aluminium conductor

Three-phase operation at 50 Hz

Design								
Design	2	Nom. cross-sectional area of conductor	mm²	25	35	50	70
	3	Shape and type of conductor		RM	RM	RM	RM
	6	Nom. cross-sectional area of screen	mm²	16	16	16	16
	9	Thickness of insulation	mm	8.0	8.0	8.0	8.0
	16	Thickness of outer sheath	mm	2.5	2.5	2.5	2.5
	17	Overall diameter of cable	mm	33	34	35	36
	19	Weight of cable	kg/km	930	1000	1100	1200
Mechanical properties	20	Minimum bending radius	mm	495	510	525	540
	21	Perm. pulling force with cable grip	N	750	1050	1500	2100
	22	Perm. pulling force with pulling head	N	750	1050	1500	2100
Electrical properties	23	d.c. resistance/unit length at 90 °C	Ω/km	1.54	1.11	0.822	0.568
	24	d.c. resistance/unit length at 20 °C	Ω/km	1.20	0.868	0.641	0.443
	25	Operating capacitance/unit length	μF/km	0.115	0.125	0.136	0.151
	26	Charging current	A/km	0.630	0.680	0.740	0.820
	28	Earth fault current	A/km	1.89	2.04	2.22	2.46
	30	Electric field strength at the conductor	kV/mm	4.10	3.90	3.70	3.50
	31	Reference diameter of cable	mm	32.1	33.2	34.4	36.1
Installation in ground	32	Current-carrying capacity	A	139	166	195	238
	33	Permissible transmission power	MVA	7.22	8.63	10.1	12.4
	34	Eff. a.c. resistance/unit length at 90 °C	Ω/km	1.55	1.13	0.836	0.582
	36	Eff. a.c. resistance/unit length at 20 °C	Ω/km	1.21	0.883	0.655	0.457
	38	Ohmic losses per cable	kW/km	30.0	31.1	31.8	32.9
	39	Fict. thermal resist. of cable for ohmic losses	Km/W	0.905	0.829	0.760	0.681
	41	Fict. thermal resist. of soil for ohmic losses	Km/W	2.653	2.631	2.607	2.575
	44	Inductance/unit length per conductor	mH/km	0.800	0.768	0.740	0.705
	45	Resistance/unit length in zero-phase seq.system	Ω/km	2.014	1.681	1.453	1.253
	46	Reactance/unit length in zero-phase seq.system	Ω/km	0.604	0.596	0.589	0.581
	47	Reduction factor		0.596	0.598	0.600	0.603
Installation in free air	32	Current-carrying capacity	A	152	184	220	274
	33	Permissible transmission power	MVA	7.90	9.56	11.4	14.2
	34	Eff. a.c. resistance/unit length at 90 °C	Ω/km	1.55	1.13	0.835	0.581
	35	Mean eff. a.c. resistance/unit length at 90 °C	Ω/km	1.55	1.12	0.832	0.578
	38	Ohmic losses per cable	kW/km	35.8	38.0	40.3	43.4
	39	Fict. thermal resist. of cable for ohmic losses	Km/W	0.906	0.830	0.761	0.682
	43	Thermal resistance of the air	Km/W	0.775	0.749	0.724	0.691
	44	Inductance/unit length per conductor	mH/km	0.709	0.682	0.657	0.630
	48	Minimum time value	s	108	144	205	260
Short-circuit	49	Rated short-time current of conductor (1s)	kA	2.35	3.29	4.70	6.58
	50	Rated short-time current of screen (1s)	kA	3.3	3.3	3.3	3.3

◉ ◉ ◉ NA2XS2Y 1×.../.. 18/30 kV (U_m=36 kV)

95 RM 16	120 RM 16	150 RM 25	185 RM 25	240 RM 25	300 RM 25	400 RM 35	500 RM 35
8.0	8.0	8.0	8.0	8.0	8.0	8.0	8.0
2.5	2.5	2.5	2.5	2.5	2.5	2.5	2.6
38	39	41	43	45	48	51	54
1350	1450	1650	1850	2050	2350	2800	3250
570	585	615	645	675	720	765	810
2850	3600	4500	5550	7200	9000	12000	15000
2850	3600	4500	5550	7200	9000	12000	15000
0.410	0.324	0.264	0.210	0.160	0.128	0.0997	0.0776
0.320	0.253	0.206	0.164	0.125	0.1000	0.0778	0.0605
0.165	0.178	0.191	0.205	0.227	0.244	0.271	0.295
0.900	0.970	1.04	1.12	1.24	1.33	1.47	1.61
2.70	2.91	3.12	3.36	3.72	3.99	4.41	4.83
3.30	3.20	3.10	3.00	2.90	2.90	2.80	2.70
37.7	39.2	40.7	42.3	44.8	46.8	50.0	53.0
283	321	355	399	460	514	572	643
14.7	16.7	18.4	20.7	23.9	26.7	29.7	33.4
0.423	0.337	0.283	0.229	0.178	0.146	0.122	0.0997
0.333	0.266	0.225	0.182	0.143	0.117	0.100	0.0826
33.9	34.7	35.7	36.4	37.7	38.5	40.0	41.2
0.619	0.572	0.521	0.482	0.431	0.396	0.338	0.304
2.546	2.520	2.494	2.467	2.427	2.396	2.348	2.306
0.678	0.656	0.632	0.615	0.591	0.575	0.545	0.529
1.129	1.061	0.830	0.788	0.748	0.722	0.559	0.541
0.576	0.571	0.320	0.317	0.313	0.310	0.195	0.192
0.605	0.606	0.456	0.457	0.459	0.460	0.362	0.364
333	383	431	493	581	660	755	863
17.3	19.9	22.4	25.6	30.2	34.3	39.2	44.8
0.423	0.337	0.284	0.230	0.180	0.148	0.126	0.104
0.420	0.334	0.279	0.226	0.175	0.143	0.120	0.0982
46.6	49.1	51.9	54.8	59.2	62.5	68.5	73.1
0.620	0.573	0.520	0.480	0.428	0.392	0.330	0.294
0.663	0.639	0.616	0.594	0.562	0.540	0.505	0.479
0.609	0.592	0.574	0.561	0.544	0.533	0.512	0.501
324	390	482	560	679	822	1117	1335
8.93	11.3	14.1	17.4	22.6	28.2	37.6	47.0
3.3	3.3	5.1	5.1	5.1	5.1	7.1	7.1

Table 5.6.9 a
PROTOTHEN X cables (XLPE insulation) with aluminium conductor

Three-phase operation at 50 Hz

Design				25	35	50	70
Design	2	Nom. cross-sectional area of conductor	mm²	25	35	50	70
	3	Shape and type of conductor		RM	RM	RM	RM
	6	Nom. cross-sectional area of screen	mm²	16	16	16	16
	9	Thickness of insulation	mm	8.0	8.0	8.0	8.0
	16	Thickness of outer sheath	mm	2.5	2.5	2.5	2.5
	17	Overall diameter of cable	mm	35	36	37	38
	19	Weight of cable	kg/km	1250	1350	1450	1550
Mechanical properties	20	Minimum bending radius	mm	525	540	555	570
	21	Perm. pulling force with cable grip	N	750	1050	1500	2100
	22	Perm. pulling force with pulling head	N	750	1050	1500	2100
Electrical properties	23	d.c. resistance/unit length at 90 °C	Ω/km	1.54	1.11	0.822	0.568
	24	d.c. resistance/unit length at 20 °C	Ω/km	1.20	0.868	0.641	0.443
	25	Operating capacitance/unit length	μF/km	0.115	0.125	0.136	0.151
	26	Charging current	A/km	0.630	0.680	0.740	0.820
	28	Earth fault current	A/km	1.89	2.04	2.22	2.46
	30	Electric field strength at the conductor	kV/mm	4.10	3.90	3.70	3.50
	31	Reference diameter of cable	mm	32.1	33.2	34.4	36.1
Installation in ground	32	Current-carrying capacity	A	125	149	175	214
	33	Permissible transmission power	MVA	6.50	7.74	9.09	11.1
	34	Eff. a.c. resistance/unit length at 90 °C	Ω/km	1.54	1.12	0.825	0.571
	36	Eff. a.c. resistance/unit length at 20 °C	Ω/km	1.20	0.871	0.644	0.446
	38	Ohmic losses per cable	kW/km	24.1	24.8	25.3	26.1
	39	Fict. thermal resist. of cable for ohmic losses	Km/W	0.911	0.836	0.769	0.692
	41	Fict. thermal resist. of soil for ohmic losses	Km/W	3.571	3.531	3.489	3.431
	44	Inductance/unit length per conductor	mH/km	0.548	0.521	0.496	0.468
	45	Resistance/unit length in zero-phase seq.system	Ω/km	2.053	1.719	1.491	1.291
	46	Reactance/unit length in zero-phase seq.system	Ω/km	0.589	0.582	0.575	0.568
	47	Reduction factor		0.561	0.563	0.566	0.569
Installation in free air	32	Current-carrying capacity	A	129	156	187	233
	33	Permissible transmission power	MVA	6.70	8.11	9.72	12.1
	34	Eff. a.c. resistance/unit length at 90 °C	Ω/km	1.54	1.12	0.825	0.571
	38	Ohmic losses per cable	kW/km	25.7	27.2	28.8	31.0
	39	Fict. thermal resist. of cable for ohmic losses	Km/W	1.115	1.029	0.950	0.862
	43	Thermal resistance of the air	Km/W	1.209	1.169	1.129	1.078
	44	Inductance/unit length per conductor	mH/km	0.548	0.521	0.496	0.468
	48	Minimum time value	s	149	200	284	359
Short-circuit	49	Rated short-time current of conductor (1s)	kA	2.35	3.29	4.70	6.58
	50	Rated short-time current of screen (1s)	kA	3.3	3.3	3.3	3.3

NA2XS(F)2Y 1×…/.. 18/30 kV (U_m=36 kV)

95 RM 16	120 RM 16	150 RM 25	185 RM 25	240 RM 25	300 RM 25	400 RM 35	500 RM 35
8.0	8.0	8.0	8.0	8.0	8.0	8.0	8.0
2.5	2.5	2.5	2.5	2.5	2.5	2.5	2.6
40	41	43	45	47	50	53	56
1700	1850	2050	2250	2500	2850	3300	3800
600	615	645	675	705	750	795	840
2850	3600	4500	5550	7200	9000	12000	15000
2850	3600	4500	5550	7200	9000	12000	15000
0.410	0.324	0.264	0.210	0.160	0.128	0.0997	0.0776
0.320	0.253	0.206	0.164	0.125	0.1000	0.0778	0.0605
0.165	0.178	0.191	0.205	0.227	0.244	0.271	0.295
0.900	0.970	1.04	1.12	1.24	1.33	1.47	1.61
2.70	2.91	3.12	3.36	3.72	3.99	4.41	4.83
3.30	3.20	3.10	3.00	2.90	2.90	2.80	2.70
37.7	39.2	40.7	42.3	44.8	46.8	50.0	53.0
255	289	323	365	423	477	543	617
13.3	15.0	16.8	19.0	22.0	24.8	28.2	32.1
0.413	0.327	0.269	0.215	0.165	0.133	0.106	0.0849
0.323	0.256	0.210	0.169	0.130	0.105	0.0846	0.0678
26.9	27.3	28.0	28.6	29.5	30.3	31.4	32.3
0.632	0.587	0.545	0.508	0.460	0.428	0.380	0.349
3.380	3.333	3.288	3.242	3.174	3.122	3.043	2.973
0.447	0.430	0.415	0.402	0.384	0.373	0.357	0.346
1.166	1.097	0.845	0.802	0.762	0.736	0.565	0.547
0.563	0.559	0.309	0.306	0.302	0.299	0.188	0.185
0.572	0.574	0.419	0.420	0.423	0.425	0.326	0.328
282	325	367	421	497	568	661	764
14.7	16.9	19.1	21.9	25.8	29.5	34.3	39.7
0.413	0.327	0.269	0.215	0.165	0.133	0.107	0.0851
32.9	34.6	36.2	38.1	40.8	43.0	46.6	49.7
0.794	0.741	0.692	0.648	0.592	0.553	0.496	0.458
1.034	0.996	0.961	0.927	0.878	0.843	0.792	0.751
0.447	0.430	0.415	0.402	0.384	0.373	0.356	0.345
451	542	665	768	928	1110	1457	1704
8.93	11.3	14.1	17.4	22.6	28.2	37.6	47.0
3.3	3.3	5.1	5.1	5.1	5.1	7.1	7.1

Table 5.6.9 b
PROTOTHEN X cables (XLPE insulation) with aluminium conductor

Three-phase operation at 50 Hz

Design							
	2	Nom. cross-sectional area of conductor	mm²	25	35	50	70
	3	Shape and type of conductor		RM	RM	RM	RM
	6	Nom. cross-sectional area of screen	mm²	16	16	16	16
	9	Thickness of insulation	mm	8.0	8.0	8.0	8.0
	16	Thickness of outer sheath	mm	2.5	2.5	2.5	2.5
	17	Overall diameter of cable	mm	35	36	37	38
	19	Weight of cable	kg/km	1250	1350	1450	1550
Mechanical properties	20	Minimum bending radius	mm	525	540	555	570
	21	Perm. pulling force with cable grip	N	750	1050	1500	2100
	22	Perm. pulling force with pulling head	N	750	1050	1500	2100
Electrical properties	23	d.c. resistance/unit length at 90 °C	Ω/km	1.54	1.11	0.822	0.568
	24	d.c. resistance/unit length at 20 °C	Ω/km	1.20	0.868	0.641	0.443
	25	Operating capacitance/unit length	μF/km	0.115	0.125	0.136	0.151
	26	Charging current	A/km	0.630	0.680	0.740	0.820
	28	Earth fault current	A/km	1.89	2.04	2.22	2.46
	30	Electric field strength at the conductor	kV/mm	4.10	3.90	3.70	3.50
	31	Reference diameter of cable	mm	32.1	33.2	34.4	36.1
Installation in ground	32	Current-carrying capacity	A	139	166	195	238
	33	Permissible transmission power	MVA	7.22	8.63	10.1	12.4
	34	Eff. a.c. resistance/unit length at 90 °C	Ω/km	1.55	1.13	0.836	0.582
	36	Eff. a.c. resistance/unit length at 20 °C	Ω/km	1.21	0.883	0.655	0.457
	38	Ohmic losses per cable	kW/km	30.0	31.1	31.8	32.9
	39	Fict. thermal resist. of cable for ohmic losses	Km/W	0.905	0.829	0.760	0.681
	41	Fict. thermal resist. of soil for ohmic losses	Km/W	2.653	2.631	2.607	2.575
	44	Inductance/unit length per conductor	mH/km	0.800	0.768	0.740	0.705
	45	Resistance/unit length in zero-phase seq.system	Ω/km	2.014	1.681	1.453	1.253
	46	Reactance/unit length in zero-phase seq.system	Ω/km	0.604	0.596	0.589	0.581
	47	Reduction factor		0.596	0.598	0.600	0.603
Installation in free air	32	Current-carrying capacity	A	152	184	220	274
	33	Permissible transmission power	MVA	7.90	9.56	11.4	14.2
	34	Eff. a.c. resistance/unit length at 90 °C	Ω/km	1.55	1.13	0.835	0.581
	35	Mean eff. a.c. resistance/unit length at 90 °C	Ω/km	1.55	1.12	0.832	0.578
	38	Ohmic losses per cable	kW/km	35.8	38.0	40.3	43.4
	39	Fict. thermal resist. of cable for ohmic losses	Km/W	0.906	0.830	0.761	0.682
	43	Thermal resistance of the air	Km/W	0.775	0.749	0.724	0.691
	44	Inductance/unit length per conductor	mH/km	0.709	0.682	0.657	0.630
	48	Minimum time value	s	108	144	205	260
Short-circuit	49	Rated short-time current of conductor (1s)	kA	2.35	3.29	4.70	6.58
	50	Rated short-time current of screen (1s)	kA	3.3	3.3	3.3	3.3

◉ ◉ ◉ NA2XS(F)2Y 1×.../.. 18/30 kV (U_m=36 kV)

95 RM 16	120 RM 16	150 RM 25	185 RM 25	240 RM 25	300 RM 25	400 RM 35	500 RM 35
8.0	8.0	8.0	8.0	8.0	8.0	8.0	8.0
2.5	2.5	2.5	2.5	2.5	2.5	2.5	2.6
40	41	43	45	47	50	53	56
1700	1850	2050	2250	2500	2850	3300	3800
600	615	645	675	705	750	795	840
2850	3600	4500	5550	7200	9000	12000	15000
2850	3600	4500	5550	7200	9000	12000	15000
0.410	0.324	0.264	0.210	0.160	0.128	0.0997	0.0776
0.320	0.253	0.206	0.164	0.125	0.1000	0.0778	0.0605
0.165	0.178	0.191	0.205	0.227	0.244	0.271	0.295
0.900	0.970	1.04	1.12	1.24	1.33	1.47	1.61
2.70	2.91	3.12	3.36	3.72	3.99	4.41	4.83
3.30	3.20	3.10	3.00	2.90	2.90	2.80	2.70
37.7	39.2	40.7	42.3	44.8	46.8	50.0	53.0
283	321	355	399	460	514	572	643
14.7	16.7	18.4	20.7	23.9	26.7	29.7	33.4
0.423	0.337	0.283	0.229	0.178	0.146	0.122	0.0997
0.333	0.266	0.225	0.182	0.143	0.117	0.100	0.0826
33.9	34.7	35.7	36.4	37.7	38.5	40.0	41.2
0.619	0.572	0.521	0.482	0.431	0.396	0.338	0.304
2.546	2.520	2.494	2.467	2.427	2.396	2.348	2.306
0.678	0.656	0.632	0.615	0.591	0.575	0.545	0.529
1.129	1.061	0.830	0.788	0.748	0.722	0.559	0.541
0.576	0.571	0.320	0.317	0.313	0.310	0.195	0.192
0.605	0.606	0.456	0.457	0.459	0.460	0.362	0.364
333	383	431	493	581	660	755	863
17.3	19.9	22.4	25.6	30.2	34.3	39.2	44.8
0.423	0.337	0.284	0.230	0.180	0.148	0.126	0.104
0.420	0.334	0.279	0.226	0.175	0.143	0.120	0.0982
46.6	49.1	51.9	54.8	59.2	62.5	68.5	73.1
0.620	0.573	0.520	0.480	0.428	0.392	0.330	0.294
0.663	0.639	0.616	0.594	0.562	0.540	0.505	0.479
0.609	0.592	0.574	0.561	0.544	0.533	0.512	0.501
324	390	482	560	679	822	1117	1335
8.93	11.3	14.1	17.4	22.6	28.2	37.6	47.0
3.3	3.3	5.1	5.1	5.1	5.1	7.1	7.1

Table 5.7.1
Cables with paper insulation with aluminium conductor

Three-phase operation at 50 Hz

Design				50	70	95	120
Design	2	Nom. cross-sectional area of conductor	mm²	50	70	95	120
	3	Shape and type of conductor		SM	SM	SM	SM
	10	Thickness of insulation conductor-conductor	mm	6.4	6.4	6.4	6.4
	11	Thickness of insulation conductor-metal sheath	mm	3.7	3.7	3.7	3.7
	12	Thickness of metal sheath	mm	1.5	1.6	1.6	1.7
	13	Diameter over metal sheath	mm	32.0	35.5	39.0	41.0
	15	Thickness of an armour tape	mm	0.5	0.8	0.8	0.8
	17	Overall diameter of cable	mm	40	45	48	51
	19	Weight of cable	kg/km	3700	4800	5400	6050
Mechanical properties	20	Minimum bending radius	mm	630	720	765	795
	21	Perm. pulling force with cable grip	N	4800	6075	6912	7803
	22	Perm. pulling force with pulling head	N	4500	6300	8550	10800
Electrical properties	23	d.c. resistance/unit length at 65 °C	Ω/km	0.757	0.523	0.378	0.299
	24	d.c. resistance/unit length at 20 °C	Ω/km	0.641	0.443	0.320	0.253
	25	Operating capacitance/unit length	μF/km	0.300	0.340	0.380	0.410
	26	Charging current	A/km	0.540	0.620	0.690	0.740
	28	Earth fault current	A/km	1.09	1.20	1.31	1.36
	31	Reference diameter of cable	mm	46.1	49.5	53.9	56.5
Installation in ground	32	Current-carrying capacity	A	132	165	200	229
	33	Permissible transmission power	MVA	2.29	2.86	3.46	3.97
	34	Eff. a.c. resistance/unit length at 65 °C	Ω/km	0.759	0.526	0.381	0.302
	38	Ohmic losses per cable	kW/km	39.7	43.0	45.7	47.5
	39	Fict. thermal resist. of cable for ohmic losses	Km/W	0.635	0.573	0.522	0.490
	41	Fict. thermal resist. of soil for ohmic losses	Km/W	0.492	0.481	0.467	0.460
	44	Inductance/unit length per conductor	mH/km	0.348	0.330	0.317	0.308
Installation in free air	32	Current-carrying capacity	A	112	140	172	198
	33	Permissible transmission power	MVA	1.94	2.42	2.98	3.43
	34	Eff. a.c. resistance/unit length at 65 °C	Ω/km	0.759	0.526	0.381	0.302
	38	Ohmic losses per cable	kW/km	28.6	30.9	33.8	35.5
	39	Fict. thermal resist. of cable for ohmic losses	Km/W	0.635	0.573	0.522	0.490
	43	Thermal resistance of the air	Km/W	0.589	0.552	0.512	0.491
	44	Inductance/unit length per conductor	mH/km	0.348	0.330	0.317	0.308
	48	Minimum time value	s	502	630	768	925
Short-circuit	49	Rated short-time current of conductor (1s)	kA	4.00	5.60	7.60	9.60

150 SM	185 SM	240 SM	300 SM
6.4	6.4	6.4	6.4
3.7	3.7	3.7	3.7
1.8	2.0	2.1	2.3
44.0	47.5	52.0	56.0
0.8	0.8	0.8	0.8
54	58	62	66
6800	7900	9100	10600
855	915	975	1050
8748	10092	11532	13068
13500	16650	21600	27000
0.243	0.194	0.148	0.118
0.206	0.164	0.125	0.1000
0.430	0.460	0.510	0.550
0.780	0.830	0.930	1.00
1.41	1.52	1.69	1.90
59.7	62.9	68.5	72.5
259	296	343	390
4.49	5.13	5.94	6.75
0.247	0.198	0.153	0.124
49.7	52.0	54.0	56.6
0.456	0.426	0.403	0.376
0.451	0.442	0.429	0.420
0.300	0.293	0.284	0.278
226	260	306	350
3.91	4.50	5.30	6.06
0.247	0.198	0.153	0.124
37.9	40.1	43.0	45.6
0.456	0.426	0.403	0.376
0.467	0.446	0.414	0.393
0.300	0.293	0.284	0.278
1109	1275	1549	1850
12.0	14.8	19.2	24.0

Table 5.7.2
Cables with paper insulation with aluminium conductor

Three-phase operation at 50 Hz

Design				35	50	70	95
Design	2	Nom. cross-sectional area of conductor	mm^2	35	50	70	95
	3	Shape and type of conductor		RM	RM	RM	RM
	9	Thickness of insulation	mm	5.5	5.5	5.5	5.5
	12	Thickness of metal sheath	mm	1.3	1.3	1.4	1.4
	13	Diameter over metal sheath	mm	20.5	22.0	24.0	25.5
	15	Thickness of an armour tape	mm	0.8	0.8	0.8	1.0
	17	Overall diameter of cable	mm	56	58	62	68
	19	Weight of cable	kg/km	6050	6550	7500	8750
Mechanical properties	20	Minimum bending radius	mm	885	915	975	1050
	21	Perm. pulling force with cable grip	N	3136	3364	3844	4624
	22	Perm. pulling force with pulling head	N	3150	4500	6300	8550
Electrical properties	23	d.c. resistance/unit length at 65 °C	Ω/km	1.03	0.757	0.523	0.378
	24	d.c. resistance/unit length at 20 °C	Ω/km	0.868	0.641	0.443	0.320
	25	Operating capacitance/unit length	μF/km	0.222	0.245	0.279	0.310
	26	Charging current	A/km	0.810	0.890	1.01	1.12
	28	Earth fault current	A/km	2.43	2.67	3.03	3.36
	30	Electric field strength at the conductor	kV/mm	3.60	3.40	3.20	3.10
	31	Reference diameter of cable	mm	60.6	63.9	67.7	71.8
Installation in ground	32	Current-carrying capacity	A	117	139	172	207
	33	Permissible transmission power	MVA	4.05	4.82	5.96	7.17
	34	Eff. a.c. resistance/unit length at 65 °C	Ω/km	1.03	0.760	0.526	0.381
	38	Ohmic losses per cable	kW/km	42.2	44.0	46.7	49.0
	39	Fict. thermal resist. of cable for ohmic losses	Km/W	0.623	0.582	0.537	0.499
	41	Fict. thermal resist. of soil for ohmic losses	Km/W	0.448	0.440	0.431	0.422
	44	Inductance/unit length per conductor	mH/km	0.476	0.454	0.427	0.409
Installation in free air	32	Current-carrying capacity	A	102	122	152	184
	33	Permissible transmission power	MVA	3.53	4.23	5.27	6.37
	34	Eff. a.c. resistance/unit length at 65 °C	Ω/km	1.03	0.760	0.526	0.381
	38	Ohmic losses per cable	kW/km	32.1	33.9	36.5	38.7
	39	Fict. thermal resist. of cable for ohmic losses	Km/W	0.623	0.582	0.537	0.499
	43	Thermal resistance of the air	Km/W	0.472	0.449	0.426	0.404
	44	Inductance/unit length per conductor	mH/km	0.476	0.454	0.427	0.409
	48	Minimum time value	s	297	423	534	671
Short-circuit	49	Rated short-time current of conductor (1s)	kA	2.69	3.85	5.39	7.31

120 RM	150 RM	185 RM	240 RM	300 RM
5.5	5.5	5.5	5.5	5.5
1.4	1.5	1.5	1.5	1.6
27.0	28.5	30.0	32.5	35.0
1.0	1.0	1.0	1.0	1.0
71	74	78	83	87
9450	10500	11400	12700	14200
1110	1155	1215	1290	1365
5041	5476	6084	6889	7569
10800	13500	16650	21600	27000
0.299	0.243	0.194	0.148	0.118
0.253	0.206	0.164	0.125	0.1000
0.338	0.367	0.398	0.445	0.483
1.23	1.33	1.44	1.61	1.75
3.69	3.99	4.32	4.83	5.25
3.00	2.90	2.90	2.80	2.70
76.2	80.1	83.7	89.7	94.2
235	266	302	353	400
8.14	9.21	10.5	12.2	13.9
0.302	0.247	0.198	0.152	0.123
50.1	52.5	54.1	57.0	59.2
0.483	0.457	0.434	0.402	0.381
0.412	0.404	0.397	0.386	0.378
0.393	0.382	0.370	0.355	0.345
211	239	274	323	368
7.31	8.28	9.49	11.2	12.7
0.302	0.247	0.198	0.152	0.123
40.4	42.3	44.5	47.7	50.1
0.483	0.457	0.434	0.402	0.381
0.384	0.367	0.353	0.331	0.317
0.393	0.382	0.370	0.355	0.345
815	992	1148	1390	1674
9.24	11.5	14.2	18.5	23.1

Table 5.7.3
Cables with paper insulation with aluminium conductor

Three-phase operation at 50 Hz

Design			mm²	35 RM	50 RM	70 RM	95 RM
	2	Nom. cross-sectional area of conductor	mm²	35	50	70	95
	3	Shape and type of conductor		RM	RM	RM	RM
	9	Thickness of insulation	mm	7.5	7.5	7.5	7.5
	12	Thickness of metal sheath	mm	1.4	1.4	1.4	1.5
	13	Diameter over metal sheath	mm	25.0	26.0	28.0	29.5
	15	Thickness of an armour tape	mm	1.0	1.0	1.0	1.0
	17	Overall diameter of cable	mm	67	69	73	76
	19	Weight of cable	kg/km	8350	8850	9650	10800
Mechanical properties	20	Minimum bending radius	mm	1035	1080	1125	1200
	21	Perm. pulling force with cable grip	N	4489	4761	5329	5776
	22	Perm. pulling force with pulling head	N	3150	4500	6300	8550
Electrical properties	23	d.c. resistance/unit length at 60 °C	Ω/km	1.01	0.744	0.514	0.372
	24	d.c. resistance/unit length at 20 °C	Ω/km	0.868	0.641	0.443	0.320
	25	Operating capacitance/unit length	µF/km	0.170	0.194	0.224	0.248
	26	Charging current	A/km	0.930	1.06	1.22	1.35
	28	Earth fault current	A/km	2.79	3.18	3.66	4.05
	30	Electric field strength at the conductor	kV/mm	4.10	4.00	3.90	3.70
	31	Reference diameter of cable	mm	74.9	75.3	77.7	81.8
Installation in ground	32	Current-carrying capacity	A	110	131	162	195
	33	Permissible transmission power	MVA	5.72	6.81	8.42	10.1
	34	Eff. a.c. resistance/unit length at 60 °C	Ω/km	1.01	0.747	0.518	0.375
	38	Ohmic losses per cable	kW/km	36.7	38.5	40.8	42.8
	39	Fict. thermal resist. of cable for ohmic losses	Km/W	0.679	0.630	0.576	0.537
	41	Fict. thermal resist. of soil for ohmic losses	Km/W	0.415	0.414	0.409	0.401
	44	Inductance/unit length per conductor	mH/km	0.531	0.496	0.462	0.442
Installation in free air	32	Current-carrying capacity	A	95	114	141	171
	33	Permissible transmission power	MVA	4.94	5.92	7.33	8.89
	34	Eff. a.c. resistance/unit length at 60 °C	Ω/km	1.01	0.747	0.517	0.375
	38	Ohmic losses per cable	kW/km	27.4	29.1	30.9	32.9
	39	Fict. thermal resist. of cable for ohmic losses	Km/W	0.679	0.630	0.576	0.537
	43	Thermal resistance of the air	Km/W	0.406	0.403	0.390	0.372
	44	Inductance/unit length per conductor	mH/km	0.531	0.496	0.462	0.442
	48	Minimum time value	s	298	422	541	678
Short-circuit	49	Rated short-time current of conductor (1s)	kA	2.55	3.65	5.11	6.93

NAEKEBA $3\times\ldots$ **18/30 kV** (U_m=36 kV)

120 RM	150 RM	185 RM	240 RM	300 RM
7.5	7.5	7.5	7.5	7.5
1.5	1.5	1.6	1.6	1.7
31.0	32.5	34.5	37.0	39.0
1.0	1.0	1.0	1.0	1.0
79	83	86	91	96
11500	12300	13600	15000	16700
1245	1290	1350	1440	1500
6241	6889	7396	8281	9216
10800	13500	16650	21600	27000
0.294	0.239	0.190	0.145	0.116
0.253	0.206	0.164	0.125	0.1000
0.269	0.290	0.313	0.348	0.376
1.46	1.58	1.70	1.89	2.05
4.38	4.74	5.10	5.67	6.15
3.60	3.50	3.40	3.30	3.20
85.2	89.1	92.7	98.9	103
223	251	286	334	379
11.6	13.0	14.9	17.4	19.7
0.298	0.243	0.195	0.150	0.122
44.4	46.0	47.9	50.3	52.5
0.507	0.480	0.456	0.424	0.400
0.394	0.387	0.381	0.370	0.363
0.424	0.411	0.398	0.382	0.369
197	223	255	301	342
10.2	11.6	13.3	15.6	17.8
0.298	0.243	0.195	0.150	0.122
34.6	36.3	38.0	40.9	42.7
0.507	0.480	0.456	0.424	0.400
0.358	0.344	0.331	0.312	0.300
0.424	0.411	0.398	0.382	0.369
815	994	1156	1396	1690
8.76	10.9	13.5	17.5	21.9

6 Explanations to the Project Planning Data of Cables

General

The different cable types and their arrangements are marked by the following *symbols* in the tables of Clause 5:

3 single-core cables in three-phase operation, bunched,

3 single-core cables in three-phase operation, flat formation

1 three-core cable in three-phase or single-phase a.c. operation,

1 three-core cable with PE or PEN conductor and a reduced cross-sectional area in three-phase operation,

1 four-core cable in three-phase operation,

1 five-core cable in three-phase operation,

1 multi-core cable (control cable), all cores are loaded with alternating current.

All calculations were carried out for a system frequency of 50 Hz. By approximation they also apply to an operation with 60 Hz, with the exception of single-core screened cables lying in a single layer.
The electrical and thermal data for the determination of the current-carrying capacity given in the tables of Clause 5 were determined by calculation. The *formulae* and *data* were taken from the following publications:

Designation	Source
Values for the material	Part 1, Clauses 18 and 19
Effective resistance or additional resistance per unit length	IEC 287 (1982)
Fictitious thermal resistance of the cable – single-core cable	Part 1, Clause 18.4.1
– three-core cable in three-phase operation	IEC 287 (1982)
– four-core cable in three-phase opration	Artbauer, J.; Winkler, F.: Belastbarkeit vieradriger 0,6/1-kV-Kabel mit Kunststoffisolierung im Drehstrombetrieb. Elektrizitätswirtschaft 83 (1984) Heft 26, S. 1104 bis 1110
Fictitious thermal resistance of the environment	Part 1, Clauses 18.4.2 and 18.4.3

The *symbols* given basically comply with the symbols in Part 1, with the exception of the symbol for the permissible tensile force F (Item 21 and 22).

For some cables with rated voltages $U_0/U = 0.6/1$ kV ($U_m = 1.2$ kV), the tables of Clause 5 do not include *thermal resistances* because at present there is no calculation method which can be generally applied.

The current-carrying capacity

▷ for a three-core cable in single-phase alternating current operation (J design) and

▷ for a three-core cable in three-phase operation and with a reduced PE or PEN conductor was determined by means of the thermal resistances of the cable and the external environment of a three-core cable of the same type in three-phase operation (O design).

The current-carrying capacity

▷ for a five-core cable in three-phase operation and

▷ for a multi-core cable (control cable, all cores are loaded with alternating current)

was determined by means of the thermal resistances of the cable and the external environment of a four-core cable of the same type in three-phase operation.

At present, the specifications for FRNC cables do not include current-carrying capacities. For this reason the thermal resistances of the cable and the environment are not mentioned.

The current-carrying capacity

▷ for a cable with the type designation code NHXHX and

▷ for a cable with the type designation code NHXCHX

was taken from a PVC-insulated and PVC-sheathed cable having the same construction.

The current-carrying capacity

▷ for a cable with the type designation code (N)2XH and

▷ for a cable with the type designation code (N)2XCH ...

was taken from an XLPE-insulated and PVC-sheathed cable having the same construction.

Explanations to the Items of the Project Planning Data

(Table 5.1.1 to 5.7.3)

Item	Designation	Symbol	Unit	Explanation	Further explanations see Part 1	Details in DIN VDE
Design						
1	Number of cores			For multi-core cables (control cables) the number of cores with the same nominal cross-sectional area of conductor is given. For single-core and multi-core cables the number of cores is mentioned in the heading.		
2	Nominal cross-sectional area of conductor	q_n	mm²	A d.c. resistance per unit length at 20 °C according to DIN VDE 0295 (see Item 24) is allocated to the nominal cross-sectional area of conductor. It is proved by measurement.	Clause 20.1, Page 320 Clause 37, Page 454 to 456	0295 0472 Part 501
3	Shape and type of conductor			The shape and type of the phase conductor are given. Shape: R Circular conductor S Sector-shaped conductor Type: E Solid conductor M Stranded conductor Example: SM Stranded sector-shaped conductor	Clause 1.2, Page 13 and 14 Clause 13.2, Page 100	0298 Part 1
4	Nominal cross-sectional area of reduced conductor	q_n	mm²	For multi-core cables with a rated voltage $U_0/U = 0.6/1$ kV, the reduced cross-sectional area of the PE or PEN conductor is given. It is allocated to the phase conductors according to the VDE specifications. Its d.c. resistance per unit length at 20 °C shall comply with DIN VDE 0295.	Clause 20.1, Page 320	0265 0266 0271 0295 0472 Part 501
5	Shape and type of reduced conductor			The shape and type of the PE or PEN conductor with a reduced cross-sectional area are given (designation code see Item 3).	Clause 1.2, Page 13 and 14 Clause 13.2, Page 100	0298 Part 1

Item	Designation	Symbol	Unit	Explanation	Further explanations see Part 1	Details in DIN VDE
6	Nominal cross-sectional area of screen	q_M	mm²	The screen always consists of copper and its cross-sectional area is allocated to the phase conductors according to the VDE specifications. The electrical effective cross-sectional area of the screen is not given because the geometrical cross-sectional area is more important with regard to the earth fault and short-circuit stresses. This stipulation deviates from the rules for concentric conductors. For cables laid in ground and having a nominal cross-sectional area of the phase conductor of 150 and 185 mm² as well as for single-core cables laid in ground and having a nominal cross-sectional area of the phase conductor of 240 mm², a nominal cross-sectional area of the screen of 16 mm² is permissible instead of 25 mm². For these designs the weight and the overall diameter slightly differ (refer to Item 17 and 19). However, this does not influence the current-carrying capacity and the electrical and thermal data.	Clause 7.2, Page 46 and 47	0271 0273
7	Nominal cross-sectional area of concentric conductor	q_n	mm²	The concentric conductor in low-voltage cables which always consists of copper is used as PE or PEN conductor; it also serves as protection against electric shock. The cross-sectional area is allocated to the phase conductors according to the VDE specifications. However, the cross-sectional area of the concentric conductor included in the type designation code always refers to the material of the phase conductors. For a cable with aluminium conductor, for example NAYCWY 3×95 SE/95 0.6/1 kV ($U_m = 1.2$ kV), the d.c. resistance per unit length at 20 °C may not be greater than the maximum permissible value for an aluminium conductor of 95 mm² according to DIN VDE 0295.	Clause 6 Page 44	0266 0271 0272 0295 0472 Part 501
8	–			Not given		

▷

Explanations to the Items of Project Planning Data (Continued)

(Table 5.1.1 to 5.7.3)

Item	Designation	Symbol	Unit	Explanation	Further explanations see Part 1	Details in DIN VDE
9	Thickness of insulation	$\delta^{1)}$	mm	The nominal value (in the case of paper-insulated cables the desired value) of the thickness of insulation is given. For polymeric-insulated cables the conductive layers are not included in the thickness of insulation. In the case of paper-insulated cables, however, the thickness of insulation also contains the conductive layers which may be up to 10 % of the total thickness.	Clause 2 Page 15 to 36	0255 0265 0266 0271 0272 0273 0472 Part 402
10	Thickness of insulation conductor - conductor	$\delta^{1)}$	mm	For belted cables the nominal value (in this case the desired value) of the thickness of insulation between the conductors (L/L) is given.	Clause 2.4, Page 35	0255
11	Thickness of insulation conductor - metal sheath	$\delta^{1)}$	mm	For belted cables the nominal value (in this case the desired value) of the thickness of insulation between the conductor and the metal sheath (L/M) is given.	Clause 2.4, Page 35	0255
12	Thickness of metal sheath	δ_M	mm	The nominal value (in this case the desired value) of the thickness of metal sheath is given.	Clause 3.4, Page 39 and 40	0255 0265 0472 Part 402
13	Diameter over metal sheath	δ_M	mm	The diameter over metal sheath is important for the selection of the accessories.	Clause 3.4, Page 39 and 40	
14	Thickness of an armour wire	$\delta^{1)}$	mm	The armour serves as mechanical protection. For multi-core polymeric-insulated cables a steel wire armour may also be used as screen. The thickness of the armour wires depends on the requirements. It shall correspond to the minimum values in the VDE specifications.	Clause 5, Page 43	0255 0271 0472 Part 402
15	Thickness of an armour tape	$\delta^{1)}$	mm	Paper-insulated cables normally have a steel tape armour with two compounded steel tapes each applied with a small pitch in such a way that the second bandage covers the gaps left by the first. The thickness is laid down in DIN VDE 0255.	Clause 5, Page 43	0255 0472 Part 402

[1] As in Part 1, no difference is made in the symbol δ

Item	Designation	Symbol	Unit	Explanation	Further explanations see Part 1	Details in DIN VDE
16	Thickness of outer sheath	δ[1]	mm	For polymeric-insulated cables the nominal thickness of the extruded outer sheath is given. For paper-insulated cables the thickness of the outer covering consisting of fibrous materials and bitumen is not specified.	Clause 3, Page 37 to 40 Clause 4, Page 41	0255 0265 0266 0271 0272 0273 0472 Part 402
17	Overall diameter of cable	d	mm	The overall diameter determined by calculation is given. Note: For polymeric-insulated cables the VDE specifications give smallest and largest diameters or recommended values.		0271 0272 0273
18	–			Not given		
19	Weight of cable	m'	kg/km	The net weight determined by calculation is given. This value is a recommended value.		

Mechanical properties

Item	Designation	Symbol	Unit	Explanation	Further explanations see Part 1	Details in DIN VDE
20	Minimum bending radius	r	mm	The permissible bending radius according to DIN VDE 0298 Part 1 is dependent on the overall diameter of the cable. With regard to installation this recommended value should not fall below the minimum value. Where a bend is to be made once only, e.g. in front of terminations, the bending radius can be decreased provided that the work is done by experts (heating to 30 °C, bending over template).	Clause 29.2, Page 400 and 401	0298 Part 1
21	Permissible pulling force with pulling grip	F	N	According to DIN VDE 0298 Part 1 the permissible pulling force when pulling the cables with a pulling grip is given.	Clause 29.4, Page 401 to 407	0298 Part 1
22	Permissible pulling force with pulling head	F	N	According to DIN VDE 0298 Part 1 the permissible pulling force when pulling the cables with a pulling head is given. The permissible pulling force with pulling head is calculated with the sum of the cross-sectional areas of conductors. Other metallic constructional elements (e.g. concentric conductor, screen) shall not be considered for the calculation. Furthermore it shall be ensured that the pulling head clamps all conductor wires equally and that a suitable sealing prevents the penetration of moisture into the conductor or into the cable end.	Clause 29.4, Page 401 to 407	0298 Part 1

$\triangleright$

[1] As in Part 1, no difference is made in the symbol δ

Explanations to the Items of Project Planning Data (Continued)

(Table 5.1.1 to 5.7.3)

Item	Designation	Symbol	Unit	Explanation	Further explanations see Part 1	Details in DIN VDE
Electrical properties						
23	d.c. resistance per unit length at ... °C	R'_ϑ	Ω/km	The d.c. resistance per unit length at the permissible operating temperature is given. This value was determined by conversion of the d.c. resistance per unit length at 20 °C (Item 24).	Clause 1, Page 11 and 12	0295 0472 Part 501
24	d.c. resistance per unit length at 20 °C	R'_{20}	Ω/km	The maximum d.c. resistance per unit length at 20 °C is given. This value was laid down again for all cables when DIN VDE 0295 was published in 1980. All electrical and thermal data listed in the tables were calculated with this d.c. resistance per unit length. For deviating conductor temperatures the d.c. resistance per unit length can be converted by means of the temperature coefficients given in Part 1.	Clause 20.1, Page 320 Clause 37, Page 454 to 456	0295 0472 Part 501
25	Operating capacitance per unit length	C'_b	$\mu F/km$	The operating capacitances per unit length were partly determined by calculation and partly by measured values. They apply to the permissible operating temperature and are recommended values. For cables with XLPE insulation or paper insulation the capacitance per unit length is almost independent of the temperature. For cables with PVC insulation the capacitance per unit length significantly deviates by influence of the temperature so that a conversion to lower operating temperatures should be made if necessary.	Clause 22, Page 331 to 334	
26	Charging current	I'_c	A/km	The charging current is a reactive current which depends on the voltage, the frequency and the capacitance of the cable. The charging current given was calculated with the operating capacitance per unit length given in Item 25, the voltage $U/\sqrt{3}$ and a system frequency of 50 Hz.	Clause 22.3 Page 334 to 335	
27	–			Not given		

Item	Designation	Symbol	Unit	Explanation	Further explanations see Part 1	Details in DIN VDE
28	Earth fault current	I_e'	A/km	A fault between a phase conductor and earth arising in an isolated neutral system is designated as earth fault. The current which results from this fault is called earth-fault current. In this table the earth-fault current is given for three-phase operation either for a multi-core cable or for a cable system consisting of three single-core cables.	Clause 22.3, Page 334 to 336	
29	Dielectric losses per cable	P_d'	kW/km	During operation cables are heated by voltage-dependent (dielectric) losses as well as by ohmic losses. For the determination of the current-carrying capacity of PVC-insulated cables the dielectric losses shall be taken into account for a rated voltage $U_0/U = 6/10$ kV and above. For XLPE- and paper-insulated cables up to $U_0/U = 18/30$ kV the dielectric losses are negligible.	Clause 22.4, Page 336	
30	Electric field strength at the conductor	E_i	kV/mm	The electric field strength on the inner conductive layer is given for the voltage $U/\sqrt{3}$ for all cables with a radial electrical field distribution.		
31	Reference diameter of cable	d	mm	A model design was taken as a basis for all cables for the determination of the current-carrying capacity (Item 32) and the thermal and electrical data. The reference diameter is the overall diameter of the cable resulting from the model design.		

Properties for laying direct in ground or installation in free air

Item	Designation	Symbol	Unit	Explanation	Further explanations see Part 1	Details in DIN VDE
32	Current-carrying capacity	I_r	A	The current-carrying capacity is the maximum permissible current under certain operating conditions. The current-carrying capacity is given for normal operation under reference operating conditions according to Table 4.2.1 for laying in ground and according to Table 4.2.10 for installation in free air.	Clause 18, Page 150 to 252	0298 Part 2

▷

Explanations to the Items of Project Planning Data (Continued)

(Table 5.1.1 to 5.7.3)

Item	Designation	Symbol	Unit	Explanation	Further explanations see Part 1	Details in DIN VDE
33	Permissible transmission power	P	MVA	The transmission power is given for three-phase operation and for the voltage U as well as for a current equal to the current-carrying capacity according to Item 32.		
34	Effective a.c. resistance per unit length at ... °C	R'_{wr}	Ω/km	The effective a.c resistance per unit length R'_{wr} is made up of the d.c. resistance per unit length and the additional resistance per unit length at single-phase a.c. operation or three-phase operation. It is given for the permissible operating temperature. The current-carrying capacity in Item 32 was determined by means of the effective resistance per unit length. In general, the additional resistance per unit length is calculated with the geometrical mean value of the cable spacings. For three single-core cables lying in a single layer, the geometrical mean value of the cable spacings is equal to 1.26 times the spacing between two adjacent cables. If the cables have a metallic covering (screen or sheath) in this arrangement, the highest additional losses will arise in an outer cable of the group. For cables laid in ground the current-carrying capacity is determined by means of the cable in the centre owing to the highest heating so that the additional resistance per unit length may be calculated with the geometrical mean value of the axial distances. For cables installed in air the mutual heating of the cables is not very important. In this case the current-carrying capacity is determined with the effective a.c. resistance per unit length of an outer cable of the group. For three single-core cables installed in air and laid in ground in bunched arrangement, the calculation of the electrical values is based on an axial distance of the cables corresponding to 1.11 times the reference diameter in Item 31. For this reason irregularities in the axial distance are considered to a certain extent.	Clause 18.3, Page 181 to 184 Clause 20, Page 320 and 321	

Item	Designation	Symbol	Unit	Explanation	Further explanations see Part 1	Details in DIN VDE
35	Mean effective a.c. resistance per unit length at ... °C	R'_{wmr}	Ω/km	The ohmic losses per cable given in Item 38 (refer also to Item 34) were determined for three single-core cables with a metallic covering, installed side by side in air, by means of the mean effective a.c. resistance per unit length at the permissible operating temperature.		
36	Effective a.c. resistance per unit length at 20 °C	R'_{w20}	Ω/km	The a.c. effective resistance per unit length at 20 °C is required for the determination of the maximum short-circuit current in a cable system (refer also to Item 34).		0102 Part 2
37	–			Not given		
38	Ohmic losses per cable	P'_i	kW/km	The ohmic losses were calculated with the current-carrying capacity according to Item 32 and the effective resistance per unit length according to Item 34 and 35 and given per cable. If the ohmic losses of single-core cables per three-phase system are required, the tabulated value shall be multiplied by the factor 3. Losses in case of a deviating current I are obtained from the following equation: $$P'_i = n \cdot I^2 \dot{R}'_{wr} \cdot 10^{-3}\,\text{kW/km}$$ or $$P'_i = n \cdot I^2 \cdot R'_{wmr} \cdot 10^{-3}\,\text{kW/km}$$ n number of loaded cores per cable (refer also to Item 34 and 35). For cables laid in ground the maximum losses occuring during the daily load cycle with a load factor of $m = 0.7$ are given. The mean ohmic losses occuring every day can be determined by multiplication with the loss factor μ: $m = 0.3\,m + 0.7\,m^2 = 0.553$ for $m = 0.7$	Clause 18.3, Page 181 to 184 Clause 20.3, Page 321 Page 200, Equ. (18.51)	
39	Fictitious thermal resistance of cable for ohmic losses	T'_{Ki}	Km/W	The fictitious thermal resistance of the cable is given in view of the ohmic losses.	Clause 18.3, Page 181 to 184 Clause 18.4.1, Page 184 to 186	
40	Fictitious thermal resistance of cable for dielectric losses	T'_{Kd}	Km/W	The fictitious thermal resistance of the cable is given in view of the dielectric losses.	Clause 18.3, Page 181 to 184 Clause 18.4.1 Page 184 to 186	

$\triangleright$

Explanations to the Items of Project Planning Data (Continued)

(Table 5.1.1 to 5.7.3)

Item	Designation	Symbol	Unit	Explanation	Further explanations see Part 1	Details in DIN VDE
41	Fictitious thermal resistance of soil for ohmic losses	T'_{xy}	Km/W	The fictitious thermal resistance of soil T'_{xy} calculated with the soil thermal resistivity of the dry area considers the cyclic course of a daily load cycle and the drying out of the soil. It was determined by means of the reference operating conditions according to Table 4.2.1. For cables with small nominal cross-sectional areas and/or a low permissible operating temperature it is possible that the soil does not dry out. This can be checked by means of equations 18.56 and 18.64, pages 201 and 202, Part 1. However, it should be taken into consideration that the soil can dry out under other operating conditions, e.g. grouping of cables, higher soil temperature or other load factor.	Clause 18.3, Page 181 to 184 Clause 18.4.3, Page 197 to 206 Clause 18.4.6, Page 218 to 230	
42	Fictitious thermal resistance of soil for dielectric losses	T'_x	Km/W	The fictitious thermal resistance of soil T'_x calculated with the soil thermal resistivity of the dry area considers the continuous operation as well as the drying out of the soil (refer also to Item 41, second paragraph). In cases where the dielectric losses shall not be neglected (refer to Item 29), the fictitious thermal resistance of soil is to be considered for the calculation of the current-carrying capacity.	Clause 18.3, Page 181 to 184 Clause 18.4.3, Page 197 to 206 Clause 18.4.6, Page 218 to 230	
43	Thermal resistance of the air	T'_{Lu}	Km/W	For cables installed in air the heat is dissipated by convection and radiation. The thermal resistance of air T'_{Lu} is given for one cable horizontally installed in free air. The cable dissipates its losses by natural convection and unhindered radiation and without influence of external heat sources (solar radiation) into its surroundings. The ambient temperature does not increase in this case.	Clause 18.3, Page 181 to 184 Clause 18.4.2, Page 186 to 196	

Item	Designation	Symbol	Unit	Explanation	Further explanations see Part 1	Details in DIN VDE
44	Inductance per unit length per conductor	L', L'_m	mH/km	The value given is a calculated value and applies to one conductor. For three single-core cables in three-phase operation laid in flat formation as well as for four-core and five-core cables, the mean inductance per unit length L'_m is given. The calculated spacing between three single-core cables in three-phase operation and in bunched arrangement (refer to Item 34) was considered.	Clause 21, Page 322 to 330	
45	Effective resistance per unit length in zero phase-sequence system	R'_{0L}	Ω/km	The effective component R'_{0L} and the reactive component X'_{0L} (refer to Item 46) of the mean zero-phase sequence system per unit length of single-core cables with metallic coverings are given. The tabulated values are calculated values. They were determined for a soil with an electrical resistivity of 100 Ω m. Ambient influences, such as pipe-lines, rails etc., were not taken into consideration.	Clause 21.4, Page 329 Clause 27.4.3, Page 385	0102 Part 2
46	Reactance per unit length in zero phase-sequence system	X'_{0L}	Ω/km	See explanation to Item 45.		
47	Reduction factor	r_{Ki}		The current reduction factor of the influencing cable was calculated with the nominal cross-sectional area of the screen and is given for a short-circuit occuring outside the cable. It is a measure for the effect of screening of a metallic cable covering which is earthed at both ends.	Clause 26, Page 349 to 361	
48	Minimum time value	τ	s	The minimum time value τ is one fifth of the time taken from the temperature curve to almost reach the permissible final temperature at constant current for installation in air. It is relevant for the calculation of the current-carrying capacity at short-time operation and intermittent operation.	Clause 18.6 Page 239 to 244	

$\triangleright$

Explanations to the Items of the Project Planning Data (Continued)

(Table 5.1.1 to 5.7.3)

Item	Designation	Symbol	Unit	Explanation	Further explanations see Part 1	Details in DIN VDE
Short-circuit						
49	Rated short-time current of conductor (1 s)	I_{thr}	kA	The rated short-time current I_{thr} of a conductor is defined for the rated short-circuit duration $t_{kr} = 1$ s (rated value of the short-circuit current-carrying capacity). Permissible short-circuit capacities I_{thz} for other times t_k up to 5 s are obtained from the following equation: $$I_{thz} = I_{thr}\sqrt{\frac{t_{kr}}{t_k}}.$$	Clause 19.3.1, Page 265 to 285 Clause 19.3.2, Page 285 to 292	
50	Rated short-time current of screen (1 s)	I_{thr}	kA	The rated short-time current of a screen is given for the rated short-circuit duration $t_{kr} = 1$ s. Permissible short-circuit capacities for other times up to 5 s can be taken from Figure 4.2.3.	Clause 19.3.1, Page 265 to 285 Clause 19.3.2, Page 285 to 292	

Accessories for Power Cables

7 Summary of Accessories

The following table shows the mounting technique, the nominal voltage and the insulation material of the cable. This facilitates the allocation of a certain accessory for a specific case of application. The relevant table with the project planning data of the accessory can be found by means of the designation code for the accessory in the summary of accessories (following pages).

Type of accessory	Nominal voltage kV	Cast resin technique	Cable Paper	Cable PVC	Cable VPE	Shrink technique	Cable Paper	Cable PVC	Cable VPE	Wrapping and casting technique	Cable Paper	Cable PVC	Cable VPE	Push-on technique	Cable Paper	Cable PVC	Cable VPE
Indoor terminations	1	PEA		×	×	SKSA		×	×								
	6	PEA, PEB		×													
	10																
	20									IKM	×			IAES 10		×	×
										EoD	×			IAE 20			×
	30													IAES 20		×	×
										EoD	×			IAES 30		×	×
Outdoor terminations	1					SKSE	×	×									
	6									FFK 10		×					
	10									FEP		×	×	FAE 10			×
										FF 10	×						
	20									FEL 20	×			FAE 20			×
										FEL-2Y			×	FAL 20			×
	30									FEL 30	×			FAE 30			×
										FEL-2Y			×				
Spreader boxes	20, 30									AS, SS	×						
Separable elbow connectors	10, 20													WS..., BWS...		×	×
Straight joints	1	PV		×	×	SKSM	×	×									
						SKSM M	×	×									
		PV		×		SKSM ST	×										
						SKSM C	×										
						SKEM	×	×									
	6	PV		×													
	10									WP, WPS 10		×	×	AMS 10			×
	20									VS	×						
										WP, WPS 20		×	×	AMS 20			×
										SMI	×						
	30									WP, WPS 30		×	×				
										SMI	×						
Branch joints	1	PA, PAK	×	×		GNKA-1	×	×									
		GMS, GMSA	×	×													
		PAD	×														
Transition joints	1					SKSM PK	×	×	×								
	10	ÜMP 10	×	×	×												
	20, 30									SM-WP	×		×				

353

Designation			Nominal voltage kV	Project planning data Table	Page
7.1 Indoor terminations	SKSA	Shrinkable spreader cap for three-core and four-core polymeric-insulated cables	1	8.1.1	358
	PEA	PROTOLIN termination (cast resin) for PVC and XLPE-insulated cables	1	8.1.2	359
	PEA	PROTOLIN termination (cast resin) for three-core PVC-insulated cables in confined connection areas	6	8.1.3	360
	PEB	PROTOLIN termination (cast resin) for three-core PVC-insulated cables	6	8.1.4	361
	IAES 10, IAES 20, IAES 30	Push-on termination with sheds for single-core PVC and XLPE-insulated cables	10, 20 and 30		
		25 to 120 mm²		8.1.5 a	362
		150 to 500 mm²		8.1.5 b	363
	IAES 10	Push-on termination with sheds for three-core	10		
		PVC-insulated cables 25 to 95 mm²		8.1.6 a	364
		120 to 300 mm²		8.1.6 b	365
		XLPE-insulated cables 50 to 95 mm²		8.1.6 c	366
		120 to 185 mm²		8.1.6 d	367
	IKM	Termination with copper housing and transparent cover for paper-insulated belted cables	10	8.1.7	368
	IAE 20	Push-on termination without sheds for single-core XLPE-insulated cables 25 to 95 mm²	20	8.1.8 a	369
		120 to 300 mm²		8.1.8 b	370
	EoD	Termination without seals for paper-insulated three-core S.L. cables	20 and 30	8.1.9	371
7.2 Outdoor terminations	SKSE	Shrinkable termination for three-core and four-core PVC and XLPE-insulated cables	1	8.2.1	373
	FFK 10	Encased termination with porcelain insulators for three-core PVC-insulated cables with flat wire armouring	6	8.2.2	374
	FAE 10, FAE 20, FAE 30	Push-on termination for single-core XLPE-insulated cables 25 to 120 mm²	10, 20 and 30	8.2.3 a	375
		150 to 500 mm²		8.2.3 b	376
	FEP	Termination with porcelain insulator for single-core and three-core PVC and XLPE-insulated cables	10	8.2.4	377

▷

Designation			Nominal voltage kV	Project planning data Table	Page
7.5 Straight joints (continued)	PV	PROTOLIN straight joint for three-core PVC-insulated cables with steel wire armouring	6	8.5.9	397
	AMS 10, AMS 20	Push-on straight joint for single-core XLPE-insulated cables	10 and 20		
		35 to 120 mm²		8.5.10 a	398
		150 to 400 mm²		8.5.10 b	399
	WPS 10, WPS 20, WPS 30	Straight joint with shrinkable tube for single-core PVC and XLPE-insulated cables	10, 20 and 30		
		25 to 120 mm²		8.5.11 a	400
		150 to 500 mm²		8.5.11 b	401
	WP	Straight joint for three-core PVC and XLPE-insulated cables	10	8.5.12	402
	WP	Straight joint for single-core PVC and XLPE-insulated cables	10, 20 and 30		
		25 to 120 mm²		8.5.13 a	403
		150 to 500 mm²		8.5.13 b	404
	VS	Straight joint for paper-insulated belted cables	10	8.5.14	405
	SMI	Straight joint for paper-insulated three-core S.L. cables	20 and 30		
		25 to 95 mm²		8.5.15 a	406
		120 to 240 mm²		8.5.15 b	407
7.6 Branch joints	GNKA-1	Shrinkable branch joint for polymeric-insulated cables	1	8.6.1	409
	PA	PROTOLIN branch joint for polymeric-insulated cables	1	8.6.2	410
	PAK	PROTOLIN branch joint for four-core PVC and XLPE-insulated cables	1	8.6.3	411
	PAD	PROTOLIN double branch joint for PVC-insulated cables	1	8.6.4	412
	GMS, GMSA	Cast-iron branch joint with fuse or automatic safety cutout for PVC and XLPE-insulated cables	1	8.6.5	413
7.7 Transition joints	SKSM PK	Shrinkable transition joint for connecting PVC and XLPE-insulated cables with three-core paper-insulated cables	1	8.7.1	415
	ÜMP 10	Transition joint for connecting PVC and XLPE-insulated cables with paper-insulated belted cables	10	8.7.2	416
	SM-WP	Transition joint for connecting three single-core XLPE-insulated cables with paper-insulated three-core S.L. cables	20 and 30	8.7.3	417

8 Project Planning Data of Accessories

8.1 Indoor Terminations

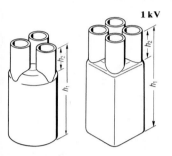

1 kV

Shrinkable spreader cap SKSA for three-core and four-core polymeric-insulated cables (see Table 8.1.1)

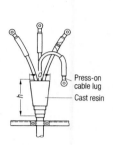

1 kV

Press-on cable lug
Cast resin

PROTOLIN termination (cast resin) PEA for PVC and XLPE-insulated cables (see Table 8.1.2)

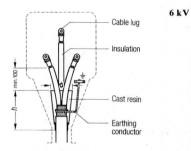

6 kV

Cable lug
Insulation
min. 100
Cast resin
Earthing conductor

PROTOLIN termination (cast resin) PEA for three-core PVC-insulated cables in confined cable connection areas (see Table 8.1.3)

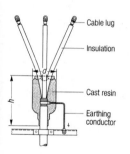

6 kV

Cable lug
Insulation
Cast resin
Earthing conductor

PROTOLIN termination (cast resin) PEB for three-core PVC-insulated cables (see Table 8.1.4)

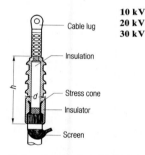

10 kV
20 kV
30 kV

Cable lug
Insulation
Stress cone
Insulator
Screen

Push-on termination with sheds IAES 10, IAES 20 and IAES 30 for single-core PVC and XLPE-insulated cables (see Table 8.1.5)

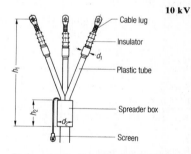

10 kV

Cable lug
Insulator
d_1
Plastic tube
Spreader box
Screen

Push-on termination with sheds IAES 10 for three-core PVC and XLPE-insulated cables (see Table 8.1.6)

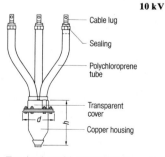

10 kV

Cable lug
Sealing
Polychloroprene tube
Transparent cover
Copper housing

Termination with copper housing and transparent cover IKM for paper-insulated belted cables (see Table 8.1.7)

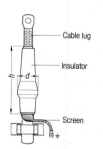

20 kV

Cable lug
Insulator
Screen

Push-on termination without sheds IAE 20 for single-core XLPE-insulated cables (see Table 8.1.8)

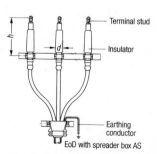

20 kV
30 kV

Terminal stud
Insulator
Earthing conductor
EoD with spreader box AS

Termination without seals EoD for paper-insulated three-core S.L. cables (see Table 8.1.9)

357

Table 8.1.1   **1 kV**
Shrinkable spreader cap SKSA
for three-core and four-core polymeric-insulated cables

Application	1	Field of application		For the protection of the crotches from dust and moisture in indoor installations, with special accessories also in outdoor installations					
Type of accessory	2	Technique		Heat shrink technique.					
	3	Construction		The spreader cap consists of cross-linked polyolefin. The outlets are coated with hot melt adhesive. For XLPE-insulated cables in outdoor installations, shrinkable tubes are necessary in order to protect the insulation from UV radiation.					
	4	Jointing and connecting		Conductors: In principle, press-on cable lugs can be used, for PVC and paper-insulated cables soldering cable lugs are also permissible. Metallic coverings: Armour or metal sheath are connected by means of a constant force spring and tinned copper braids.					
Allocation of cables	5	Design		PVC and XLPE-insulated cables 0.6/1 kV ($U_m = 1.2$ kV)					
	6	Number of cores		3	3	3	4	4	4	4
	7	Nominal cross-sectional area of conductor	mm^2	4–35	50–150	185–300	4–35	35–70	70–150	185–300
Dimensions, weight	10	Size SKSA ...		3/35	3/150	3/300	4/35	4/70	4/150	4/300
	11	Height h_1[1]	mm	103	180	200	96	165	217	223
		Height h_2[1]	mm	28	40	40	25	40	44	51
	16	Weight	kg	0.04	0.11	0.16	0.04	0.08	0.11	0.16
Tests	17	Specifications		Based on DIN VDE 0278 Part 1						
Special properties	20	Installation position		Optional						
Accessories	22	Auxiliary devices		Suitable tool when using press-on cable lugs. Hot-air blower or gas burner for shrinking						

[1] Dimensions after shrinking

Table 8.1.2
PROTOLIN termination (cast resin) PEA
for PVC and XLPE-insulated cables

Application	1	Field of application	For the protection of the crotches from dust and moisture in indoor installations. Preferably used in switching and distribution systems				
Type of accessory	2	Technique	Cast resin technique. The mechanical strength at the crotch is considerably increased by the cast resin.				
	3	Construction	Cast resin body.				
	4	Jointing and connecting	In principle, press-on cable lugs can be used, for PVC-insulated cables, however, soldering cable lugs are also permissible.				
Allocation of cables	5	Design	NYY, NAYY, NYCY, NYCWY, NAYCWY 0.6/1 kV (U_m = 1.2 kV) NA2XY 0.6/1 kV (U_m = 1.2 kV)				
	6	Number of cores	1–5	1–5	1–5	1–5	1–5
	7	Nominal cross-sectional area of conductor	for NYY, NAYY and NA2XY				
		single-core mm²	4–240	300–500	–	–	
		three-core mm²	1.5–35	50–95	120–150	185–240	
		four-core mm²	1.5–25	35–70	95–120	150–185	240–300
		five-core mm²	1.5–16	25	–	–	
		three-core mm²	for NYCY, NYCWY and NAYCWY 1.5–25	35–70	95–120	150	185, 240
		four-core mm²	1.5–16	25–70	95–120	150	185, 240
Dimensions, weight	10	Size PEA ..	2	3	4	5	6
	11	Height h mm	150	210	200	220	270
	14	Diameter d mm	60	80	90	100	120
	15	Filling volume l	0.14	0.42	0.52	0.65	1.3
	16	Weight kg	0.7	1.4	1.4	1.8	2.6
Tests	17	Specifications	Based on DIN VDE 0278 Part 1				
Special properties	19	Permissible peak short-circuit current kA	40	40	40	40	40
	20	Installation position	Optional. The termination shall be in a vertical position during casting and curing.				
Accessories	22	Auxiliary devices	Suitable tool when using press-on cable lugs				

Table 8.1.3   **6 kV**
PROTOLIN termination (cast resin) PEA
for three-core PVC-insulated cables in confined connection areas

Application	1	Field of application	For the protection of the crotches from dust and moisture in indoor installations. Preferably used in confined connection areas, e.g. in motor junction boxes			
Type of accessory	2	Technique	Cast resin technique. The mechanical and electrical strength at the crotch is considerably increased by the cast resin.			
	3	Construction	Cast resin body. If the minimum distance of 100 mm (refer to the figure on page 357) is not maintained, shrinkable tubes shall be shrunk on over the core ends and the cable lugs such that the creepage distance of 100 mm is achieved again.			
	4	Jointing and connecting	Conductor: Press-on or soldering cable lugs Armour: By soldering on an earthing conductor			
Allocation of cables	5	Design	NYFGY 3.6/6 kV (U_m = 7.2 kV)			
	7	Nominal cross-sectional area of conductor mm²		25, 35	50 – 95	120 – 240
Dimensions, weight	10	Size PEA ..		3	4	5
	11	Height h mm		210	200	220
	14	Diameter d mm		80	90	100
	15	Filling volume l		0.42	0.52	0.65
	16	Weight kg		1.5	1.5	1.9
Tests	17	Specifications	DIN VDE 0278 Part 1 and 4			
Special properties	19	Permissible peak short-circuit current kA		63	63	63
	20	Installation position	Optional. The termination shall be in a vertical position during casting and curing.			
Accessories	22	Auxiliary devices	Suitable tool when using press-on cable lugs			

Table 8.1.4
PROTOLIN termination (cast resin) PEB
for three-core PVC-insulated cables

Application	1	Field of application	For the protection of the crotches from dust and moisture in indoor installations. Preferably used in power plants and industry			
Type of accessory	2	Technique	Cast resin technique. The mechanical and electrical strength at the crotch is considerably increased by the cast resin.			
	3	Construction	Cast resin body, if necessary protective caps against condensation			
	4	Jointing and connecting	Conductor: Press-on or soldering cable lugs Armour: By means of a metal tube with cleat			
Allocation of cables	5	Design	NYFGY 3.6/6 kV (U_m = 7.2 kV)			
	7	Nominal cross-sectional area of conductor mm^2	25 – 50	70 – 120	150, 185	240, 300
Dimensions, weight	10	Size PEB ..	1	2	3	4
	11	Height h mm	210	225	245	270
	14	Diameter d mm	82	90	115	125
	15	Filling volume l	0.72	0.95	1.55	2.0
	16	Weight kg	2.1	2.7	3.7	4.5
Tests	17	Specifications	DIN VDE 0278 Part 1 and 4			
Special properties	19	Permissible peak short-circuit current[1] kA	63	63	63	63
	20	Installation position	Optional. The termination shall be in a vertical position during casting and curing.			
Accessories	22	Auxiliary devices	Suitable tool when using press-on cable lugs			

[1] For a nominal cross-sectional area of conductor of 95 mm^2 and above, peak short-circuit currents up to 125 kA are permissible provided that special accessories are used

Table 8.1.5 a
10, 20 and 30 kV

Push-on terminations with sheds IAES 10, IAES 20 and IAES 30
for single-core PVC and XLPE-insulated cables (up to 120 mm²)

Application	1	Field of application	The good surface properties of the insulator permit the operation under rigorous conditions such as those caused by dust, condensation or moisture. In open and metal-clad switchboards, switching and distribution systems, transformers. The push-on termination IAES 10 is suitable for installation in 10 kV motor terminal boxes 1XB8911 because of its small dimensions.					
Type of accessory	2	Technique	Push-on technique					
	3	Construction	The termination consists of silicone rubber and has an integrated stress cone.					
	4	Jointing and connecting	Press-on cable lugs					
Allocation of cables	5	Design[1]	NYSY, N2XS2Y, NA2XS2Y, NA2XS(F)2Y 6/10 kV (U_m = 12 kV) N2XS2Y, NA2XS2Y, NA2XS(F)2Y 12/20 kV (U_m = 24 kV) N2XS2Y, NA2XS2Y, NA2XS(F)2Y 18/30 kV (U_m = 36 kV)					
	7	Nominal cross-sectional area of conductor mm²	25	35	50	70	95	120
	9	Nominal cross-sectional area of screen mm²	16	16	16	16	16	16
Dimensions, weight (10 kV)	10	Size IAES 10/...	25 RM	35 RM	50 RM	70 RM	95 RM	120 RM
	11	Height h mm	175	175	175	175	175	175
	14	Diameter d mm	40	40	40	40	40	40
	16	Weight kg	0.8	0.8	0.8	0.8	0.8	0.8
Dimensions, weight (20 kV)	10	Size IAES 20/...	25 RM	35 RM	50 RM	70 RM	95 RM	120 RM
	11	Height h mm	253	253	253	253	253	253
	14	Diameter d mm	62	62	62	62	62	65
	16	Weight kg	1.3	1.3	1.6	1.6	1.6	1.6
Dimensions, weight (30 kV)	10	Size IAES 30/...	–	–	50 RM	70 RM	95 RM	120 RM
	11	Height h mm	–	–	318	318	318	318
	14	Diameter d mm	–	–	96	96	96	96
	16	Weight kg	–	–	3.1	3.1	3.1	3.1
Tests	17	Specifications	DIN VDE 0278 Part 1 and 4					
Special properties	19	Permissible peak short-circuit current kA	125	125	125	125	125	125
	20	Installation position	Optional					
Accessories	22	Auxiliary devices	Stripping tool for the outer semi-conductive layer of XLPE-insulated cables. Suitable tool for the press-on cable lugs					

[1] The termination can also be used for cables with PVC outer sheath

Table 8.1.5b

10, 20 and 30 kV

**Push-on termination with sheds IAES 10, IAES 20 and IAES 30
for single-core PVC and XLPE-insulated cables (150 mm² and above)**

Application	1	Field of application		The good surface properties of the insulator permit the operation under rigorous conditions such as those caused by dust, condensation or moisture. In open and metal-clad switchboards, switching and distribution systems, transformers. The push-on termination IAES 10 is suitable for installation in 10 kV motor terminal boxes 1XB8911 because of its small dimensions.					
Type of accessory	2	Technique		Push-on technique					
	3	Construction		The termination consists of silicone rubber and has an integrated stress cone.					
	4	Jointing and connecting		Press-on cable lugs					
Allocation of cables	5	Design[1]		NYSY, N2XS2Y, NA2XS2Y, NA2XS/F)2Y 6/10 kV (U_m = 12 kV) N2XS2Y, NA2XS2Y, NA2XS(F)2Y 12/20 kV (U_m = 24 kV) N2XS2Y, NA2XS2Y, NA2XS(F)2Y 18/30 kV (U_m = 36 kV)					
	7	Nominal cross-sectional area of conductor	mm²	150	185	240	300	400	500
	9	Nominal cross-sectional area of screen	mm²	25	25	25	25	35	35
Dimensions, weight (10 kV)	10	Size IAES 10/ . . .		150 RM	185 RM	240 RM	300 RM	400 RM	500 RM
	11	Height h	mm	175	175	175	175	175	175
	14	Diameter d	mm	40	52	52	52	52	52
	16	Weight	kg	0.8	0.8	0.8	0.8	1.2	1.2
Dimensions, weight (20 kV)	10	Size IAES 20/ . . .		150 RM	185 RM	240 RM	300 RM	400 RM	500 RM
	11	Height h	mm	253	253	253	253	318	318
	14	Diameter d	mm	65	71	71	71	96	96
	16	Weight	kg	1.6	1.4	1.4	1.4	3.1	3.1
Dimensions, weight (30 kV)	10	Size IAES 30/ . . .		150 RM	185 RM	240 RM	300 RM	400 RM	500 RM
	11	Height h	mm	318	318	318	318	318	318
	14	Diameter d	mm	96	96	96	96	96	96
	16	Weight	kg	3.1	3.1	3.1	3.1	3.1	3.1
Tests	17	Specifications		DIN VDE 0278 Part 1 and 4					
Special properties	19	Permissible peak short-circuit current	kA	125	125	125	125	125	125
	20	Installation position		Optional					
Accessories	22	Auxiliary devices		Stripping tool for the outer semi-conductive layer of XLPE-insulated cables. Suitable tool for the press-on cable lugs					

[1] The termination can also be used for cables with PVC outer sheath

363

Table 8.1.6 a
Push-on termination with sheds IAES 10
for three-core PVC-insulated cables (up to 95 mm²)

Application	1	Field of application	The good surface properties of the insulator permit the operation under rigorous conditions such as those caused by dust, condensation or moisture. In open and metal-clad switchboards, switching and distribution systems, transformers. The push-on termination IAES 10 is suitable for installation in 10 kV motor terminal boxes 1XB8911 because of its small dimensions.				
Type of accessory	2	Technique	Push-on technique				
	3	Construction	The termination consists of three insulators and the spreader box. The insulator consists of silicone rubber and has an integrated stress cone. The cores are wrapped with carbon crepe paper and additionally protected by means of a push-on plastic sleeve.				
	4	Jointing and connecting	Press-on or soldering cable lugs				
Allocation of cables	5	Design	NYSEY 6/10 kV ($U_m = 12$ kV)				
	7	Nominal cross-sectional area of conductor mm²	25	35	50	70	95
	8	Type and shape of conductor	RM	RM	RM	RM	RM
	9	Nominal cross-sectional area of screen mm²	16	16	16	16	16
Dimensions, weight	10	Size IAES 10/ ...	35 RM	50 RM	70 RM	95 RM	120 RM
	11	Height h_1 mm Height h_2 mm	550 130	550 130	550 130	550 130	550 130
	14	Diameter d_1 mm Diameter d_2 mm	40 75	40 75	40 75	40 75	40 75
	16	Weight kg	3.0	3.0	3.0	3.0	4.0
Tests	17	Specifications	DIN VDE 0278 Part 1 and 4				
Special properties	19	Permissible peak short-circuit current kA	40	40	40	40	40
	20	Installation position	Optional. The termination shall be in a vertical position during casting and curing.				
	21	Treatment of crotch	PVC pipe filled with cast resin. For peak short-circuit currents up to 80 kA, a larger PVC pipe shall be used; up to 125 kA a metal funnel is necessary.				
Accessories	22	Auxiliary devices	Stripping tool for the outer semi-conductive layer of XLPE-insulated cables. Suitable tool for the press-on cable lugs				

Table 8.1.6 b
Push-on termination with sheds IAES 10
for three-core PVC-insulated cables (120 mm² and above)

Application	1	Field of application	The good surface properties of the insulator permit the operation under rigorous conditions such as those caused by dust, condensation or moisture. In open and metal-clad switchboards, switching and distribution systems, transformers. The push-on termination IAES 10 is suitable for installation in 10 kV motor terminal boxes 1XB8911 because of its small dimensions.				
Type of accessory	2	Technique	Push-on technique				
	3	Construction	The termination consists of three insulators and the spreader box. The insulator consists of silicone rubber and has an integrated stress cone. The cores are wrapped with carbon crepe paper and additionally protected by means of a push-on plastic sleeve.				
	4	Jointing and connecting	Press-on or soldering cable lugs				
Allocation of cables	5	Design	NYSEY 6/10 kV ($U_m = 12$ kV)				
	7	Nominal cross-sectional area of conductor mm²	120	150	185	240	300
	8	Type and shape of conductor	RM	RM	RM	RM	RM
	9	Nominal cross-sectional area of screen mm²	16	25	25	25	25
Dimensions, weight	10	Size IAES 10/ . . .	150 RM	185 RM	185 RM	300 RM	400 RM
	11	Height h_1 mm Height h_2 mm	550 130	550 140	550 140	550 140	550 140
	14	Diameter d_1 mm Diameter d_2 mm	40 75	52 90	52 90	52 90	52 110
	16	Weight kg	4.0	4.0	4.0	6.1	8.5
Tests	17	Specifications	DIN VDE 0278 Part 1 and 4				
Special properties	19	Permissible peak short-circuit current kA	40	40	40	40	40
	20	Installation position	Optional. The termination shall be in a vertical position during casting and curing.				
	21	Treatment of crotch	PVC pipe filled with cast resin. For peak short-circuit currents up to 80 kA, a larger PVC pipe shall be used; up to 125 kA a metal funnel is necessary.				
Accessories	22	Auxiliary devices	Stripping tool for the outer semi-conductive layer of XLPE-insulated cables. Suitable tool for the press-on cable lugs				

Table 8.1.6 c   **10 kV**
Push-on termination with sheds IAES 10
for three-core XLPE-insulated cables (up to 95 mm^2)

Application	1	Field of application	The good surface properties of the insulator permit the operation under rigorous conditions such as those caused by dust, condensation or moisture. In open and metal-clad switchboards, switching and distribution systems, transformers. The push-on termination IAES 10 is suitable for installation in 10 kV motor terminal boxes 1XB8911 because of its small dimensions.		
Type of accessory	2	Technique	Push-on technique		
	3	Construction	The termination consists of three insulators and the spreader box. The insulator consists of silicone rubber and has an integrated stress cone. The cores are wrapped with carbon crepe paper and additionally protected by means of a push-on plastic sleeve.		
	4	Jointing and connecting	Press-on cable lugs		
Allocation of cables	5	Design[1]	NA2XS2Y 6/10 kV ($U_m = 12$ kV)		
	7	Nominal cross-sectional area of conductor mm^2	50	70	95
	8	Type and shape of conductor	SE	SE	SE
	9	Nominal cross-cross-sectional of screen mm^2	16	16	16
Dimensions, weight	10	Size IAES 10/ . . .	50 SE	70 SE	95 SE
	11	Height h_1 mm Height h_2 mm	550 130	550 130	550 130
	14	Diameter d_1 mm Diameter d_2 mm	40 75	40 75	40 75
	16	Weight kg	3.0	3.0	3.0
Tests	17	Specifications	DIN VDE 0278 Part 1 and 4		
Special properties	19	Permissible peak short-circuit current kA	40	40	40
	20	Installation position	Optional. The termination shall be in a vertical position during casting and curing.		
	21	Treatment of crotch	PVC pipe filled with cast resin. For peak short-circuit currents up to 80 kA, a larger PVC pipe shall be used; up to 125 kA a metal funnel is necessary.		
Accessories	22	Auxiliary devices	Stripping tool for the outer semi-conductive layer of XLPE-insulated cables. Suitable tool for the press-on cable lugs		

[1] The termination can also be used for cables with PVC outer sheath

Table 8.1.6 d   **10 kV**
Push-on termination with sheds IAES 10
for three-core XLPE-insulated cables (120 mm² and above)

Application	1	Field of application	The good surface properties of the insulator permit the operation under rigorous conditions such as those caused by dust, condensation or moisture. In open and metal-clad switchboards, switching and distribution systems, transformers. The push-on termination IAES 10 is suitable for installation in 10 kV motor terminal boxes 1XB8911 because of its small dimensions.		
Type of accessory	2	Technique	Push-on technique		
	3	Construction	The termination consists of three insulators and the spreader box. The insulator consists of silicone rubber and has an integrated stress cone. The cores are wrapped with carbon crepe paper and additionally protected by means of a push-on plastic sleeve.		
	4	Jointing and connecting	Press-on cable lugs		
Allocation of cables	5	Design[1]	NA2XS2Y 6/10 kV ($U_m = 12$ kV)		
	7	Nominal cross-sectional area of conductor mm²	120	150	185
	8	Type and shape of conductor	SE	SE	SE
	9	Nominal cross-cross-sectional of screen mm²	16	25	25
Dimensions, weight	10	Size IAES 10/ . . .	120 SE	150 SE	185 SE
	11	Height h_1 mm Height h_2 mm	550 130	550 130	550 130
	14	Diameter d_1 mm Diameter d_2 mm	40 75	40 75	40 90
	16	Weight kg	4.0	4.0	4.0
Tests	17	Specifications	DIN VDE 0278 Part 1 and 4		
Special properties	19	Permissible peak short-circuit current kA	40	40	40
	20	Installation position	Optional. The termination shall be in a vertical position during casting and curing.		
	21	Treatment of crotch	PVC pipe filled with cast resin. For peak short-circuit currents up to 80 kA, a larger PVC pipe shall be used; up to 125 kA a metal funnel is necessary.		
Accessories	22	Auxiliary devices	Stripping tool for the outer semi-conductive layer of XLPE-insulated cables. Suitable tool for the press-on cable lugs		

[1] The termination can also be used for cables with PVC outer sheath

Table 8.1.7   **10 kV**
Termination with copper housing and transparent cover IKM
for paper-insulated belted cables

Application	1	Field of application		These terminations are suitable for installation in open and metal-clad switchboards.							
Type of accessory	2	Technique		Wrapping and casting technique							
	3	Construction		A copper housing with a transparent cover of transparent plastics and compound-resistant tubes of polychloroprene pushed on over the free core ends terminate the end of the cable. The ends of the tubes shall be sealed by sleeve cleats. The compound level can be regulated by means of a filler hole.							
	4	Jointing and connecting		Press-on cable lugs and sealings							
Allocation of cables	5	Design		NKBA, NAKBA 6/10 ($U_m = 12$ kV)							
	7	Nominal cross-sectional area of conductor	mm^2	35	50	70	95	120	150	185	240
Dimensions, weight	10	Size	IKM ..	30	30	35	40	40	40	50	50
	11	Height h	mm	190	190	190	215	215	215	265	265
	14	Diameter d	mm	130	130	130	155	155	155	170	170
	15	Filling volume	l	1.0	1.0	1.0	2.0	2.0	2.0	3.0	3.0
	16	Weight	kg	3.0	3.0	3.5	5.0	5.0	5.0	6.5	6.5
Tests	17	Specifications		DIN VDE 0278 Part 1 and 4							
Special properties	19	Permissible peak-short-circuit current[1]	kA	40	40	40	40	40	40	40	40
	20	Installation position		The termination shall always be installed vertically.							
Accessories	22	Auxiliary devices		Suitable tool for the press-on cable lugs							

[1] Peak short-circuit currents up to 80 kA are permissible provided that suitable special accessories are used

Table 8.1.8 a
Push-on termination without sheds IAE 20
for single-core XLPE-insulated cables (up to 95 mm^2)

Application	1	Field of application		The good surface properties of the smooth insulator permit the operation also in case of occasional condensation. In extremely confined connection areas				
Type of accessory	2	Technique		Push-on technique				
	3	Construction		The termination consists of silicone rubber and has an integrated stress cone.				
	4	Jointing and connecting		Press-on cable lugs				
Allocation of cables	5	Design[1]		N2XS2Y, NA2XS2Y, NA2XS(F)2Y 12/20 kV (U_m = 24 kV)				
	7	Nominal cross-sectional area of conductor	mm^2	25	35	50	70	95
	9	Nominal cross-sectional area of screen	mm^2	16	16	16	16	16
Dimensions, weight	10	Size	IAE 20/..	25	35	50	70	95
	11	Height h	mm	253	253	253	253	253
	14	Diameter d	mm	57	57	62	62	62
	16	Weight	kg	1.3	1.3	1.6	1.6	1.6
Tests	17	Specifications		DIN VDE 0278 Part 1 and 4				
Special properties	19	Permissible peak short-circuit current	kA	125	125	125	125	125
	20	Installation position		Optional				
Accessories	22	Auxiliary devices		Stripping tool for the outer semi-conductive layer. Suitable tool for the press-on cable lugs				

[1] The termination can also be used for cables with PVC outer sheath

Table 8.1.8 b
Push-on termination without sheds IAE 20
for single-core XLPE-insulated cables (120 mm² and above)

Application	1	Field of application		The good surface properties of the smooth insulator permit the operation also in case of occasional condensation. In extremely confined connection areas				
Type of accessory	2	Technique		Push-on technique				
	3	Construction		The termination consists of silicone rubber and has an integrated stress cone.				
	4	Jointing and connecting		Press-on cable lugs				
Allocation of cables	5	Design[1]		N2XS2Y, NA2XS2Y, NA2XS(F)2Y 12/20 kV (U_m = 24 kV)				
	7	Nominal cross-sectional area of conductor	mm²	120	150	185	240	300
	9	Nominal cross-sectional area of screen	mm²	16	25	25	25	25
Dimensions, weight	10	Size IAE 20/ ...		120	150	185	240	300
	11	Height h	mm	253	253	253	253	253
	14	Diameter d	mm	65	65	71	71	71
	16	Weight	kg	1.6	1.6	1.4	1.4	1.4
Tests	17	Specifications		DIN VDE 0278 Part 1 and 4				
Special properties	19	Permissible peak short-circuit current	kA	125	125	125	125	125
	20	Installation position		Optional				
Accessories	22	Auxiliary devices		Stripping tool for the outer semi-conductive layer. Suitable tool for the press-on cable lugs				

[1] The termination can also be used for cables with PVC outer sheath

Table 8.1.9
Termination without seals EoD
for paper-insulated three-core S.L. cables

Application	1	Field of application	This termination shall be used in indoor installations without condensation. Protective caps against condensation are recommended where condensation is likely to occur.				
Type of accessory	2	Technique	Wrapping and casting technique				
	3	Construction	The electrical strength is ensured by a wrapped stress cone of metallized paper in conjunction with the filling compound. The transparent cast resin insulator allows the level of the compound to be checked. Soldered top and bottom fittings prevent the penetration of moisture. The termination EoD shall only be used together with the spreader box AS or SS.				
	4	Jointing and connecting	Conductor: For copper conductors the terminal studs are soldered; for aluminium conductors they are pressed. Lead sheath: By soldering on an earthing conductor				
Allocation of cables	5	Design	NEKEBA, NAEKEBA 12/20 kV (U_m = 24 kV)			NEKEBA, NAEKEBA 18/30 kV (U_m = 36 kV)	
	7	Nominal cross-sectional area of conductor　　mm²	25–120	150, 185	240	35–70	95–240
Dimensions, weight	10	Size　　EoD ..	1	1	2	2	2
	11	Height h　　mm	350	350	430	430	430
	14	Diameter d　　mm	67	67	91	91	91
	15	Filling volume[1]　　l	2	2	5	5	5
	16	Weight　　kg	12	13	20	20	21
Tests	17	Specifications	DIN VDE 0278 Part 1 and 4				
Special properties	19	Permissible peak short-circuit current[2]　　kA	40	40	40	40	40
	20	Installation position	The termination shall always be installed vertically.				
	21	Treatment of crotch	Spreader box AS or SS (details are given in Tables 8.3.1 and 8.3.2)				
Accessories	22	Auxiliary devices	Suitable tool for cables with aluminium conductors				

[1] Per set
[2] In conjunction with the spreader box AS; when using the spreader box SS a peak short-circuit current of 80 kA is permissible and by means of additional fixing cleats and core supporting struts 125 kA are permissible

8.2 Outdoor Terminations

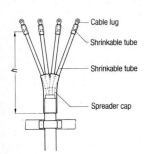

1 kV

Cable lug

Shrinkable tube

Shrinkable tube

Spreader cap

Shrinkable termination SKSE for three-core and four-core PVC and XLPE-insulated cables (see Table 8.2.1)

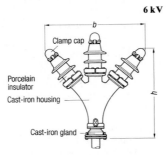

6 kV

Clamp cap

Porcelain insulator

Cast-iron housing

Cast-iron gland

Encased termination FFK 10 with porcelain insulators for three-core PVC-insulated cables with flat wire armouring (see Table 8.2.2)

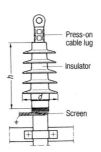

10 kV
20 kV
30 kV

Press-on cable lug

Insulator

Screen

Push-on termination FAE 10, FAE 20 and FAE 30 for single-core XLPE-insulated cables (see Table 8.2.3)

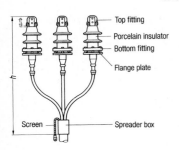

10 kV

Top fitting

Porcelain insulator

Bottom fitting

Flange plate

Screen

Spreader box

Termination FEP with porcelain insulator for single-core and three-core PVC and XLPE insulated cables (see Table 8.2.4)

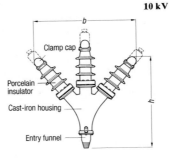

10 kV

Clamp cap

Porcelain insulator

Cast-iron housing

Entry funnel

Encased termination FF 10 with porcelain insulators for paper-insulated belted cables (see Table 8.2.5)

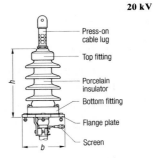

20 kV

Press-on cable lug

Top fitting

Porcelain insulator

Bottom fitting

Flange plate

Screen

Termination with porcelain insulator FAL 20 for single-core XLPE-insulated cables (see Table 8.2.6)

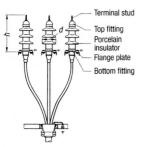

20 kV
30 kV

Terminal stud

Top fitting

Porcelain insulator

Flange plate

Bottom fitting

Termination with porcelain insulator FEL for paper-insulated three-core S.L. cables (see Table 8.2.7)

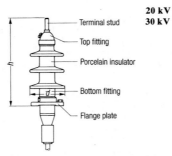

20 kV
30 kV

Terminal stud

Top fitting

Porcelain insulator

Bottom fitting

Flange plate

Termination without porcelain insulator FEL-2Y for single-core XLPE-insulated cables (see Table 8.2.8)

372

Table 8.2.1
Shrinkable termination SKSE
for three-core and four-core PVC and XLPE-insulated cables

Application	1	Field of application		For the protection of the crotches from dust and moisture in outdoor installations						
Type of accessory	2	Technique		Heat shrink technique						
	3	Construction		The shrinkable termination SKSE consists of a shrinkable spreader cap which is coated with hot melt adhesive, the shrinkable tubes for the protection of the core ends and the short, coated tube ends which are shrunk on over the bare conductors as sealings. XLPE-insulated cores shall basically be protected against UV radiation by shrinkable tubes.						
	4	Jointing and connecting		Conductor: In principle, press-on cable lugs can be used, for PVC-insulated cables soldering cable lugs are also permissible. Lead sheath: By soldering on an earthing conductor						
Allocation of cables	5	Design		PVC and XLPE-insulated cables 0.6/1 kV (U_m = 1.2 kV)						
	6	Number of cores		3	3	3	4	4	4	4
	7	Nominal cross-sectional area of conductor	mm²	4–35	50–150	185–300	4–35	35–70	70–150	185–300
Dimensions, weight	10	Size	SKSE ...	3/35	3/150	3/300	4/35	4/70	4/150	4/300
	11	Height h	mm	300	450	500	300	470	450	500
	16	Weight	kg	0.08	0.2	0.3	0.1	0.22	0.22	0.35
Tests	17	Specifications		Based on DIN VDE 0278 Part 1						
Special properties	19	Permissible peak short-circuit current	kA	40	40	40	40	40	40	40
	20	Installation position		Optional						
Accessories	22	Auxiliary devices		Suitable tool when using press-on cable lugs. Hot-air blower or gas burner for shrinking						

Table 8.2.2
Encased termination FFK 10 with porcelain insulators
for three-core PVC-insulated cables with flat wire armouring

Application	1	Field of application	Termination for three-core polymeric-insulated cables in industry and switchgear
Type of accessory	2	Technique	Casting technique
	3	Construction	This termination consists of a cast-iron housing with porcelain insulators. A clamp cap connects the conductors and the fittings at the top. The seal to the cable sheath is made by means of a gland.
	4	Jointing and connecting	Conductors: Clamp cap Armour: Armour and housing are connected by an earthing conductor
Allocation of cables	5	Design	NYFGY 3.6/6 kV ($U_m = 7.2$ kV)
	7	Nominal cross-sectional area of conductor mm^2	25 – 240
Dimensions, weight	11	Height h mm	750
	13	Width b mm	820
	15	Filling volume l	12
	16	Weight kg	53
Tests	17	Specifications	DIN VDE 0278 Part 1 and 5
Special properties	18	Supporting function	Provided by the cast-iron housing
	19	Permissible peak short-circuit current kA	125
	20	Installation position	Vertical

Table 8.2.3 a
Push-on termination FAE 10, FAE 20 and FAE 30
for single-core XLPE-insulated cables (up to 120 mm²)

Application	1	Field of application		The push-on termination meets all requirements for outdoor installation.					
Type of accessory	2	Technique		Push-on technique					
	3	Construction		The insulator with large sheds consists of non-tracking silicone rubber. The downward screen wires are embedded in a silicone sealing compound in order to prevent the penetration of moisture into the termination. The sealing at the end of the conductor is ensured by the cable lug with sealing collar and sealing lips at the insulator. The termination has an integrated stress cone.					
	4	Jointing and connecting		Special press-on cable lugs with sealing collars					
Allocation of cables	5	Design[1]		N2XS2Y, NA2XS2Y, NA2XS(F)2Y 6/10 kV (U_m = 12 kV) N2XS2Y, NA2XS2Y, NA2XS(F)2Y 12/20 kV (U_m = 24 kV) N2XS2Y, NA2XS2Y, NA2XS(F)2Y 18/30 kV (U_m = 36 kV)					
	7	Nominal cross-sectional area of conductor	mm²	25	35	50	70	95	120
	9	Nominal cross-sectional area of screen	mm²	16	16	16	16	16	16
Dimensions, weight (10 kV)	10	Size FAE 10/ ...		25 RM	35 RM	50 RM	70 RM	95 RM	120 RM
	11	Height h	mm	270	270	270	270	270	270
	14	Diameter d	mm	105	105	105	105	105	105
	16	Weight	kg	3.1	3.1	3.1	3.3	3.3	3.3
Dimensions, weight (20 kV)	10	Size FAE 20/ ...		25 RM	35 RM	50 RM	70 RM	95 RM	120 RM
	11	Height h	mm	370	370	370	370	370	370
	14	Diameter d	mm	135	135	135	135	135	135
	16	Weight	kg	4.3	4.3	4.3	4.5	4.5	4.5
Dimensions, weight (30 kV)	10	Size FAE 30/ ...		–	35 RM	50 RM	70 RM	95 RM	120 RM
	11	Height h	mm	–	465	465	465	465	465
	14	Diameter d	mm	–	145	145	145	145	145
	16	Weight	kg	–	5.5	5.5	5.7	5.7	5.7
Tests	17	Specifications		DIN VDE 0278 Part 1, 5 and 100					
Special properties	19	Permissible peak short-circuit current[2]	kA	40	40	40	40	40	40
	20	Installation position		The installation position may not exceed 30° from the vertical position.					
Accessories	22	Auxiliary devices		Stripping tool for the outer semi-conductive layer. Suitable tool for the press-on cable lugs					

[1] The termination can also be used for cables with PVC outer sheath
[2] A peak short-circuit current of 80 kA is permissible provided that the termination is connected to a fixed rigid busbar

Table 8.2.3 b
Push-on termination FAE 10, FAE 20 and FAE 30
for single-core XLPE-insulated cables (150 mm^2 and above)

10, 20 and 30 kV

Application	1	Field of application	The push-on termination meets all requirements for outdoor installation.					
Type of accessory	2	Technique	Push-on technique					
	3	Construction	The insulator with large sheds consists of non-tracking silicone rubber. The downward screen wires are embedded in a silicone sealing compound in order to prevent the penetration of moisture into the termination. The sealing at the end of the conductor is ensured by the cable lug with sealing collar and sealing lips at the insulator. The termination has an integrated stress cone.					
	4	Jointing and connecting	Special press-on cable lugs with sealing collars					
Allocation of cables	5	Design[1]	N2XS2Y, NA2XS2Y, NA2XS(F)2Y 6/10 kV (U_m = 12 kV) N2XS2Y, NA2XS2Y, NA2XS(F)2Y 12/20 kV (U_m = 24 kV) N2XS2Y, NA2XS2Y, NA2XS(F)2Y 18/30 kV (U_m = 36 kV)					
	7	Nominal cross-sectional area of conductor mm^2	150	185	240	300	400	500
	9	Nominal cross-sectional area of screen mm^2	25	25	25	25	35	35
Dimensions, weight (10 kV)	10	Size FAE 10/ . . .	150 RM	185 RM	240 RM	300 RM	400 RM	500 RM
	11	Height h mm	270	270	240	240	240	240
	14	Diameter d mm	105	105	121	121	121	121
	16	Weight kg	3.5	3.5	3.5	3.5	3.5	3.5
Dimensions, weight (20 kV)	10	Size FAE 20/ . . .	150 RM	185 RM	240 RM	300 RM	–	–
	11	Height h mm	370	370	370	370	–	–
	14	Diameter d mm	135	135	135	135	–	–
	16	Weight kg	4.7	4.7	4.7	4.7	–	–
Dimensions, weight (30 kV)	10	Size FAE 30/ . . .	150 RM	185 RM	240 RM	–	–	–
	11	Height h mm	465	465	465	–	–	–
	14	Diameter d mm	145	145	145	–	–	–
	16	Weight kg	5.9	5.9	5.9	–	–	–
Tests	17	Specifications	DIN VDE 0278 Part 1, 5 and 100					
Special properties	19	Permissible peak short-circuit current[2] kA	40	40	40	40	40	40
	20		The installation position may not exceed 30° from the vertical position.					
Accessories	22	Auxiliary devices	Stripping tool for the outer semi-conductive layer. Suitable tool for the press-on cable lugs					

[1] The termination can also be used for cables with PVC outer sheath
[2] A peak short-circuit current of 80 kA is permissible provided that the termination is connected to a fixed rigid busbar

Table 8.2.4
Termination FEP with porcelain insulator
for single-core and three-core PVC and XLPE-insulated cables

Application	1	Field of application		Mainly used as pole-mounted termination in power supply and industry						
Type of accessory	2	Technique		Wrapping and casting technique						
	3	Construction		This termination is a porcelain insulator flanged to a bottom fitting. A conductive tape with additional compression is fitted to the stripped-off conductive layer. A brass tube pushed on over the cable end serves both for accommodating the inner seal wrapping and for soldering on the entry funnel.						
	4	Jointing and connecting		Copper or aluminium conductors are clamped at the top fitting. Both busbars and cable lugs can be connected to the angled connector brought out of the side.						
Allocation of cables	5	Design[1]		NYSY, NYSEY 6/10 kV ($U_m = 12$ kV)			N2XS2Y,NA2XS2Y, NA2XS(F)2Y 6/10 kV ($U_m = 12$ kV)		NA2XS2Y 6/10 kV ($U_m = 12$ kV)	
	6	Number of cores		1 or 3	1 or 3	1 or 3	1	1	3	3
	7	Nominal cross-sectional area of conductor	mm²	25–120	150–240	300	25–120	150–300	50–120	150, 185
	8	Shape and type of conductor		RM	RM	RM	RM	RM	SE	SE
	9	Nominal cross-sectional area of screen	mm²	16	25	25	16	25	16	25
Dimensions, weight	11	Height h	mm	390	390	390	390	390	390	390
	15	Filling volume	l	6	6	6	6	6	6	6
	16	Weight[2]	kg	50	50	50	50	50	50	50
Tests	17	Specifications		DIN VDE 0278 Part 1 and 5						
Special properties	18	Supporting function		Provided by the porcelain insulator and the flange plate						
	19	Permissible peak short-circuit current	kA	40[3]	40[3]	40[3]	125	125	40	40
	20	Installation position		Vertical						
	21	Treatment of crotch		PVC pipe filled with cast resin for three-core cables						
Accessories	22	Auxiliary devices		Stripping tool for the outer semi-conductive layer of XLPE-insulated cables						

[1] The termination can also be used for cables with PVC outer sheath
[2] Per set (three terminations)
[3] Permissible peak short-circuit current for three-core cables; for single-core cables a peak short-circuit current up to 125 kA is permissible

Table 8.2.5
Encased termination FF10 with porcelain insulators
for paper-insulated belted cables

10 kV

Application	1	Field of application		Mainly in distribution systems	
Type of accessory	2	Technique		Casting technique	
	3	Construction		The termination consists of a cast-iron housing with porcelain insulators. A clamp cap connects the conductors and the top fittings.	
	4	Jointing and connecting		Conductor: Clamp cap Metallic coverings: Lead sheath and armour are sealed by soldering with the termination funnel.	
Allocation of cables	5	Design		NKBA 6/10 kV (U_m = 12 kV)	
	7	Nominal cross-sectional area of conductor	mm^2	35–185	240–300
Dimensions, weight	10	Size	FF . . .	10	10/I
	11	Height h	mm	750	820
	13	Width b	mm	820	860
	15	Filling volume	l	12	14
	16	Weight	kg	50	60
Tests	17	Specifications		DIN VDE 0278 Part 1 and 5	
Special properties	18	Supporting function		Provided by the cast-iron housing	
	19	Permissible peak short-circuit current	kA	125	125
	20	Installation position		Vertical	

Table 8.2.6
Termination with porcelain insulator FAL 20
for single-core XLPE-insulated cables

Application	1	Field of application	This termination is mainly used as pole-mounted termination with supporting function.
Type of accessory	2	Technique	Push-on technique
	3	Construction	A push-on stress cone of silicone rubber is fitted in the porcelain insulator for sealing and controlling the electrical field. The insulator is terminated by means of a top and bottom fitting, the void between core insulation and porcelain insulator is not filled with filling compound.
	4	Jointing and connecting	Press-on cable lugs
Allocation of cables	5	Design	N2XS2Y, NA2XS2Y, NA2XS(F)2Y 12/20 kV ($U_m = 24$ kV)
	7	Nominal cross-sectional area of conductor mm²	25–500
	9	Nominal cross-sectional area of screen mm²	16–35
Dimensions, weight	11	Height h mm	380
	13	Width b mm	245
	16	Weight kg	40.5
Tests	17	Specifications	DIN VDE 0278 Part 5
Special properties	18	Supporting function	Provided by the porcelain insulator and the flange plate
	19	Permissible peak short-circuit current kA	125
	20	Installation position	Optional
Accessories	22	Auxiliary devices	Stripping tool for the outer semi-conductive layer. Suitable tool for the press-on cable lugs

[1] The termination can also be used for cables with PVC outer sheath

Table 8.2.7 **20 and 30 kV**
Termination with porcelain insulator FEL 20 and FEL 30
for paper-insulated three-core S.L. cables

Application	1	Field of application	This termination is mainly used as pole-mounted termination in power supply.	
Type of accessory	2	Technique	Wrapping and casting technique	
	3	Construction	The termination consists of a porcelain insulator; the stress cone consists of wrapped metallized paper. The ends of the porcelain insulator are metallized, the metal parts are soldered on in the works so that no seals are needed. The termination with porcelain insulator FEL may only be used together with the spreader box AS or SS.	
	4	Jointing and connecting	Conductor: Copper conductors are connected by means of copper press-on terminal studs. Aluminium conductors can only be pressed with Al/Cu press-on terminal studs. Lead sheath: By soldering with the metal part at the bottom	
Allocation of cables	5	Design	NEKEBA, NAEKEBA 12/20 kV (U_m = 24 kV)	NEKEBA, NAEKEBA 18/30 kV (U_m = 36 kV)
	7	Nominal cross-sectional area of conductor mm^2	25–240	35–240
Dimensions, weight	10	Size FEL ..	20	30
	11	Height h mm	475	565
	14	Diameter d mm	185	210
	15	Filling volume[1] l	5	10
	16	Weight kg	33	47
Tests	17	Specifications	DIN VDE 0278 Part 1 and 5	
Special properties	18	Supporting function	Provided by the porcelain insulator and the flange plate	
	19	Permissible peak short-circuit current[2] kA	40	40
	20	Installation position	Vertical	
	21	Treatment of crotch	Spreader box AS or SS (see Table 8.3.1 or 8.3.2)	
Accessories	22	Auxiliary devices	Suitable compression tool for the terminal studs	

[1] Per set
[2] In conjunction with the spreader box AS; when using the spreader box SS a peak short-circuit current of 80 kA
is permissible and by means of additional fixing cleats and core supporting struts 125 kA are permissible

Table 8.2.8
Termination with porcelain insulator FEL-2Y
for single-core XLPE-insulated cables

Application	1	Field of application		Mainly used as pole-mounted termination in power supply, but also for single-phase a.c. power supply, e.g. railways.				
Type of accessory	2	Technique		Wrapping and casting technique				
	3	Construction		This termination is a porcelain insulator, the wrapped stress cone consists of conductive and self-welding insulating tapes. A brass tube pushed on over the cable ends serves both for accommodating the inner seal wrappings and for soldering on the entry funnel. The porcelain insulator is metallized at the ends, the metal parts are soldered on in the works so that no seals are needed.				
	4	Jointing and connecting		Conductor: Copper conductors are connected by means of copper press-on terminal studs, aluminium conductors can only be pressed with Al/Cu press-on terminal studs. Screen: By soldering with the entry funnel				
Allocation of cables	5	Design[1]		N2XS2Y, NA2XS2Y, NA2XS(F)2Y 12/20 kV ($U_m = 24$ kV)			N2XS2Y, NA2XS2Y, NA2XS(F)2Y 18/30 kV ($U_m = 36$ kV)	
	7	Nominal cross-sectional area of conductor	mm^2	25–120	150–300	400, 500	35–120	150–300	400, 500
	9	Nominal cross-sectional area of screen	mm^2	16	25	35	16	25	35
Dimensions, weight	10	Size FEL-2Y ..		20	20	30	30	30	30
	11	Height h_1	mm	475	475	565	565	565	565
	14	Diameter d_1	mm	185	185	210	210	210	210
	15	Filling volume[2]	l	4	4	6	6	6	6
	16	Weight[2]	kg	37	37	54	54	54	54
Tests	17	Specifications		DIN VDE 0278 Part 1 and 5				
Special properties	18	Supporting function		Provided by the porcelain insulator and the flange plate				
	19	Permissible peak short-circuit current	kA	125	125	125	125	125	125
	20	Installation position		Vertical				
Accessories	22	Auxiliary devices		Stripping tool for the outer semi-conductive layer. Suitable compression tool for the terminal studs				

[1] The termination can also be used for cables with PVC outer sheath
[2] Per set

8.2 Spreader Boxes

20 kV
30 kV

20 kV
30 kV

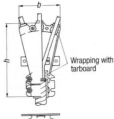

Wrapping with
tarboard

Spreader box AS for paper-insulated three-core
S.L. cables in indoor and outdoor installations
up to $I_s = 40$ kA (see Table 8.3.1)

Spreader box SS for paper-insulated three-core
S.L. cables in indoor and outdoor installations
up to $I_s = 125$ kA (see Table 8.3.2)

Table 8.3.1
Spreader box AS
for paper-insulated three-core S.L. cables in indoor and outdoor installations
up to $I_s = 40$ kA

20 and 30 kV

Application	1	Field of application		The spreader box AS protects the crotch of three-core S.L. cables from mechanical damage in case of short-circuit. It can be used in indoor and outdoor installations.				
Type of accessory	2	Technique		Casting technique				
	3	Construction		The spreader box consists of cast-iron and is casted with filling compound.				
Allocation of cables	5	Design		NEKEBA, NAEKEBA 12/20 kV ($U_m = 24$ kV)			NEKEBA, NAEKEBA 18/30 kV ($U_m = 36$ kV)	
	7	Nominal cross-sectional area of conductor	mm^2	25, 35	50–120	150–240	35–70	95–240
Dimensions, weight	10	Size	AS ...	60	80	110	80	110
	11	Height h	mm	120	138	170	138	170
	13	Width b	mm	144	174	222	174	222
	14	Diameter d	mm	60	80	110	80	110
	15	Filling volume	l	0.5	1.0	2.0	1.0	2.0
	16	Weight	kg	4.5	6.5	11.0	6.5	11.0
Special properties	19	Permissible peak short-circuit current[2]	kA	40	40	40	40	40
	20	Installation position		The spreader box can only be fixed horizontally.				

383

Table 8.3.2
Spreader box SS
for paper-insulated three-core S.L. cables in indoor and outdoor installations
up to $I_s = 125$ kA

				NEKEBA, NAEKEBA 12/20 kV ($U_m = 24$ kV)		NEKEBA, NAEKEBA 18/30 kV ($U_m = 36$ kV)	
Application	1	Field of application		The spreader box SS protects the crotch of three-core S.L. cables from mechanical damage in case of short-circuit. It can be used in indoor and outdoor installations.			
Type of accessory	2	Technique		Wrapping technique (in outdoor installations casting technique is additionally required)			
	3	Construction		The spreader box consists of cast-iron. The spreader box which is not filled with compound can be installed in any position. The cable and the single cores are wrapped with tarboard at the entry points. In outdoor installations the cable is sealed at the crotch by casting filling compound into the box. For this purpose, the wrapping at the entry point is lengthened by 60 to 80 mm.			
Allocation of cables	5	Design		NEKEBA, NAEKEBA 12/20 kV ($U_m = 24$ kV)		NEKEBA, NAEKEBA 18/30 kV ($U_m = 36$ kV)	
	7	Nominal cross-sectional area of conductor	mm²	25–95	120–240	35	50–240
Dimensions, weight	10	Size	SS ...	75	100	75	100
	11	Height h	mm	360	500	360	500
	13	Width b	mm	195	265	195	265
	16	Weight	kg	10	18	10	18
Special properties	19	Permissible peak short-circuit current	kA	125	125	125	125
	20	Installation position		Optional			

8.4 Ellbow Connectors

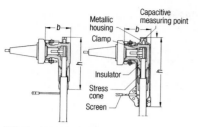

Elbow connectors for single-core PVC
and XLPE-insulated cables without metallic
protection against electric shock WS 200 and 400
(left figure) and elbow connectors safe to touch
BWS 200 and 400 (right figure)
(see Table 8.4.1 a and b)

Table 8.4.1 a
Separable elbow connectors WS 200 and BWS 200
for single-core PVC and XLPE-insulated cables

Application	1	Field of application	Separable elbow connectors WS and BWS are used for connecting medium-voltage cables in switchgear, to transformers and to electrical machines. They are mainly used in SF_6-insulated metal-clad load-break switchgear. The separable elbow connectors WS without metallic protection against electric shock should be preferably used for metal-clad connection areas, the separable elbow connectors BWS with metallic protection against electric shock are recommended for open connection areas.			
Type of accessory	2	Technique	Push-on technique			
	3	Construction	The separable elbow connector consists of a stud clamp connecting the conductor of the cable with the bushing stem, the insulating part of silicone rubber with the stress cone for covering the cut-back of the cable core and the field-limiting outer coating. The angled open part of the plug matches to the standardized conical part of the bushing. After pushing-on, the plug is fixed in its final position by fastening elements.			
	4	Jointing and connecting	Conductor: The conductor is clamped by two screws inside the separable elbow connectors. Screen: Press-on cable lug			
Allocation of cables	5	Design[1]	NYSY, N2XS2Y, NA2XS2Y, NA2XS(F)2Y 6/10 kV (U_m = 12 kV) N2XS2Y, NA2XS2Y, NA2XS(F)2Y 12/20 kV (U_m = 24 kV)			
	7	Nominal cross-sectional area of conductor mm^2	25	35	50	70
	9	Nominal cross-sectional area of screen mm^2	16	16	16	16
Dimensions, weight	10	Size WS 200/ . . .	10/25	10/35	10/50	10/70
	11	Height h mm	236	236	236	236
	13	Width b mm	105	105	105	105
	16	Weight kg	0.7	0.7	0.7	0.7
Dimensions, weight	10	Size BWS 200/ . . .	10/25	10/35	10/50	10/70
	11	Height h mm	312	312	312	312
	13	Width b mm	113	113	113	113
	16	Weight kg	1.1	1.1	1.1	1.1
Tests	17	Specifications	Based on DIN VDE 0278			
Special properties	19	Permissible peak short-circuit current[2] kA	40	40	40	40
	20	Installation position	Optional			
Accessories	22	Auxiliary devices	Stripping tool for the outer semi-conductive layer of XLPE-insulated cables. Suitable tool for the press-on cable lug for fixing the screen			

[1] The separable elbow connector can also be used for cables with PVC outer sheath
[2] The value applies to the separable elbow connector WS; for the separable elbow connector BWS a peak short-circuit current up to 80 kA is permissible

Table 8.4.1 b
Separable elbow connectors WS 400 and BWS 400
for single-core PVC and XLPE-insulated cables

Application	1	Field of application		Separable elbow connectors WS and BWS are used for connecting medium-voltage cables in switchgear, to transformers and to electrical machines. They are mainly used in SF_6-insulated metal-clad load-break switchgear. The separable elbow connectors WS without metallic protection against electric shock should be preferably used for metal-clad connection areas, the separable elbow connectors BWS with metallic protection against electric shock are recommended for open connection areas.				
Type of accessory	2	Technique		Push-on technique				
	3	Construction		The separable elbow connector consists of a stud clamp connecting the conductor of the cable with the bushing stem, the insulating part of silicone rubber with the stress cone for covering the cut-back of the cable core and the field-limiting outer coating. The angled open part of the plug matches to the standardized conical part of the bushing. After pushing-on, the plug is fixed in its final position by fastening elements.				
	4	Jointing and connecting		Conductor: The conductor is clamped by two screws inside the separable elbow connectors. Screen: Press-on cable lug				
Allocation of cables	5	Design[1]		NYSY, N2XS2Y, NA2XS2Y, NA2XS(F)2Y 6/10 kV (U_m = 12 kV) N2XS2Y, NA2XS2Y, NA2XS(F)2Y 12/20 kV (U_m = 24 kV)				
	7	Nominal cross-sectional area of conductor	mm^2	70	95	120	150	185
	9	Nominal cross-sectional area of screen	mm^2	16	16	16	25	25
Dimensions, weight	10	Size WS 400/ ...		20/70	20/95	20/120	20/150	20/185
	11	Height h	mm	255	255	255	255	255
	13	Width b	mm	155	155	155	155	155
	16	Weight	kg	0.9	0.9	0.9	0.9	0.9
Dimensions, weight	10	Size BWS 400/ ...		20/70	20/95	20/120	20/150	20/185
	11	Height h	mm	335	335	335	335	335
	13	Width b	mm	168	168	168	168	168
	16	Weight	kg	1.9	1.9	1.9	1.9	1.9
Tests	17	Specifications		Based on DIN VDE 0278				
Special properties	19	Permissible peak short-circuit current[2]	kA	40	40	40	40	40
	20	Installation position		Optional				
Accessories	22	Auxiliary devices		Stripping tool for the outer semi-conductive layer of XLPE-insulated cables. Suitable tool for the press-on cable lug for fixing the screen				

[1] The separable elbow connector can also be used for cables with PVC outer sheath
[2] The value applies to the separable elbow connector WS; for the separable elbow connector BWS a peak short-circuit current up to 80 kA is permissible

8.5 Straight Joints

1 kV

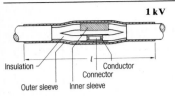

Shrinkable straight joint SKSM according to DIN 47632 for four-core PVC and XLPE-insulated cables (see Table 8.5.1)

1 kV

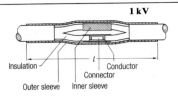

Shrinkable multi-range straight joint SKSM M for four-core PVC and XLPE-insulated cables (see Table 8.5.2)

1 kV

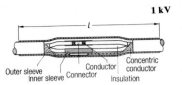

Shrinkable straight joint SKSM C for three-core PVC-insulated cables with concentric conductor (see Table 8.5.3)

1 kV

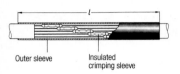

Shrinkable straight joint SKSM ST for PVC-insulated control cables (see Table 8.5.4)

1 kV

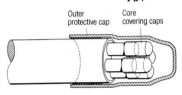

Shrinkable voltage-resistant end cap SKEM for three-core and four-core PVC and XLPE-insulated cables (see Table 8.5.5)

1 kV

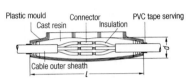

PROTOLIN straight joint PV for PVC and XLPE-insulated cables (see Table 8.5.6)

1 kV

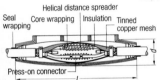

PROTOLIN straight joint PV for underground mining for PVC-insulated cables (see Table 8.5.7)

1 kV

PROTOLIN straight joint PV for PVC-insulated cables with lead sheath (see Table 8.5.8)

6 kV

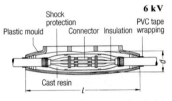

PROTOLIN straight joint PV for three-core PVC-insulated cables with steel wire armouring (see Table 8.5.9)

10 kV
20 kV

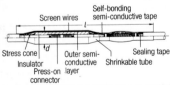

Push-on straight joint AMS 10 and AMS 20 for single-core XLPE-insulated cables see Table 8.5.10)

10 kV
20 kV
30 kV

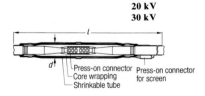

Straight joint with shrinkable tube WPS for single-core PVC and XLPE-insulated cables (see Table 8.5.11)

10 kV

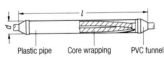

Straight joint WP for three-core PVC and XLPE-insulated cells (see Table 8.5.12)

10 kV
20 kV
30 kV

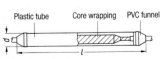

Straight joint WP for single-core PVC and XLPE-insulated cables (see Table 8.5.13)

10 kV

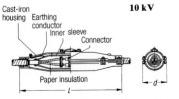

Straight joint VS for paper-insulated belted cables (see Table 8.5.14)

20 kV
30 kV

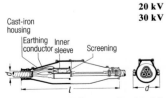

Straight joint SMI for paper-insulated three-core S.L. cables (see Table 8.5.15)

Table 8.5.1
Shrinkable straight joint SKSM according to DIN 47 632
for four-core PVC and XLPE-insulated cables

Application	1	Field of application		Polymeric-insulated low-voltage cables are connected to each other by means of the straight joint SKSM. The cable connection can immediately be laid in ground or in water after the straight joint has cooled down.					
Type of accessory	2	Technique		Heat shrink technique					
	3	Construction		The shrinkable straight joint consists of the inner sleeves which are shrunk on over each core and the common outer sleeve (protective tube). The inner surfaces of the sleeves are coated with thermoplastic hot-melt adhesive. The adhesive melts by the thermal effect during shrinking. For this reason the adjacent parts are bonded longitudinally watertight.					
	4	Jointing and connecting		In principle, screw connectors or press-on connectors can be used, for PVC-insulated cables soldering connectors are also permissible.					
Allocation of cables	5	Design		NYY, NAYY, NA2XY 0.6/1 kV ($U_m = 1.2$ kV)					
	7	Nominal cross-sectional area of conductor	mm^2	1.5–6	6–16	25, 35	50, 70	95–150	185–240
Dimensions, weight	10	Size	SKSM ...	25/5[1]	35/4	55/4	65/4	80/4	110/4
	12	Length l[2]	mm	250	460	665	700	760	900
	14	Diameter d	mm	The diameter depends on the spreading of the cores, however, it is considerably smaller than the diameter of comparable joints in cast resin technique.					
	16	Weight	kg	0.04	0.16	0.36	0.46	0.61	0.86
Tests	17	Specifications		DIN 47 632, DIN VDE 0278 Part 1 and 3					
Special properties	20	Installation position		Optional					
Accessories	22	Auxiliary devices		Suitable tool when using press-on connectors. Hot-air blower or gas burner for shrinking					

[1] Based on DIN 47 632
[2] Length after shrinking

Table 8.5.2

1 kV

Shrinkable multi-range straight joint SKSM M
for four-core PVC and XLPE-insulated cables

Application	1	Field of application	Polymeric-insulated low-voltage cables are connected to each other by means of the straight joint SKSM M. The cable connection can be immediately laid in ground or in water after the straight joint has cooled down.		
Type of accessory	2	Technique	Heat shrink technique		
	3	Construction	The shrinkable straight joint consists of the inner sleeves which are shrunk on over each core and the common outer sleeve (protective tube). The inner surfaces of the sleeves are coated with thermoplastic hot-melt adhesive. The adhesive melts by the thermal effect during shrinking. For this reason the adjacent parts are bonded longitudinally watertight.		
	4	Jointing and connecting	In principle, screw connectors or press-on connectors can be used, for PVC-insulated cables soldering connectors are also permissible.		
Allocation of cables	5	Design	NYY, NAYY, NA2XY 0.6/1 kV (U_m = 1.2 kV)		
	7	Nominal cross-sectional area of conductor mm^2	6–25	35–150	185–300
Dimensions, weight	10	Size SKSM M ...	4/25	4/150	4/300
	12	Length $l^{1)}$ mm	600	760	1000
	14	Diameter d mm	The diameter depends on the spreading of the cores, however, it is considerably smaller than the diameter of comparable joints in cast resin technique.		
	16	Weight kg	0.16	0.61	0.86
Tests	17	Specifications	DIN VDE 0278 Part 1 and 3		
Special properties	20	Installation position	Optional		
Accessories	22	Auxiliary devices	Suitable tool when using press-on connectors. Hot-air blower or gas burner for shrinking		

[1] Length before shrinking

Table 8.5.3
Shrinkable straight joint SKSM C
for three-core PVC-insulated cables with concentric conductor

Application	1	Field of application	Polymeric-insulated low-voltage cables are connected to each other by means of the straight joint SKSM C. The cable connection can immediately be laid in ground or in water after the straight joint has cooled down.				
Type of accessory	2	Technique	Shrink-on technique				
	3	Construction	The shrinkable straight joint consists of the inner sleeves which are shrunk on over each core and the common outer sleeve (protective tube). The inner surfaces of the sleeves are coated with thermoplastic hot-melt adhesive. The adhesive melts by the thermal effect during shrinking. For this reason the adjacent parts are bonded longitudinally watertight.				
	4	Jointing and connecting	In principle, screw connectors or press-on connectors can be used, for PVC-insulated cables soldering connectors are also permissible. Concentric conductor: The concentric conductors are connected with press-on connectors and additionally covered by a tinned copper mesh.				
Allocation of cables	5	Design	NYCY, NYCWY, NAYCWY 0.6/1 kV ($U_m = 1.2$ kV)				
	7	Nominal cross-sectional area of conductor mm²	6–16	25, 35	50, 70	95–150	185–240
Dimensions, weight	10	Size SKSM C...	3/16	3/35	3/70	3/150	3/240
	12	Length l[1) mm	460	665	700	760	900
	14	Diameter d mm	The diameter depends on the spreading of the cores, however, it is considerably smaller than the diameter of comparable joints in cast resin technique.				
	16	Weight kg	0.175	0.365	0.455	0.585	0.805
Tests	17	Specifications	DIN VDE 0278 Part 1 and 3				
Special properties	20	Installation position	Optional				
Accessories	22	Auxiliary devices	Suitable tool when using press-on connectors. Hot-air blower or gas burner for shrinking				

[1) Length before shrinking

Table 8.5.4
Shrinkable straight joint SKSM ST
for PVC-insulated control cables

Application	1	Field of application	The shrinkable straight joint SKSM ST is suitable for all polymeric-insulated control cables. After installation of the joint the cable can immediately be connected to the system.						
Type of accessory	2	Technique	Heat shrink technique						
	3	Construction	The shrinkable straight joint consists of the inner sleeves which are shrunk on over each core and the common outer sleeve (protective tube). The inner surfaces of the sleeves are coated with thermoplastic hot-melt adhesive. The adhesive melts by the thermal effect during shrinking. For this reason the adjacent parts are bonded longitudinally watertight.						
	4	Jointing and connecting	Insulated crimping sleeves						
Allocation of cables	5	Design	NYY 0.6/1 kV ($U_m = 1.2$ kV)						
	6	Number of cores	4–7	8–14	15–21	22–40	41–61	62–85	86–100
	7	Nominal cross-sectional area of conductor mm^2	1.5	1.5	1.5	1.5	1.5	1.5	1.5
Dimensions, weight	10	Size SKSM ST ...	4–7	8–14	15–21	22–40	41–61	62–85	86–100
	12	Length $l^{1)}$ mm	300	300	350	350	400	400	450
	14	Diameter d	The diameter depends on the spreading of the cores, however, it is considerably smaller than the diameter of comparable joints in cast resin technique.						
	16	Weight kg	0.09	0.11	0.14	0.20	0.25	0.30	0.35
Tests	17	Specifications	The straight joints SKSM ST are checked by the manufacturer.						
Special properties	20	Installation position	Optional						
Accessories	22	Auxiliary devices	Suitable tool when using crimping sleeves. Hot-air blower or gas burner for shrinking						

[1] Length before shrinking

Table 8.5.5
Shrinkable voltage-resistant end cap SKEM
for three-core and four-core PVC and XLPE-insulated cables

1 kV

Application	1	Field of application	Dead-end cables can be sealed with shrinkable voltage-resistant end caps so that the cable route may be under voltage, if required.					
Type of accessory	2	Technique	Heat shrink technique					
	3	Construction	The voltage-resistant end cap consists of three or four small and one big shrinkable cap which is marked with a voltage symbol. The single cores are insulated and sealed by the small caps whereas the big cap serves as outer protection of the cable.					
Allocation of cables	5	Design	PVC and XLPE-insulated cables 0.6/1 kV ($U_m = 1.2$ kV)					
	6	Number of cores	3	3	3	4	4	4
	7	Nominal cross-sectional area of conductor mm^2	6–25	35–150	185–300	6–25	35–150	185–300
Dimensions, weight	10	Size SKEM ...	3/25	3/150	3/300	4/25	4/150	4/300
	14	Diameter d	After shrinking on the outer cap the cable diameter is slightly enlarged.					
	16	Weight kg	0.08	0.15	0.30	0.10	0.20	0.40
Tests	17	Specifications	The voltage-resistant end cap analogously meets the requirements of DIN VDE 0278.					
Special properties	20	Installation position	Optional					
Accessories	22	Auxiliary devices	Hot-air blower or gas burner for shrinking					

Table 8.5.6
PROTOLIN straight joint PV
for PVC and XLPE-insulated cables

Application	1	Field of application	The straight joint PV is suitable for all PVC and XLPE-insulated cables. It can be laid in ground and in water without additional measures.
Type of accessory	2	Technique	Cast resin technique
	3	Construction	Cast resin body
	4	Jointing and connecting	Press-on connectors can be used, for PVC-insulated cables soldering connectors are also permissible.
Allocation of cables	5	Design	NYY, NAYY, NYCY, NYCWY, NAYCWY 0.6/1 kV ($U_m = 1.2$ kV) NA2XY 0.6/1 kV ($U_m = 1.2$ kV)
	6	Number of cores	1, 3, 4 and 5

Nominal cross-sectional area of conductor (7):

		1	2	3	4	5	6	7	8
NYY									
single-core	mm²	4–10	16–35	50–150	185–300	400, 500	–	–	–
three-core	mm²	–	1.5–6	10–25	35, 50	70–120	150	185, 240	–
four-core	mm²	–	1.5–6	10, 16	25, 35	50–95	120, 150	185	240, 300
five-core	mm²	–	1.5–4	6, 10	16, 25	–	–	–	–
NYCWY, NYCY									
three-core	mm²	–	1.5/1.5–6/6	10/10, 16/16	25/16 35/16	50/25– 95/50	120/70 150/70	185/95	240/120
four-core	mm²	–	1.5/1.5, 2.5/2.5	4/4– 10/10	16/16, 25/16	35/16 50/25	70/35– 120/70	150/70	185/95, 240/120
NAYY, NA2XY									
single-core	mm²	–	–	–	50	70–185	240, 300	–	–
four-core	mm²	–	–	–	–	25–70	95–150	185	–
NAYCWY									
three-core	mm²	–	–	–	–	50/50, 70/70	95/95– 150/150	185/185	–

Dimensions, weight:

			1	2	3	4	5	6	7	8
10	Size	PV ..	1	2	3	4	5	6	7	8
12	Length l	mm	112	148	220	310	422	540	580	680
14	Diameter d	mm	30	40	46	60	80	94	110	130
15	Filling volume	l	0.05	0.1	0.275	0.7	1.6	2.8	4.0	6.5
16	Weight	kg	0.5	0.5	0.7	1.8	3.3	5.5	7.5	11.0

Tests	17	Specifications	DIN VDE 0278 Part 1 and 3
Special properties	20	Installation position	Optional. The straight joint shall be in a horizontal position during casting and curing
Accessories	22	Auxiliary devices	Suitable tool when using press-on connectors

Table 8.5.7
PROTOLIN straight joint PV
for underground mining for PVC-insulated cables

Application	1	Field of application		The straight joint PV is suitable for all mining areas.						
Type of accessory	2	Technique		Cast resin technique						
	3	Construction		Cast resin body; wrapping of self-bonding insulating tape; helical distance spreader for the single cores.						
	4	Terminal components		Only press-on connectors are permitted.						
Allocation of cables	5	Design		NYY, NYCWY, NYCY 0.6/1 kV (U_m = 1.2 kV)						
	7	Nominal cross-sectional area of conductor four-core	mm²	**NYY** 1.5–6	10, 16	25, 35	50–95	120, 150	185	240, 300
		three-core	mm²	**NYCWY, NYCY** 1.5/1.5–6/6	10/10, 16/16	25/25, 35/35	50/50–95/95	120/120, 150/150	185/95	240/120
Dimensions, weight	10	Size	PV ..	2	3	4	5	6	7	8
	12	Length l	mm	148	220	310	422	540	580	680
	14	Diameter d		40	46	60	80	94	110	130
	15	Filling volume	l	0.1	0.275	0.7	1.6	2.8	4.0	6.5
	16	Weight	kg	0.6	0.9	1.8	4.2	7.4	9.8	14.5
Tests	17	Specifications		DIN VDE 0278 Part 1 and 3 The requirements of the German mining authorities are fulfilled.						
Special properties	20	Installation position		Optional. The straight joint shall be in a horizontal position during casting and curing. When observing special rules for installation it may also be installed vertically.						
Accessories	22	Auxiliary devices		Suitable tool when using press-on connectors						

Table 8.5.8 1 kV
PROTOLIN straight joint PV
for PVC-insulated cables with lead sheath

Application	1	Field of application		The straight joint PV is suitable for all multi-core cables with lead sheath. It is mainly used for filling stations and in chemical and heavy industry.			
Type of accessory	2	Technique		Cast resin technique			
	3	Construction		Cast resin body. Earthing conductor for connecting the lead sheaths and the sheath wires, if available. In case of heavy and frequent stress by chemical essences, a lead plate shall be soldered on from lead sheath to lead sheath during installation.			
	4	Jointing and connecting		Press-on or soldering connectors			
Allocation of cables	5	Design		NYKY 0.6/1 kV ($U_m = 1.2$ kV)			
	6	Number of cores		3–5			
	7	Nominal cross-sectional area of conductor					
		three-core	mm^2	1.5–4	6–10	–	–
		four-core	mm^2	1.5, 2.5	4–25	35–70	95, 120
		five-core	mm^2	–	1.5–6	–	–
Dimensions, weight	10	Size	PV ..	3	4	5	6
	12	Length l	mm	220	310	422	540
	14	Diameter d	mm	46	60	80	95
	15	Filling volume	l	0.3	0.7	1.6	2.8
	16	Weight	kg	0.7	1.8	3.3	5.5
Tests	17	Specifications		DIN VDE 0278 Part 1 and 3			
Special properties	20	Installation position		Optional. The straight joint shall be in a horizontal position during casting and curing.			
Accessories	22	Auxiliary devices		Suitable tool when using press-on connectors			

Table 8.5.9
PROTOLIN straight joint PV
for PVC-insulated cables with steel wire armouring

Application	1	Field of application	The straight joint PV is fitted to the cable NYFGY 3.6/6 kV (U_m = 7.2 kV). It is mainly used in power plants and in industry.			
Type of accessory	2	Technique	Cast resin technique			
	3	Construction	Cast resin body; a shock protection encloses the connecting point.			
	4	Jointing and connecting	Conductors: Press-on or soldering connectors Armour: The armour and the shock protection are wrapped with a tinned copper wire, thickness 1 mm, and soldered together.			
Allocation of cables	5	Design	NYFGY 3.6/6 kV (U_m = 7.2 kV)			
	7	Nominal cross-sectional area of conductor	25	35–70	95–150	185–240
Dimensions, weight	10	Size PV ..	5	6	7	8
	12	Length l mm	422	540	580	680
	14	Diameter d mm	80	95	110	130
	15	Filling volume l	1.6	2.8	4.0	6.5
	16	Weight kg	4.2	6.9	9.0	14.0
Tests	17	Specifications	DIN VDE 0278 Part 1 and 2			
Special properties	19	Permissible peak short-circuit current kA	125	125	125	125
	20	Installation position	Optional. The straight joint shall be in a horizontal position during casting and curing.			
Accessories	22	Auxiliary devices	Suitable tool when using press-on connectors			

Table 8.5.10 a
Push-on straight joint AMS 10 and AMS 20
for single-core XLPE-insulated cables (up to 120 mm²)

Application	1	Field of application		The push-on straight joint is preferably used because of its extremely short installation time (approx. 1 hour). It is mainly used in power supply networks, but also in industry and in power plants.			
Type of accessory	2	Technique		Push-on technique, heat shrink technique			
	3	Construction		The insulation is made by an insulating body of silicone rubber which is electrically shielded from outside influences with integrated stress control elements. A tinned copper mesh connecting the screen wires is fitted over the insulation. The tinned copper mesh also serves as protection against electric shock. The outer protection is ensured by a thick-walled shrinkable tube which is coated with hot-melt adhesive.			
	4	Jointing and connecting		Conductors: Special press-on connectors Screen: Press-on connectors			
Allocation of cables	5	Design		N2XS2Y, NA2XS2Y, NA2XS(F)2Y 6/10 kV ($U_m = 12$ kV) N2XS2Y, NA2XS2Y, NA2XS(F)2Y 12/20 kV ($U_m = 24$ kV)			
	7	Nominal cross-sectional area of conductor	mm²	35	50	70	95	120
	9	Nominal cross-sectional area of screen	mm²	16	16	16	16	16
Dimensions, weight (10 kV)	10	Size AMS 10/ . . .		35	50	70	95	120
	12	Length l[2]	mm	750	750	750	750	750
	14	Diameter d	mm	50	50	50	55	55
	16	Weight	kg	1.7	1.7	1.7	1.7	1.7
Dimensions, weight (20 kV)	10	Size AMS 20/ . . .		35	50	70	95	120
	12	Length l[2]	mm	750	750	750	750	750
	14	Diameter d	mm	70	70	70	70	75
	16	Weight	kg	2.2	2.2	2.2	2.2	2.2
Tests	17	Specifications		DIN VDE 0278 Part 1 and 2			
Special properties	19	Permissible peak short-circuit current	kA	125	125	125	125	125
	20	Installation position		Optional			
Accessories	22	Auxiliary devices		Stripping tool for the outer semi-conductive layer. Suitable tool for the press-on connectors. Hot-air blower or gas burner for shrinking			

[1] The straight joint can also be used for cables with PVC outer sheath
[2] Length before shrinking

Table 8.5.10 b
Push-on straight joint AMS 10 and AMS 20
for single-core XLPE-insulated cables (150 mm² and above)

Application	1	Field of application	The push-on straight joint is preferably used because of its extremely short installation time (approx. 1 hour). It is mainly used in power supply networks, but also in industry and in power plants.				
Type of accessory	2	Technique	Push-on technique, heat shrink technique				
	3	Construction	The insulation is made by an insulating body of silicone rubber which is electrically shielded from outside influences with integrated stress control elements. A tinned copper mesh connecting the screen wires is fitted over the insulation. The tinned copper mesh also serves as protection against electric shock. The outer protection is ensured by a thick-walled shrinkable tube which is coated with hot-melt adhesive.				
	4	Jointing and connecting	Conductors: Special press-on connectors Screen: Press-on connectors				
Allocation of cables	5	Design	N2XS2Y, NA2XS2Y, NA2XS(F)2Y 6/10 kV (U_m = 12 kV) N2XS2Y, NA2XS2Y, NA2XS(F)2Y 12/20 kV (U_m = 24 kV)				
	7	Nominal cross-sectional area of conductor mm²	150	185	240	300	400
	9	Nominal cross-sectional area of screen mm²	25	25	25	25	35
Dimensions, weight (10 kV)	10	Size AMS 10/ . . .	150	185	240	300	–
	12	Length l[2] mm	750	750	750	750	–
	14	Diameter d mm	55	60	60	60	–
	16	Weight kg	1.7	1.9	1.9	1.9	–
Dimensions, weight (20 kV)	10	Size AMS 20/ . . .	150	185	240	300	400
	12	Length l[2] mm	750	1000	1000	1000	1000
	14	Diameter d mm	75	80	80	80	80
	16	Weight kg	2.2	2.6	2.6	2.6	2.6
Tests	17	Specifications	DIN VDE 0278 Part 1 and 2				
Special properties	19	Permissible peak short-circuit current kA	125	125	125	125	125
	20	Installation position	Optional				
Accessories	22	Auxiliary devices	Stripping tool for the outer semi-conductive layer. Suitable tool for the press-on connectors. Hot-air blower or gas burner for shrinking				

[1] The straight joint can also be used for cables with PVC outer sheath
[2] Length before shrinking

Table 8.5.11 a **10, 20 and 30 kV**
Straight joint with shrinkable tube WPS 10, WPS 20 and WPS 30
for single-core PVC and XLPE-insulated cables (up to 120 mm²)

Application	1	Field of application	The straight joint WPS is mainly used in power supply networks, but also in industry and power plants for their station service system.					
Type of accessory	2	Technique	Wrapping and heat shrink technique					
	3	Construction	The core wrapping consists of self-bonding high-voltage insulating tape, the electrical screening of conductive contact tape and a tinned copper mesh. The tinned copper mesh serves as protection against electric shock. The outer protection is ensured by a thick-walled shrinkable tube which is coated with hot-melt adhesive.					
	4	Jointing and connecting	Press-on connector					
Allocation of cables	5	Design[1]	NYSY $\quad$ 6/10 kV (U_m = 12 kV) N2XS2Y, NA2XS2Y, NA2XS(F)2Y $\quad$ 6/10 kV (U_m = 12 kV) N2XS2Y, NA2XS2Y, NA2XS(F)2Y 12/20 kV (U_m = 24 kV) N2XS2Y, NA2XS2Y, NA2XS(F)2Y 18/30 kV (U_m = 36 kV)					
	7	Nominal cross-sectional area of conductor $\quad$ mm²	25	35	50	70	95	120
	9	Nominal cross-sectional area of screen $\quad$ mm²	16	16	16	16	16	16
Dimensions, weight (10 kV)	10	Size[2] $\quad$ WPS 10/ . . .	55/19	55/19	55/19	55/19	55/19	55/19
	12	Length l[3] $\quad$ mm	750	750	750	750	750	750
	14	Diameter d $\quad$ mm	30	30	30	35	35	40
	16	Weight $\quad$ kg	1.9	1.9	1.9	1.9	1.9	2.0
Dimensions, weight (20 kV)	10	Size[2] $\quad$ WPS 20/ . . .	72/30	72/30	72/30	72/30	72/30	72/30
	12	Length l[3] $\quad$ mm	940	940	940	940	940	940
	14	Diameter d $\quad$ mm	35	35	35	40	40	40
	16	Weight $\quad$ kg	2.4	2.4	2.4	2.6	2.6	2.6
Dimensions, weight (30 kV)	10	Size[2] $\quad$ WPS 30/ . . .	–	72/30	72/30	80/22	80/22	80/22
	12	Length l[3] $\quad$ mm	–	940	940	1000	1000	1000
	14	Diameter d $\quad$ mm	–	40	40	45	45	45
	16	Weight $\quad$ kg	–	2.7	2.7	2.7	2.9	2.9
Tests	17	Specifications	DIN VDE 0278 Part 1 and 2					
Special properties	19	Permissible peak short-circuit current $\quad$ kA	125	125	125	125	125	125
	20	Installation position	Optional					
Accessories	22	Auxiliary devices	Stripping tool for the outer semi-conductive layer. Suitable tool for the press-on connectors. Hot-air blower or gas burner for shrinking					

[1] The straight joint can also be used for cables with PVC outer sheath
[2] Size of the shrinkable tube SFH
[3] Length before shrinking

Table 8.5.11 b
Straight joint with shrinkable tube WPS 10, WPS 20 and WPS 30
for single-core PVC and XLPE-insulated cables (150 mm² and above)

Application	1	Field of application		The straight joint WPS is mainly used in power supply networks, but also in industry and power plants for their station service system.					
Type of accessory	2	Technique		Wrapping and shrink-on technique					
	3	Construction		The core wrapping consists of self-bonding high-voltage insulating tape, the electrical screening of conductive contact tape and a tinned copper mesh. The tinned copper mesh serves as protection against electric shock. The outer protection is ensured by a thick-walled shrinkable tube which is coated with hot-melt adhesive.					
	4	Jointing and connecting		Press-on connectors					
Allocation of cables	5	Design[1]		NYSY 6/10 kV ($U_m = 12$ kV) N2XS2Y, NA2XS2Y, NA2XS(F)2Y 6/10 kV ($U_m = 12$ kV) N2XS2Y, NA2XS2Y, NA2XS(F)2Y 12/20 kV ($U_m = 24$ kV) N2XS2Y, NA2XS2Y, NA2XS(F)2Y 18/30 kV ($U_m = 36$ kV)					
	7	Nominal cross-sectional area of conductor	mm²	150	185	240	300	400	500
	9	Nominal cross-sectional area of screen	mm²	25	25	25	25	35	35
Dimensions, weight (10 kV)	10	Size[2] WPS 10/ . . .		55/19	55/19	72/30	72/30	72/30	72/30
	12	Length l[3]	mm	750	750	940	940	940	940
	14	Diameter d	mm	40	40	45	45	50	50
	16	Weight	kg	2.0	2.0	2.1	2.1	2.5	2.5
Dimensions, weight (20 kV)	10	Size[2] WPS 20/ . . .		72/30	72/30	80/22	80/22	80/22	102/39
	12	Length l[3]	mm	940	940	1000	1000	1000	1000
	14	Diameter d	mm	45	45	45	50	55	60
	16	Weight	kg	2.7	2.7	2.8	2.8	3.2	3.2
Dimensions, weight (30 kV)	10	Size[2] WPS 30/ . . .		80/22	80/22	80/22	102/39	102/39	102/39
	12	Length l[3]	mm	1000	1000	1000	1000	1000	1000
	14	Diameter d	mm	50	50	55	60	60	60
	16	Weight	kg	2.9	3.0	3.0	3.0	3.0	3.8
Tests	17	Specifications		DIN VDE 0278 Part 1 and 2					
Special properties	19	Permissible peak short-circuit current	kA	125	125	125	125	125	125
	20	Installation position		Optional					
Accessories	22	Auxiliary devices		Stripping tool for the outer semi-conductive layer. Suitable tool for the press-on connectors. Hot-air blower or gas burner for shrinking					

[1] The straight joint can also be used for cables with PVC outer sheath
[2] Size of the shrinkable tube
[3] Length before shrinking

Table 8.5.12   **10 kV**
Straight joint WP
for three-core PVC and XLPE-insulated cables

Application	1	Field of application		The straight joint WP is mainly used in power supply networks, but also in industry and power plants for their station service system.		
Type of accessory	2	Technique		Wrapping and cast resin technique		
	3	Construction		The core wrapping consists of self-bonding high-voltage insulating tape, the electrical screening of conductive contact tape and a tinned copper mesh. The tinned copper mesh serves as protection against electric shock. The outer protection is ensured by a plastic pipe filled with cast resin.		
	4	Jointing and connecting		Press-on connectors		
Allocation of cables	5	Design[1]		NYSEY $\quad$ 6/10 kV (U_m = 12 kV) NA2XS2Y $\quad$ 6/10 kV (U_m = 12 kV)		
	7	Nominal cross-sectional area of conductor	mm^2	25–70	95, 120	150–300
	9	Nominal cross-sectional area of screen	mm^2	16	16	25
Dimensions, weight	10	Size	WP . .	5	6	7
	12	Length l	mm	1100	1100	1100
	14	Diameter d	mm	90	110	125
	15	Filling volume	l	5.2	7.8	9.2
	16	Weight	kg	11.7	17.5	20
Tests	17	Specifications		DIN VDE 0278 Part 1 and 2		
Special properties	19	Permissible peak short-circuit current	kA	125	125	125
	20	Installation position		Optional. The straight joint shall be in a horizontal position during casting and curing.		
Accessories	22	Auxiliary devices		Stripping tool for the outer semi-conductive layer of XLPE-insulated cables. Suitable tool for the press-on connectors		

[1] The straight joint can also be used for cables with PVC outer sheath

Table 8.5.13 a
Straight joint WP
for single-core PVC and XLPE-insulated cables (up to 120 mm²)

Application	1	Field of application		The straight joint WP is mainly used in power supply networks, but also in industry and power plants for their station service system.					
Type of accessory	2	Technique		Wrapping and cast resin technique					
	3	Construction		The core wrapping consists of self-bonding high-voltage insulating tape, the electrical screening of conductive contact tape and a tinned copper mesh. The tinned copper mesh serves as protection against electric shock. The outer protection is ensured by a plastic pipe filled with cast resin.					
	4	Jointing and connecting		Press-on connectors					
Allocation of cables	5	Design[1]		NYSY $\quad$ 6/10 kV (U_m = 12 kV) N2XS2Y, NA2XS2Y, NA2XS(F)2Y $\quad$ 6/10 kV (U_m = 12 kV) N2XS2Y, NA2XS2Y, NA2XS(F)2Y $\quad$ 12/20 kV (U_m = 24 kV) N2XS2Y, NA2XS2Y, NA2XS(F)2Y $\quad$ 18/30 kV (U_m = 36 kV)					
	7	Nominal cross-sectional area of conductor	mm²	25	35	50	70	95	120
	9	Nominal cross-sectional area of screen	mm²	16	16	16	16	16	16
Dimensions, weight (10 kV)	10	Size	WP ..	10	10	10	10	10	11
	12	Length l	mm	615	615	615	615	615	615
	14	Diameter d	mm	63	63	63	63	63	75
	15	Filling volume	l	1.2	1.2	1.2	1.2	1.2	1.8
	16	Weight	kg	3.9	3.9	3.9	3.9	3.9	5.3
Dimensions, weight (20 kV)	10	Size	WP ..	1	1	1	1	1	1
	12	Length l	mm	840	840	840	840	840	840
	14	Diameter d	mm	75	75	75	75	75	75
	15	Filling volume	l	2.08	2.08	2.08	2.08	2.08	2.08
	16	Weight	kg	6.1	6.1	6.1	6.3	6.3	6.3
Dimensions, weight (30 kV)	10	Size	WP ..	–	3	3	3	4	4
	12	Length l	mm	–	840	840	840	840	840
	14	Diameter d	mm	–	75	75	75	90	90
	15	Filling volume	l	–	2.08	2.08	2.08	3.2	3.2
	16	Weight	kg	–	6.3	6.3	6.3	8.6	8.6
Tests	17	Specifications		DIN VDE 0278 Part 1 and 2					
Special properties	19	Permissible peak short-circuit current	kA	125	125	125	125	125	125
	20	Installation position		Optional					
Accessories	22	Auxiliary devices		Stripping tool for the outer semi-conductive layer of XLPE-insulated cables. Suitable tool for the press-on connectors					

[1] The straight joint can also be used for cables with PVC outer sheath

Table 8.5.13 b
10, 20 and 30 kV
Straight joint WP
for single-core PVC and XLPE-insulated cables (150 mm² and above)

Application	1	Field of application		The straight joint WP is mainly used in power supply networks, but also in industry and power plants for their station service system.					
Type of accessory	2	Technique		Wrapping and cast resin technique					
	3	Construction		The core wrapping consists of self-bonding high-voltage insulating tape, the electrical screening of conductive contact tape and a tinned copper mesh. The tinned copper mesh serves as protection against electric shock. The outer protection is ensured by a plastic pipe filled with cast resin.					
	4	Jointing and connecting		Press-on connectors					
Allocation of cables	5	Design[1]		NYSY $\qquad$ 6/10 kV (U_m = 12 kV) N2XS2Y, NA2XS2Y, NA2XS(F)2Y $\;$ 6/10 kV (U_m = 12 kV) N2XS2Y, NA2XS2Y, NA2XS(F)2Y $\;$ 12/20 kV (U_m = 24 kV) N2XS2Y, NA2XS2Y, NA2XS(F)2Y $\;$ 18/30 kV (U_m = 36 kV)					
	7	Nominal cross-sectional area of conductor	mm²	150	185	240	300	400	500
	9	Nominal cross-sectional area of screen	mm²	25	25	25	25	35	35
Dimensions, weight (10 kV)	10	Size	WP ..	11	11	11	11	12	12
	12	Length l	mm	615	615	615	615	840	840
	14	Diameter d	mm	75	75	75	75	90	90
	15	Filling volume	l	1.8	1.8	1.8	1.8	3.2	3.2
	16	Weight	kg	5.3	5.3	5.3	5.3	8.6	8.6
Dimensions, weight (20 kV)	10	Size	WP ..	2	2	2	2	22	22
	12	Length l	mm	840	840	840	840	1000	1000
	14	Diameter d	mm	90	90	90	90	110	110
	15	Filling volume	l	3.2	3.2	3.2	3.2	4.8	4.8
	16	Weight	kg	8.6	8.6	8.6	8.6	10.0	10.0
Dimensions, weight (30 kV)	10	Size	WP ..	4	4	4	4	44	44
	12	Length l	mm	840	840	840	840	1000	1000
	14	Diameter d	mm	90	90	90	90	110	110
	15	Filling volume	l	3.2	3.2	3.2	3.2	4.8	4.8
	16	Weight	kg	8.6	8.6	8.6	8.6	10.0	10.0
Tests	17	Specifications		DIN VDE 0278 Part 1 and 2					
Special properties	19	Permissible peak short-circuit current	kA	125	125	125	125	125	125
	20	Installation position		Optional					
Accessories	22	Auxiliary devices		Stripping tool for the outer semi-conductive layer of XLPE-insulated cables. Suitable tool for the press-on connectors					

[1] The straight joint can also be used for cables with PVC outer sheath

Table 8.5.14
Straight joint VS
for paper-insulated belted cables

Application	1	Field of application		Three-core belted cables are connected by means of the straight joint VS.				
Type of accessory	2	Technique		Wrapping and casting technique				
	3	Construction		The connecting points are wrapped with mass-impregnated paper and the cores are spaced by supporting bandages. The lead sheaths are connected by the inner joint. A conductive connection to the cast-iron joint housing is provided by an earthing conductor. The lead inner joint and the cast-iron protective joint are filled with filling compound.				
	4	Jointing and connecting		Conductor: Copper conductors are connected by means of press-on or soldering connectors. Aluminium conductors can be soldered, pressed or welded. Lead sheath: Connection by a lead inner joint				
Allocation of cables	5	Design		NKBA, NAKBA 6/10 kV ($U_m = 12$ kV)				
	7	Nominal cross-sectional area of conductor	mm^2	25	35–70	95	120, 150	185–300
Dimensions, weight	10	Size	VS ...	VI 34	VI 43	VI 53	VI 64	VI 10/87
	12	Length l	mm	655	765	880	1000	1264
	14	Diameter d	mm	140	203	222	236	265
	15	Filling volume	l	5	7	10	14	23
	16	Weight	kg	19.5	28.5	37.0	51.0	69.0
Tests	17	Specifications		DIN VDE 0278 Part 1 and 2				
Special properties	19	Permissible peak short-circuit current	kA	125	125	125	125	125
	20	Installation position		Optional. The straight joint shall be in a horizontal position during casting and curing.				
Accessories	22	Auxiliary devices		Suitable tool when using press-on connectors				

Table 8.5.15 a
Straight joint SMI
for paper-insulated three-core S.L. cables (up to 95 mm²)

Application	1	Field of application	The straight joint SMI is used for connecting three-core S.L. cables.				
Type of accessory	2	Technique	Wrapping and casting technique				
	3	Construction	The connecting points are wrapped with mass-impregnated paper. The core wrapping has a metallic screening and each core is additionally terminated by means of a lead inner joint. The lead inner joint is filled with cable impregnating compound and the outer cast-iron protective joint with black insulating compound.				
	4	Jointing and connecting	Conductor: Copper conductors are connected by soldering sleeves; aluminium conductors should preferably be welded. Lead sheath: Connection by lead inner joints				
Allocation of cables	5	Design	NEKEBA, NAEKEBA 12/20 kV ($U_m = 24$ kV) NEKEBA, NAEKEBA 18/30 kV ($U_m = 36$ kV)				
	7	Nominal cross-sectional area of conductor mm²	25	35	50	70	95
Dimensions, weight (20 kV)	10	Size SMI 20/ . . .	1175	1175	1175	1175	1175
	12	Length l mm	1175	1175	1175	1175	1175
	14	Diameter d mm	330	330	330	330	330
	15	Filling volume Inner joint l Protective joint l	5 35	5 35	5 35	5 35	5 35
	16	Weight kg	110	110	110	110	110
Dimensions, weight (30 kV)	10	Size SMI 30/ . . .	–	1500	1500	1500	1500
	12	Length l mm	–	1500	1500	1500	1500
	14	Diameter d mm	–	330	330	330	330
	15	Filling volume Inner joints l Protective joint l	– –	7 43	7 43	7 43	7 43
	16	Weight kg	–	150	150	150	150
Tests	17	Specifications	DIN VDE 0278 Part 1 and 2				
Special properties	19	Permissible peak short-circuit current kA	125	125	125	125	125
	20	Installation position	Optional				
Accessories	22	Auxiliary devices	Suitable tool for welding the conductors				

Table 8.5.15 b
Straight joint SMI
for paper-insulated three-core S.L. cables (120 mm² and above)

Application	1	Field of application		The straight joint SMI is used for connecting three-core S.L. cables.		
Type of accessory	2	Technique		Wrapping and casting technique		
	3	Construction		The connecting points are wrapped with mass-impregnated paper. The core wrapping has a metallic screening and each core is additionally terminated by means of a lead inner joint. The lead inner joint is filled with cable impregnating compound and the outer cast-iron protective joint with black insulating compound.		
	4	Jointing and connecting		Conductor: Copper conductors are connected by soldering sleeves; aluminium conductors should preferably be welded. Lead sheath: Connection by lead inner joints		
Allocation of cables	5	Design		NEKEBA, NAEKEBA 12/20 kV ($U_m = 24$ kV) NEKEBA, NAEKEBA 18/30 kV ($U_m = 24$ kV)		
	7	Nominal cross-sectional area of conductor	mm²	120	150	185	240
Dimensions, weight (20 kV)	10	Size SMI 20/ . . .		1500	1500	1500	1500
	12	Length l	mm	1500	1500	1500	1500
	14	Diameter d	mm	330	330	330	330
	15	Filling volume Inner joints	l	5	5	7	7
		Protective joint	l	45	45	43	43
	16	Weight	kg	140	140	145	145
Dimensions, weight (30 kV)	10	Size SMI 30/ . . .		1500	1500	1500	1500
	12	Length l	mm	1500	1500	1500	1500
	14	Diameter d	mm	330	330	330	330
	15	Filling volume Inner joints	l	7	7	7	7
		Protective joint	l	43	43	43	43
	16	Weight	kg	150	150	150	150
Tests	17	Specifications		DIN VDE 0278 Part 1 and 2		
Special properties	19	Permissible peak short-circuit current	kA	125	125	125	125
	20	Installation position		Optional		
Accessories	22	Auxiliary devices		Suitable tool for welding the conductors		

8.6 Branch Joints

1 kV

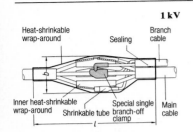

Shrinkable branch joint GNKA-1
for polymeric-insulated cables
(see Table 8.6.1)

1 kV

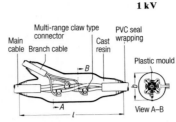

PROTOLIN branch joint PA for
polymeric-insulated cables
(see Table 8.6.2)

1 kV

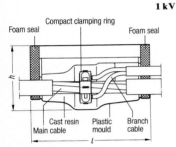

PROTOLIN branch joint PAK for
four-core PVC and XLPE-insulated
cables (see Table 8.6.3)

1 kV

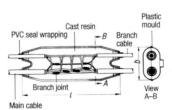

PROTOLIN double branch joint
PAD for PVC-insulated cables
(see Table 8.6.4)

1 kV

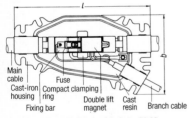

Cast-iron branch joint with fuse GMS
or with automatic safety cutout GMSA for
PVC and XLPE-insulated
cables (see Table 8.6.5)

Table 8.6.1
Shrinkable branch joint GNKA-1
for polymeric-insulated cables

Application	1	Field of application		The shrinkable branch joint is suitable for polymeric-insulated low-voltage cables, e.g. as service joint in distribution networks. It can be installed without cutting the cable because of the special design of the special single branch-off clamps, if necessary even under voltage. The branch-off area of the cable is protected against moisture and mechanical influences by means of the branch joint.		
Type of accessory	2	Technique		Heat shrink technique		
	3	Construction		The shrinkable branch joint consists of insulated special single branch-off clamps and the shrinkable inner wrap-arounds which are coated with hot-melt adhesive, a protective layer and a shrinkable wrap-around as protection against external influences. The inner side of this shrinkable wrap-around is coated with hot-melt adhesive.		
	4	Jointing and connecting		Special single branch-off clamps		
Allocation of cables	5	Design		NYY, NAYY $\quad$ 0.6/1 kV (U_m = 1.2 kV) NA2XY $\quad\quad\quad$ 0.6/1 kV (U_m = 1.2 kV) NYCWY, NAYCWY 0.6/1 kV (U_m = 1.2 kV)		
	7	Nominal cross-sectional area of conductor		for NYY, NAYY, NA2XY		
		Main cable	mm^2	95–185		–
		Branch cable	mm^2	25–95		–
				for NYCWY, NAYCWY		
		Main cable	mm^2	–		95–185
		Branch cable	mm^2	–		25–95
Dimensions, weight	10	Size $\quad$ GNKA-1/..		4		3
	12	Length l	mm	620		620
	13	Width b	mm	160		160
	16	Weight	kg	3.0		3.0
Tests	17	Specifications		DIN VDE 0278 Part 1 and 3		
Special properties	20	Installation position		Optional		
Accessories	22	Auxiliary devices		A hexagon socket screw key, size 7, is required for the special single branch-off clamps; if the joint shall be installed under voltage, an insulated key is necessary. Hot-air blower or gas burner for shrinking		

Table 8.6.2
PROTOLIN branch joint PA
for polymeric-insulated cables

Application	1	Field of application	For branching off service cables, e.g. branch cables in distribution networks. The branch-off area of the cable is protected against moisture and mechanical influences by means of the branch joint.				
Type of accessory	2	Technique	Cast resin technique				
	3	Construction	The branch joint PA consists of a two-part plastic mould with lateral branch. The mould closed with keyway and featherkey is filled with cast resin.				
	4	Jointing and connecting	The conductors are connected by multi-range claw-type connectors (single claw-type connectors). Small cross-sectional areas are soldered on to the main cable.				
Allocation of cables	5	Design	NYY, NAYY $\quad$ 0.6/1 kV (U_m = 1.2 kV) NYCWY, NAYCWY 0.6/1 kV (U_m = 1.2 kV) NA2XY $\quad$ 0.6/1 kV (U_m = 1.2 kV)				
	6	Number of cores	3 or 4	3 or 4	3 or 4	3	3
	7	Nominal cross-sectional area of conductor					
		Main cable $\quad$ mm^2	1.5–10	16–35	50–150	1.5/1.5–10/10	16/16
		Branch cable $\quad$ mm^2	1.5–10	16–35	50	1.5/1.5–10/10	16/16
Dimensions, Weight	10	Size $\quad$ PA ..	1	2	3	1	2
	12	Length l $\quad$ mm	220	280	440	220	280
	14	Diameter d $\quad$ mm	70	100	120	70	100
	15	Filling volume $\quad$ l	0.47	1.25	3.35	0.47	1.25
	16	Weight $\quad$ kg	1.3	2.7	6	1.3	2.7
Tests	17	Specifications	DIN VDE 0278 Part 1 and 3				
Special properties	20	Installation position	Optional. The branch joint shall be in horizontal position during casting and curing.				

Table 8.6.3
PROTOLIN branch joint PAK
for four-core PVC and XLPE-insulated cables

Application	1	Field of application	Branch joints PAK shall be used for branching off service cables from PVC and XLPE-insulated main cables in distribution networks if cable compact clamping rings are used. The branch-off area of the cable is protected against moisture and mechanical influences by means of the branch joint.		
Type of accessory	2	Technique	Cast resin technique		
	3	Construction	The branch is arranged parallel to the main cable. Pre-fabricated foam seals can be matched to the cable diameter. The ready-fitted mould is filled with cast resin and closed with a cover.		
	4	Jointing and connecting	The conductors are connected by a compact clamping ring.		
Allocation of cables	5	Design	NYY, NAYY 0.6/1 kV (U_m = 1.2 kV) NA2XY 0.6/1 kV (U_m = 1.2 kV)		
	7	Nominal cross-sectional area of conductor			
		Main cable mm^2	70–185	70–150	70–185
		Branch cable mm^2	25–50	25–50	25–95
	8	Shape and type of conductor	SE	SM	SE or SM
Dimensions, weight	10	Size PAK . . .	81	81	82
	11	Height h mm	185	185	210
	12	Length l mm	325	325	350
	15	Filling volume l	2.1	2.1	3.4
	16	Weight kg	3.5	3.5	4.5
Tests	17	Specifications	DIN VDE 0278 Part 1 and 3		
Special properties	20	Installation position	Optional		
Accessories	22	Auxiliary devices	Hexagon socket screw key. The branch joint shall be in a horizontal position during casting and curing.		

Table 8.6.4
PROTOLIN double branch joint PAD
for PVC-insulated cables

Application	1	Field of application	The double branch joint is suitable for branching off multi-core cables. It is preferably used for street lighting cables. It can also be used for airport beacons.			
Type of accessory	2	Technique	Cast resin technique			
	3	Construction	PVC-insulated cables up to a maximum of 2×25 mm^2 or 4×16 mm^2 can be connected by means of the symmetrical mould with keyway and featherkey. It is filled with cast resin. The cast resin prevents the penetration of moisture into the connector area.			
	4	Jointing and connecting	Conductor cross-sectional areas of 1.5 to 4 mm^2 are soldered, for larger cross-sectional areas branch terminals shall be used.			
Allocation of cables	5	Design	NYY, NYCWY, NYCY 0.6/1 kV ($U_\mathrm{m} = 1.2$ kV)			
	6	Number of cores	3	3	4	4
	7	Nominal cross-sectional area of conductor	For NYY			
		Main cable mm^2	1.5–4	6–16	1.5, 2.5	4–16
		Branch cable mm^2	1.5–4	6–16	1.5, 2.5	4–16
			For NYCWY, NYCY			
		Main cable mm^2	1.5/1.5, 2.5/2.5	4/4–16/16	1.5/1.5, 2.5/2.5	4/4–16/16
		Branch cable mm^2	1.5/1.5, 2.5/2.5	4/4–16/16	1.5/1.5, 2.5/2.5	4/4–16/16
Dimensions, weight	10	Size PAD ..	1	2	1	2
	11	Height h mm	80	100	80	100
	12	Length l mm	235	302	235	302
	15	Filling volume l	0.5	1.2	0.5	1.2
	16	Weight kg	1.1	2.3	1.1	2.3
Tests	17	Specifications	DIN VDE 0278 Part 1 and 3			
Special properties	20	Installation position	Optional. The double branch joint PAD shall be in a horizontal position during casting and curing.			

Table 8.6.5
Cast-iron branch joint with fuse GMS or with automatic safety cutout GMSA
for PVC and XLPE-insulated cables

Application	1	Field of application	The branch joint provides the protection by fuse for the branch cable, e.g. for operating street lighting, signs, phone box lighting etc., without installing a separate cable network. The safety cutout of the branch joint GMSA can be switched on and off by means of advanced double lift magnets at any point. For distance switching a 12 V or 24 V d.c. voltage source is necessary besides the permanently-wired tracker wires.	
Type of accessory	2	Technique	Cast resin technique	
	3	Construction	The branch joint consists of a cast iron housing, a fixing bar with socket for a fuse, a compact clamping ring and a double lift magnet for the design GMSA. The cast iron housing is filled with cast resin up to the fuse socket. After curing, the area of the fuses is sealed by a transparent cap which is cast-en bloc. Fuses or automatic safety cutouts up to a maximum of 63 A can be put in place.	
	4	Jointing and connecting	For the branch standard cable clamping rings with insulating parts can be used. The connection from the clamping ring to the fuses consists of pre-fabricated cores of 16 mm².	
Allocation of cables	5	Design Main cable Branch cable	NYY, NAYY, NA2XY 0.6/1 kV (U_m = 1.2 kV) NYY, NYCWY, NYCY 0.6/1 kV (U_m = 1.2 kV)	
	7	Nominal cross-sectional area of conductor		
		Main cable mm²	70–150	70–150
		Branch cable mm²	1.5–16	1.5–16
Dimensions, weight	10	Size	GMS 1	GMSA 1
	12	Length l mm	395	395
	13	Width b mm	230	230
	15	Filling volume l	4.68	4.68
	16	Weight kg	23	23
Tests	17	Specifications	DIN VDE 0278 Part 1 and 3	
Special properties	19	Installation position	Optional. The branch joint shall be in a horizontal position during casting and curing.	
Accessories	22	Auxiliary devices	Hexagon socket screw key	

8.7 Transition joints

1 kV

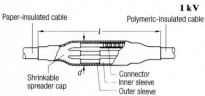

Shrinkable transition joint SKSM PK
for connecting PVC and XLPE-insulated cables
with three-core paper-insulated cables
(see Table 8.7.1)

10 kV

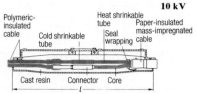

Transition joint MP 10 for connecting PVC and
XLPE-insulated cables with paper-insulated belted
cables (see Table 8.7.2)

20 kV
30 kV

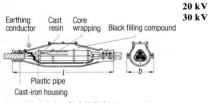

Transition joint SM-WP for connecting
three single-core XLPE-insulated cables with
paper-insulated three-core S.L.
cables (see Table 8.7.3)

Table 8.7.1
Shrinkable transition joint SKSM PK
for connecting PVC and XLPE-insulated cables with three-core paper-insulated cables

Application	1	Field of application	For connecting paper-insulated cables with polymeric-insulated cables			
Type of accessory	2	Technique	Heat shrink technique			
	3	Construction	The shrinkable transition joint consists of the inner sleeves which are shrunk on over each core and the common outer sleeve (protective sleeve). The inner sides of the joints are coated with hot-melt adhesive. The adhesive melts because of the thermal effect during shrinking. For this reason the parts laying side by side are bonded longitudinally watertight.			
	4	Jointing and connecting	Conductor: In principle, screw-clamp connectors or press-on connectors can be used, for PVC-insulated cables soldering connectors are also permissible.			
Allocation of cables	5	Design	NKBA, NAKBA, NAKLEY 0.6/1 kV ($U_m = 1.2$ kV) NYY, NAYY $\quad\quad\quad\quad$ 0.6/1 kV ($U_m = 1.2$ kV) NA2XY $\quad\quad\quad\quad\quad$ 0.6/1 kV ($U_m = 1.2$ kV)			
	7	Nominal cross-sectional area of conductor $\quad$ mm^2	25, 35	50, 70	95–150	185, 240
Dimensions, weight	10	Size $\quad$ SKSM PK ...	55/4	65/4	80/4	110/4
	12	Length $l^{1)}$ $\quad\quad$ mm	665	700	760	900
	14	Diameter d	The diameter depends on the spreading of the cores, however, it is considerably smaller than the diameter of comparable joints in cast resin technique.			
	16	Weight $\quad\quad\quad$ kg	0.36	0.46	0.61	0.86
Tests	17	Specifications	DIN VDE 0278 Part 1 and 3			
Special properties	20	Installation position	Optional			
	21	Treatment of crotch	The paper-insulated cable is sealed against bleeding of compound by means of a shrinkable spreader cap.			
Accessories	22	Auxiliary devices	Suitable tool when using press-on connectors. Hot-air blower or gas burner for shrinking			

$^{1)}$ Length after shrinking

Table 8.7.2   **6 kV**
Transition joint ÜMP 10
for connecting PVC and XLPE-insulated cables with paper-insulated belted cables

Application	1	Field of application	For connecting paper-insulated belted cables with polymeric-insulated cables, preferably used in power supply and industry.
Type of accessory	2	Technique	Cast resin technique
	3	Construction	The joint consists of a longitudinally split cast-iron housing which is filled with cast resin. The cores are sealed by self-welding and compound-resistant insulating tapes in conjunction with heat and cold shrinkable tubes in order to prevent the compound from bleeding. The joint shall lie somewhat deeper than the cable in order to prevent the compound from migrating at the cable ends. In this case the mass effect on the polymeric cable side is avoided by means of press-on connectors with barrier. The cast-iron housing serves as protection against electric shock and provides a screening for the connection point towards the outside.
	4	Jointing and connecting	The copper or aluminium conductors of the cables shall be connected by press-on connectors with barrier.
Allocation of cables	5	Design Mass-impregated cable Polymeric-insulated cable[1]	NKBA, NAKBA $\qquad$ 6/10 kV $(U_\mathrm{m}) = 12$ kV NYSY, NYSEY $\qquad$ 6/10 kV $(U_\mathrm{m}) = 12$ kV N2XS2Y, NA2XS2Y, NA2XS(F)2Y 6/10 kV $(U_\mathrm{m}) = 12$ kV
	7	Nominal cross-sectional area of conductor mm^2	35 - 240
	9	Nominal cross-sectional area of screen mm^2	16 - 25
Dimensions, weight	12	Length l mm	1000
	13	Width b mm	226
	15	Filling volume l	12.5
	16	Weight kg	60
Tests	17	Specifications	DIN VDE 0278 Part 1 and 2
Special properties	19	Permissible peak short-circuit current kA	125
	20	Installation position	Horizontal. The joint shall lie deeper than the cable in order to prevent the compound from migrating at the ends of belted cables.
Accessories	22	Auxiliary devices	Stripping tool for the outer semi-conductive layer of XLPE-insulated cables. Suitable tool for the press-on connectors

[1] The transition joint can also be used for cables with PVC outer sheath

Table 8.7.3
Transition joint SM-WP   **20 and 30 kV**
for connecting three single-core XLPE-insulated cables with paper-insulated three-core S.L. cables

Application	1	Field of application	For connecting paper-insulated three-core S.L. cables with XLPE-insulated cables. Owing to a special design the screens can be separated for sheath measurements on XLPE-insulated cables.			
Type of accessory	2	Technique	Wrapping, cast resin, casting technique			
	3	Construction	Each core has a screened core wrapping which is protected against mechanical damage by a PVC pipe filled with cast resin or by a shrinkable tube. The outer protective joint is casted with filling compound. The joint shall lie somewhat deeper than the cable in order to prevent the compound from migrating at the cable ends. In this case the mass effect on the polymeric cable side is avoided by means of the press-on connector with barrier.			
	4	Jointing and connecting	The copper or aluminium conductors of the cables are connected by press-on connectors with barrier. Aluminium conductors can also be welded (Cadweld Procedure).			
Allocation of cables	5	Design[1]	NEKEBA, NAEKEBA N2XS2Y, NA2XS2Y, NA2XS(F)2Y 12/20 kV (U_m = 24 kV)		NEKEBA, NAEKEBA N2XS2Y, NA2XS2Y, NA2XS(F)2Y 18/30 kV (U_m = 36 kV)	
	7	Nominal cross-sectional area of conductor mm²	25 - 150	185 - 300	35 - 70	95 - 240
Dimensions, weight	10	Size SM-WP ..	1500-8	1800-9	1500-8	1800-9
	12	Length *l* mm	1500	1800	1500	1800
	13	Width *b* mm	330	316	330	316
	15	Filling volume Cast resin l Filling compound l	7.44 45	9.6 65	7.44 45	9.6 65
	16	Weight kg	180	200	180	200
Tests	17	Specifications	DIN VDE 0278 Part 1 and 2			
Special properties	19	Permissible peak short-circuit current kA	125	125	125	125
	20	Installation position	Horizontal. The joint shall lie deeper than the cable in order to prevent the compound from migrating at the ends of three-core S.L. cables.			
Accessories	22	Auxiliary devices	Stripping tool for the outer semi-conductive layer of XLPE-insulated cables. Suitable tool for the press-on connectors			

[1] The transition joint can also be used for cables with PVC outer sheath

9 Explanations to the Project Planning Data of Accessories

Explanations to the Items of the Project Planning Data
(Table 8.1.1 to 8.7.3)

Item	Designation	Symbol	Unit	Explanation	Further explanations see Part 1
1	Field of application			This item gives information on the system-dependent application preferred and, if there is no indication in the designation code for the accessory, on the function, e.g. – protection of the cable fillers from dirt and water, – mechanical protection in case of short-circuit	32.1
2	Technique			The technique for the construction and installation of the accessory is given.	32.4
3	Construction			List of the main constructional elements, e.g. insulators, housings etc.	32.3
4	Terminal components			Suitable terminal components for the conductor and the metallic coverings are given.	30.2 30.3
5	Design			Designation code of the corresponding cable type	13.2
6	Number of cores			Clause 6, Item 1	
7	Nominal cross-sectional area of conductor	q_n	mm^2	Clause 6, Item 2	20.1
8	Shape and type of conductor			Clause 6, Item 3	1.2 13.7
9	Nominal cross-sectional area of screen	q_n	mm^2	Clause 6, Item 6	7.2
10	Size			Designation of the accessory with type designation code and, if available, – information on the voltage – information on the cross-sectional area of conductors – or serial number	

Item	Designation	Symbol	Unit	Explanation	Further explanations see Part 1
11	Height	h	mm	The most important installation measures are given in order to determine the suitability for installation in confined connection areas and in order to check the required space for joints. Depending on the design the data in the tables shall be observed.	
12	Length	l	mm		
13	Width	b	mm		
14	Diameter	d	mm		
15	Filling volume		l	Required filling volume of the poured compounds such as cast resin and filling compounds	32.4.1 32.4.2
16	Weight		kg	Weight of the unit pack	
17	Specifications			Information on the relevant parts in DIN VDE 0278	32.2
18	Supporting function			For outdoor terminations if mechanically fixed housings are available.	
19	Permissible peak short-circuit current	I_s	kA	The peak value is given. For single-core cables this value is valid for fixing at the stud and at the bottom of the accessory. For three-core cables the strength is ensured by spreader boxes and core supporting struts.	19.4.4
20	Installation position			For terminations usually vertical, for joints horizontal, for accessories with solid dielectric optional, with the exception of outdoor terminations with sheds.	
21	Treatment of crotch			For three-core cables the design of the crotch sealing is given.	
22	Auxiliary devices			Information on necessary special tools, such as crimping tools, stripping tools, heating appliances etc.	

419